Microbial Endocrinology

Mark Lyte · Primrose P.E. Freestone
Editors

Microbial Endocrinology

Interkingdom Signaling in Infectious Disease and Health

 Springer

Editors
Mark Lyte
Texas Tech University
Health Sciences Center
Lubbock, TX
USA
mark.lyte@ttuhsc.edu

Primrose P.E. Freestone
Department of Infection, Immunity
and Inflammation
University of Leicester School of Medicine
Leicester
UK
ppef1@le.ac.uk

ISBN 978-1-4419-5575-3   e-ISBN 978-1-4419-5576-0
DOI 10.1007/978-1-4419-5576-0
Springer New York Dordrecht Heidelberg London

Library of Congress Control Number: 2010922891

Printed on acid-free paper

Springer is part of Springer Science+Business Media (www.springer.com)

About the Editors

Prof. Mark Lyte, Ph.D., M.S., MT(ASCP) is a Professor in the Department of Pharmacy Practice, School of Pharmacy and adjunct Professor in the Departments of Internal Medicine and Microbiology and Immunology, School of Medicine of Texas Tech University Health Sciences Center. Professor Lyte obtained his Ph.D. and M.S. degrees from the Weizmann Institute of Science in Israel. His undergraduate degree in the clinical laboratory sciences and work as a board-certified medical technologist in the hospital setting have influenced his translational approach throughout his career. Early work investigating the ability of stress to modulate immunity during infection led Professor Lyte to examine the role of bacterial recognition of stress-related neuroendocrine hormones in the infective process. This research ultimately led him in 1992 to establish the field of microbial endocrinology. In addition to having served on scientific review panels for the National Institutes of Health and other worldwide agencies, Professor Lyte has been awarded the Joseph Susman Memorial Award for Surgical Infectious Disease Research by the Surgical Infection Society and was named a finalist for the NIH Director's Pioneer Award.

Dr. Primrose Freestone, B.Sc. (Hons), Ph.D., PGCE is a Lecturer in Clinical Microbiology, in the Department of Infection, Immunity and Inflammation, School of Medicine, University of Leicester. She has been a close collaborator of Professor Mark Lyte for over 10 years and is a foundation contributor in the field of microbial endocrinology. A microbial biochemist by training, Dr Freestone was the first to identify tyrosine phosphorylation as a regulatory mechanism in bacteria, and to elucidate the mechanisms by which stress hormones induce bacterial growth. She is also a highly active inventor in the fields of microbial diagnostics and dietary bioactives and has been holder of several prestigious Fellowships to develop entrepreneurial activities in the area of life science biotechnology.

ML would like to dedicate the book to his wife Phyllis who has supported him through this journey since the first day and to his two sons, Joshua and Jeremy, for their dedication to acquiring knowledge.

PF would like to dedicate her part of the book to her family. In particular, her aunt Betty, cousin Wendy, and above all, her mother, Rosie, to whom she owes a great debt of gratitude for her love, support and her tasty and no prompting supply of snacks when working late!

Preface

Microbial endocrinology represents a newly emerging interdisciplinary field that is formed by the intersection of the fields of neurobiology and microbiology. It is the intent of this book to introduce a new perspective to the current understanding not only of the factors that mediate the ability of microbes to cause disease, but also of the mechanisms that maintain normal homeostasis. The discovery that microbes can both synthesize and directly respond to neuroendocrine hormones provides for a new framework with which to investigate how microorganisms interface not only with vertebrates, but also with invertebrates and even plants. As illustrated by the breadth of expertise of the contributors, the reader will learn that the neuroendocrine hormones that one most commonly associates with mammals, are actually found throughout the plant, insect and microbial communities to an extent that will undoubtedly surprise many, and most importantly, highlight how interactions between microbes and neuroendocrine hormones can influence the pathophysiology of infectious disease. This book will lead its readers through a journey to the rather startling realization that through eons of residence within our bodies, bacteria have evolved what appear to be specific detection systems for our neuroendocrine hormones, and that they not only seem to be able to "sense" when we are stressed, but also to utilize and effectively co-op products of our neurophysiological response to stress for their own usage.

It was in 1992 that one of us (ML) gave the first talk on the subject of neuroendocrine–bacterial interactions at that year's annual meeting of the American Society for Microbiology General Meeting in New Orleans to an audience of only two people (and one was ML's technician). That over the succeeding years the concept of microbial endocrinology met with much skepticism is an understatement. However, it was 25 years later from that first talk, in 2007, that both of us were pleased to act as organizers of the first American Society of Microbiology microbial endocrinology symposium that was held during the 2007 annual society meeting in Toronto. Thankfully, the wheel of scientific skepticism has now come full circle, and we are very pleased to be able to present the very first book on microbial endocrinology.

Lubbock, TX
Leicester, UK
February 2010

Prof. Mark Lyte
Dr. Primrose P.E. Freestone

Contents

Contributors

Noura Al-Dayan
Department of Infection, Immunity and Inflammation
University of Leicester, School of Medicine
University Road, Leicester LE1 9HN, UK
na71@leicester.ac.uk

Dr. John C. Alverdy
Professor, Director, Minimally Invasive/Bariatric Surgery,
Department of Surgery, University of Chicago, 5841
South Maryland Avenue, MC 6090, Chicago, IL 60637, USA
jalverdy@surgery.bsd.uchicago.edu

Dr. Michael T. Bailey
Division of Oral Biology, College of Dentistry, The Ohio State University,
4161 Postle Hall, 305 W. 12th Avenue, Columbus, OH 43210, USA
bailey.494@osu.edu

Dr. Bradley L. Bearson
USDA-ARS-National Laboratory for Agriculture and the Environment,
Agroecosystem Management Research Unit, 2110 University Drive,
NSRIC-2103, Ames, IA 50011-3310, USA
brad.bearson@ars.usda.gov

Prof. David R. Brown
Professor, Department of Veterinary and Biomedical Sciences,
University of Minnesota, 295 Animal Science/ Veterinary Medicine,
1988 Fitch Avenue, St. Paul, MN 55108-6010, USA
brown013@umn.edu

Dr. Karl V. Clemons
California Institute for Medical Research, 2260 Clove Drive, San Jose,
CA 95128, USA
clemons@cimr.org

Dr. Scot E. Dowd
Director, Research and Testing Laboratory, 4321 Marsha Sharp Freeway,
Lubbock, TX 79407, USA
sdowd@pathogenresearch.org

Dr. Primrose P.E. Freestone
Department of Infection, Immunity and Inflammation, School of Medicine,
University of Leicester, University Road, Leicester LE1 9HN, UK
ppefl@le.ac.uk

Dr. Benedict T. Green
Research Pharmacologist, Agricultural Research Service,
United States Department of Agriculture, 1150 E. 1400 N. Logan,
UT 84341, USA
ben.green@ars.usda.gov

Dr. Richard Haigh
Department of Genetics, University of Leicester, University Road,
Leicester LE1 7RH, UK
rxh@le.ac.uk

Aruna Jahoor
Department of Cell Biology and Biochemistry, Texas Tech University
Health Sciences Center, 3601 4th Street, MS 6540, Lubbock, TX 79430, USA
aruna.jahoor@ttuhsc.edu

Prof. Mark Lyte
Department of Pharmacy Practice, School of Pharmacy,
Texas Tech University Health Sciences Center, 3601 4th Street, MS 8162,
Lubbock, TX 79430-8162, USA
mark.lyte@ttuhsc.edu

Dr. Cristiano G. Moreira
Department of Microbiology, UT Southwestern Medical Center,
5323 Harry Hines Blvd., Dallas, TX 75390-9048, USA
cristiano.moreira@utsouthwestern.edu

Dr. Anthony Roberts, BSc, BDS, FDS RCPS, FDS (Rest Dent) RCPS, PhD, FHEA
Senior Lecturer and Honorary Consultant in Restorative Dentistry,
School of Dentistry, The University of Manchester, Higher Cambridge Street,
Manchester, M15 6FH, UK
anthony.roberts@manchester.ac.uk

Dr. Kathleen S. Romanowski
Department of Surgery, University of Chicago Medical Center,
5841 South Maryland Avenue, MC 6040, Chicago, IL 60637, USA
kathleen.romanowski@uchospitals.edu

Prof. Victoria V. Roshchina
Institute of Cell Biophysics, Russian Academy of Sciences Pushchino,
Moscow Region 142290, Russia
roshchinavic@mail.ru

Dr. Kendra Rumbaugh
Assistant Professor, Department of Surgery, Texas Tech University Health
Sciences Center, 3601 4th Street, MS 8312, Lubbock, TX 79430, USA
kendra.rumbaugh@ttuhsc.edu

Dr. Sara Sandrini
Department of Infection, Immunity and Inflammation, University of Leicester,
School of Medicine, University Road, Leicester LE1 9HN, UK
sms26@le.ac.uk

Dr. Jata Shankar
California Institute for Medical Research, 2260 Clove Drive, San Jose,
CA 95126, USA
jata@stanford.edu

Dr. Neil Shearer
Campylobacter Research Group, Institute of Food Research,
Colney Lane, Norwich, Norfolk, NR4 7UA, UK
neil.shearer@bbsrc.ac.uk

Dr. Vanessa Sperandio
Associate Professor of Microbiology, Department of Microbiology,
UT Southwestern Medical Center, 5323 Harry Hines Blvd., Dallas,
TX 75390-9048, USA
vanessa.sperandio@utsouthwestern.edu

Dr. David A. Stevens
Division of Infectious Diseases, Department of Medicine, Santa Clara Valley
Medical Center, Rm. 6C097, 751 S. Bascom Avenue, San Jose,
CA 95128-2699, USA
stevens@stanford.edu

Prof. Mark P. Stevens
Division of Microbiology, Institute for Animal Health, Compton,
Newbury, Berkshire, RG20 7NN, UK
mark-p.stevens@bbsrc.ac.uk

Dr. Cordula Stover
Department of Infection, Immunity and Inflammation, School of Medicine,
University of Leicester, University Road, Leicester LE1 9HN, UK
cms13@leicester.ac.uk

Dr. Nicholas J. Walton
Institute of Food Research, Norwich Research Park, Colney,
Norwich NR4 7UA, UK
nichjwalton@aol.com

Dr. Simon C. Williams
Associate Professor, Department of Cell Biology & Biochemistry,
Texas Tech University Health Sciences Center, 3601 4th Street,
MS 8326, Lubbock, TX 79430, USA
simon.williams@ttuhsc.edu

Dr. Olga Zaborina
Research Associate (Associate Professor), Laboratory for Surgical Infection
Research and Therapeutics, Department of Surgery, University of Chicago,
Medical Center 5841 S. Maryland Avenue, MC 5032, Chicago, IL 60637, USA
ozaborin@surgery.bsd.uchicago.edu

Dr. Alexander Zaborin
Laboratory for Surgical Infection Research and Therapeutics,
Department of Surgery, University of Chicago 5841 S. Maryland Avenue,
MC 5032, Chicago, IL 60637, USA
azaborin@surgery.bsd.uchicago.edu

Chapter 1
Microbial Endocrinology: *A Personal Journey*

Mark Lyte

1.1 Introduction

The development of the field of microbial endocrinology has now spanned nearly 18 years from the time I first proposed its creation in 1992 (Lyte 1992; Lyte and Ernst 1992; Lyte 1993). During the intervening time, this interdisciplinary field has experienced two of the characteristics of a typical microbial growth curve: a long lag phase during which acceptance of articles into mainstream journals was problematic to say the least, followed by an early log phase of growth characterized by increasing awareness that the intersection of microbiology, endocrinology, and neurophysiology offers a unique way to understand the mechanisms underlying health and disease. This book, I am happy to report, comes at the start of that early log phase with the multitude of possibilities for future rapid growth.

As is more common than many are often wont to admit, the development of a new discipline does not occur in a vacuum. And, if one chooses to look hard enough, one can usually find reports dating back many decades which document many of the experimental findings that help form the founding tenets of the new discipline. Given the history of use of neuroactive substances in the treatment of human disease, which provided for ample opportunity for the interaction of microorganisms with neuroendocrine hormones, it is not surprising that such is the case in microbial endocrinology (Lyte 2004). It is a testament to how the prevailing notions of what separates "us" from "them" can influence scientific inquiry that scientists of a bygone era did not fully recognize the ability of a lowly bacterium to produce and recognize substances that are more commonly thought of as defining a mammal (i.e., vertebrate nervous system) could prove critical to both health and disease. Indeed, J.A. Shapiro (2007) may have put it best by titling, in part, his recent article covering his nearly 40-year career observing the unique growth patterns of bacteria: "Bacteria are small, not stupid."

M. Lyte (✉)
Department of Pharmacy Practice, School of Pharmacy, Texas Tech University
Health Sciences Center, 3601 4th Street, MS 8162, Lubbock, TX 79430, USA
e-mail: mark.lyte@ttuhsc.edu

M. Lyte and P.P.E. Freestone (eds.), *Microbial Endocrinology*,
Interkingdom Signaling in Infectious Disease and Health,
DOI 10.1007/978-1-4419-5576-0_1, © Springer Science+Business Media, LLC 2010

1.2   From Psychoneuroimmunology to Microbial Endocrinology

1.2.1   Theoretical Reflections

During the late 1980s to the early 1990s, I, as well as many others, were involved in the examination of the ability of stress to affect immune responsiveness (Peterson et al. 1991). The field of psychoneuroimmunology (PNI), founded by Robert Ader and Nicholas Cohen in 1975 (Ader and Cohen 1975; Ader et al. 1995), was just emerging from its infancy into mainstream thinking. That most ambiguous of biological terms, stress was taking center stage not only in scientific thought but also in the public's perception of immediate, potentially controllable factors that determined health and well-being. Both in the scientific and public spheres, stress has for many decades been negatively associated with health in general. The demonstration that psychological stress could impact the generation of an immune response (Ader et al. 1995), coupled with reports which showed neural innervation of immune organs such as the spleen (Felten et al. 1990), led to the realization that two seemingly disparate disciplines, one immune and one neural, interacted with each other and that interaction was critical in homeostasis and disease. While the need for such interdisciplinary research is well-recognized today as intrinsic to the study of health (witness the priority of interdisciplinary funding initiatives from the National Institutes of Health), the obviousness of such an approach was less evident in the 1970s to the early 1990s. I can vividly recall the heated discussions at conferences among leaders in the immunology field arguing against the inclusion of neural or endocrinological factors (and certainly not something such as psychological, which was not as scientifically rigorous as "hard" science) in the study of immune responsiveness. The advent of increasingly accessible molecular biological tools was beginning to make inroads into deciphering the mechanisms governing the generation of an immune response. While immunological pathways had in the past been deciphered through the study of cell to cell interactions, molecular biological tools afforded a new way to examine such pathways, and cellular immunology began to yield to molecular immunology (with the attendant changes in departmental names). The argument by many of these immunological leaders was that with these new tools we were just beginning to understand the complexity of the immune system, and to add on to that the complexity of the neural and endocrinological systems, let alone the even more unknowable psychological factors, would be scientifically "unwise" (actually, more descriptive terms were used at the time) and impede progress. Once we understood the molecular mechanisms governing the immune system, as was the mainstream consensus at the time, only then should we tackle any interactions between different disciplines.

The recounting of the beginnings of PNI is relative to the origins of microbial endocrinology for a number of reasons. First, and foremost, the realization that an interdisciplinary approach was needed if a fuller understanding of the mechanisms that govern immune responsiveness in the host was to be achieved. No one biological system that operates in isolation of another may, on the face of it, be self-evident today; such was not the case even a quarter of a century ago. Since immunological phenomena, such as the production of antibodies, could occur in a completely in vitro setting (e.g., Mishell–Dutton culture) where no brain or endocrinological organs are present, why should the products of such systems, i.e., neuroendocrine hormones, be needed for an immune response? Thus, the predominant reasoning was that the immune system was a free-standing biological system that could operate in the absence of any other system. The recognition of neuroimmune inter-actions as being critical to the development and maintenance of immune responsiveness in an individual can best be seen in the emergence of PNI and the associated neuroimmunology-related field over the past two decades (Irwin 2008).

In many ways, microbial endocrinology has gone, and continues to go, through similar growing pains as that experienced by PNI. Cannot bacteria grow and be studied in vitro in the absence of any nervous or endocrinological components? Is such a question no different contextually from that which immunologists once asked of the relevance of neurohormones to the study of immunology? One of the "dirty little secrets" of the time in immunology was that the ability to demonstrate in vitro immunological phenomena, such as the generation of antibodies in a Mishell–Dutton system, and hence the independence of immunology from other biological systems, was that multiple lots of a key media component needed to be first screened to find the one "magical" lot that worked best. Once that lot was identified, multi-liter shipments would be ordered and stored for future use. That key media ingredient was fetal bovine serum, which by itself is a rich compendium of neuroendocrine hormones. The realization that endocrine components were nec-essary to even immunological phenomena, such as antibody formation, underscored the need to study the role of such neuroendocrine influences in the individual.

That the implications of such a connection between media components and sus-tainability of a biological reaction was not fully recognized at the time is immediately applicable to microbiology and is best illustrated by the response engendered the first time the microbial endocrinology concept was presented at a scientific meeting. At the 1992 American Society of Microbiology 92nd General Meeting in New Orleans, I gave a 10 min slide presentation entitled "Modulation of gram-negative bacterial growth by catecholamines" (Lyte and Freestone 2009; Mullard 2009). By the time I presented as last speaker in the session, there were only two people in the audience and, the two session chairs, one of which was a well-known chair of a large microbiology department. After speaking for about 2 min about the presence of neuroendocrine hormones in bacteria and the need for an interdisciplinary approach to understanding the pathogenesis of infectious disease, one of the audience members left leaving only a solitary person in a room meant for a few hundred people. That audience member happened to be my laboratory technician, Sharon Ernst, who was a co-author on my second microbial endocrinology-related paper. At the finish of my talk, one of the chairs (not knowing I was lecturing to my own technician) evidently felt duty bound due to the presence of an audience member to ask a question, which (to paraphrase) was "why would anyone want to grow bacteria in a serum containing medium containing hormones when such good rich media exist such as tryptic soy broth and brain heart infusion." My answer (again paraphrasing from memory) was simple and still encapsulates one of the underlying tenets that have driven the creation of microbial endocrinology: "…because we do not have tryptic soy broth and brain heart infusion media floating through our veins and arteries and until we use media that reflects the same environment that bacteria must survive in, then we will never fully understand the mechanisms underlying the ability of infectious agents to cause disease."

1.2.2 Experimental Observations Leading to Microbial Endocrinology

The involvement of PNI in the creation of microbial endocrinology went far beyond the theoretical aspects described above. By 1992 I had obtained my first NIH grant which embodied a PNI approach examining the mechanisms by which stress could affect susceptibility to infectious disease. Although stress had been well recognized to affect susceptibility to infections for nearly 100 years (Peterson et al. 1991), I sought to identify relevant immune-based mechanisms through the use of the ethologically-relevant stress of social conflict (Fig. 1.1), instead of the more artificial stressors such as restraint stress or electric shock, which did not

Fig. 1.1 Social conflict in mice is conducted by the simple placement of a group-housed mouse also known as an "intruder" (*black*, C57BL/6J male) into the cage of a singly-housed mouse, also known as the "resident" (*white*, CF-1 male). The resident will engage the intruder ultimately resulting in the "defeat" of the intruder as shown by the limp forepaws and angled ears. Once the intruder assumes the defeat posture, the resident then disengages and at this point the intruder is removed. The social conflict procedure is done under reversed day-night light cycle using low level red light for illumination. For a fuller description of social conflict procedure see Lyte et al. (1990b) and Miczek et al. (2001)

reflect any sort of stress that an animal would have any evolutionary experience (Miczek et al. 2001). Among my early findings was that social conflict stress induced an *increase* in those immune functions, notably phagocytosis, that constitutes a first line of defense against infection (Lyte et al. 1990b). From an evolutionary perspective, the finding of increased immune responsiveness against infection made perfect sense. If an animal is wounded, then bacterial infection would almost certainly be encountered. It made little sense from the animal's perspective to have immune responsiveness decreased at a time that it was presented with an infectious challenge to its survival. What would be needed during this time of acute stress would be heightened immune activity, which was what the social conflict study had shown.

However, this surprising result presented a paradox. If immune responsiveness is increased during time of acute, ethologically-relevant stress, then why is the animal more susceptible to an infectious challenge? Most of the literature over the last century had indeed shown that stressed animals did exhibit increased susceptibility to infectious disease challenge (Peterson et al. 1991). With that in mind, I conducted a series of experiments during 1991–1992 in which social conflict stressed animals were challenged with oral pathogens such as *Yersinia enterocolitica*. The results of those experiments showed the surprising result of increased mortality in stressed animals as compared to home cage controls (Fig. 1.2). Should not these animals, which showed greater than a 500% increase in phagocytic capacity

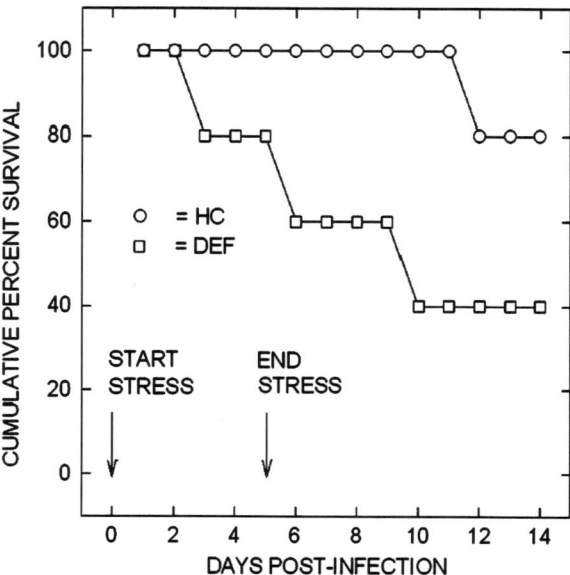

Fig. 1.2 Animals were per orally challenged with *Y. enterocolitica* immediately prior to social conflict stress (DEF, defeated, *squares*) or only handling and transport into procedure room (HC, home cage controls, *circles*). The stress or handling was conducted once per day for 5 days and percent survival followed for 14 days

(Lyte et al. 1990a, b), also display increased resistance to infectious challenge and not the increased mortality (Fig. 1.2)?

It was these sets of experiments during 1991–1992 that led me to reconsider the whole concept of stress and susceptibility to infectious disease not from the perspective of the animal but from that of the infecting bacterium. For a number of reasons, the infecting organism is as highly stressed, if not more so, than the stressed host. First, most infectious agents, such as food-borne pathogens, have survived food preservation and cooking steps that result in a damaged cellular state. Upon entrance into the host, the infecting bacterium must survive the host's physical defenses such as stomach acid and then survive and proliferate within the gastrointestinal tract amid the trillions of indigenous bacteria, which rigorously maintain ecological balance among various species through means including, for example, the elaboration of bacteriocins (Riley and Wertz 2002). Central among the factors that influence the ability of any infecting microbe to survive in a host is the capacity to recognize its environment and then employ that information to initiate pathogenic processes (i.e., adherence onto epithelium) and proliferate. The central question then became, what host-derived signals would be available to an infecting bacterium that could be used to the bacterium's own advantage and ultimately survival within the host? It was at this point that I made the decision to eliminate (for the time being) the role of immunology in addressing the effect of stress on the pathogenesis of infectious disease and instead to concentrate on the role of stress on the infecting bacterium within the hostile environment of the host. In other words, were there direct effects of the *host's* stress response on the bacterium?

Critical to the above line of reasoning was an overlooked phenomenon of infectious disease as experienced in nature (real world) as opposed to the laboratory. That aspect specifically concerns the dose of infectious organisms that are needed to effect overt disease in the host. It is well established in food microbiology that the number of infecting organisms needed to cause food-related gastrointestinal infection can be as low as ten bacteria per gram of food (Willshaw et al. 1994). However, in the laboratory, the challenge of animals with infectious bacteria can well go as high as 10^{10-11} bacteria or colony forming units (CFU) per ml. Further adding to this discrepancy between real world and laboratory infectious doses is that, on average, a mouse weighs 20–25 g while a human weighs 70 kg, meaning that the dosage a laboratory animal receives is many-fold greater than what is experienced by an individual. Over the last century, a number of investigators have raised the issue of whether non-ecologically relevant doses of infectious organisms can provide complete understanding of the mechanisms that underlie the pathogenesis of infectious disease in vivo (Smith 1996). In a similar fashion, this same question can also be raised regarding in vitro studies, which utilize high ($>10^4$ CFU per ml) bacterial inoculums. Not unlike the question of how a *single* individual may respond to a new environment as compared to how a large *group* of individuals may respond to the same new environment, the survival behavior of low numbers of bacteria within the new environment of the gastrointestinal tract may radically differ from that of large numbers of bacteria. This social aspect of bacterial behavior represents the newly emerging field of sociomicrobiology (Parsek and Greenberg 2005; West et al. 2006). Specifically, the environmental signals that *single* or low numbers of bacteria may look for markedly differ from that sought by high numbers of bacteria. And in addition to the above point of low, not high, numbers of bacteria that contaminate food, this also applies to the vast majority of infections in general in which infecting doses of bacteria are small ($<10^4$ CFU) in number.

Thus, from the outset, one of the guiding principles in microbial endocrinology has been the use of low bacterial numbers ($1–10^3$ CFU per ml) coupled with a medium that is reflective of the in vivo milieu. Other guiding principles, such as the combination of neuroendocrine hormones and bacteria under study should be matched such that each is found to occur in the same anatomical region in vivo, have also been formulated. In addition to the chapters contained in the present book, the reader is further directed to a recent comprehensive review, which thoroughly discusses the methodological aspects of conducting microbial endocrinology-related experiments (Freestone and Lyte 2008).

The choice of the initial neuroendocrine hormones for the first experiment was based on the stress response itself and the well-known increase in catecholamines (Gruchow 1979; Woolf et al. 1992). Further, the stress-induced release of catecholamines had been one of the primary mechanisms that had been proposed in PNI-related research to account for the ability of stress to suppress immune responsiveness, and hence increase susceptibility to an infectious challenge (Ader and Cohen 1993; Webster Marketon and Glaser 2008). As has been recognized for many decades, the induction and sustained release of the catecholamines, especially norepinephrine, occurs during many forms of stress extending from psychological to surgical (Fink 2000). The Gram-negative bacterium, *Y. enterocolitica*, was chosen as the first bacterium to test whether a neuroendocrine hormone, namely norepinephrine, could have *direct* effects on growth. The results of this initial experiment in 1991, which was carried out in liquid culture using small 60 mm Petri dishes, combined a low inoculum of *Y. enterocolitica* (33 CFU per ml of a serum-supplemented minimal medium) with norepinephrine, epinephrine, or diluent (Fig. 1.3). In many ways, this experiment, which was the proverbial "shot in the dark," is the one that has led through many years to the creation of this current book. As shown in Fig. 1.3, there is a very small amount of visual growth evident in both the control and epinephrine supplemented plates (indicated by arrows). However, in the norepinephrine supplemented culture, there is dense growth throughout. To this day, I still remember my excitement at seeing these results. And from that day on, I effectively ceased looking at PNI-related phenomena and instead turned my research direction to the study of neuroendocrine–bacterial interactions and the creation of the field of microbial endocrinology.

1.2.3   Gaining Acceptance of Microbial Endocrinology

As can often be the case in any endeavor that seeks to introduce a paradigm-shift in thinking, the introduction of neuroendocrine–bacterial interactions as a hitherto unrecognized mechanism in the pathogenesis of infectious disease was met not only with initial skepticism, but also downright hostility. At a mid-1990s meeting

GROWTH OF Y. ENTEROCOLITICA

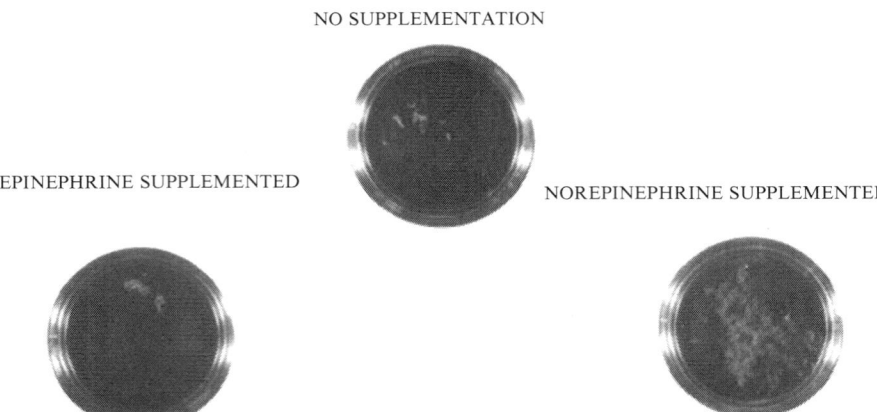

NO SUPPLEMENTATION

EPINEPHRINE SUPPLEMENTED

NOREPINEPHRINE SUPPLEMENTED

Fig. 1.3 The experiment that launched the field of microbial endocrinology. *Yersinia enterocolitica* culture plates in 1991 showing that bacterial growth in serum-based medium was enhanced in the presence of the neuroendocrine stress hormone norepinephrine, but not epinephrine or control diluent

in Toronto that focused on the role of neuroendocrine mediators and immunity in drug addiction, I gave a microbial endocrinology-based lecture as part of a session on stress and its relationship to drug addiction sequelae such as increased prevalence of infectious disease in drug addicts. At the conclusion of my talk before I could take any questions, the session chair addressed the audience and said that my ideas were so radical that they should not be taken seriously, and the audience should in essence forget what I just presented. More than one member of that audience has approached me over the years to recount that episode and the shock of the audience being told to disregard what they had just heard as well to ask why I did not get mad (which I did not). Such opposition, although admittedly more restrained, was also encountered during the early years in terms of gaining acceptance into the scientific literature. I have been told by more than one individual that the integration of microbial endocrinology into mainstream infectious disease research would have been accelerated if I had chosen to publish in more microbiology-oriented journals. However, my choice to publish in journals not typically read by microbiologists was dictated not by choice, but instead by necessity. My early attempts to publish in basic microbiology-based journals were universally met with rejection. Undoubtedly, while one may take the convenient road of blaming the reviewer for failure to consider a highly interdisciplinary approach where it is often not possible to address all the questions regarding each of the fields, I shall instead take a fair share of the blame since it is also the responsibility of the author to educate the reader of the need to go beyond traditional thinking.

With that said, I have also come to recognize that one of the defining reasons that these early papers were rejected from basic microbiology-centric journals was the reliance on phenomenology rather than mechanisms. My own training in the clinical laboratory sciences and subsequent work in hospital laboratories before entering graduate school in 1977, ingrained in me a powerful sense of the clinical side of microbiology. And that side is one that is grounded in growth, for without evident growth and sufficient numbers of bacteria, little can be done, even today, to diagnose suspected bacterial disease. Thus, it seemed to me at the time (and still does today) that the ability to show growth-related effects of neuroendocrine hormones on bacteria would have profound implications for the study of the host factors, which influence susceptibility to infectious disease. However, I was surprised that this was generally not the case. A similar refrain ran through those early reviews that the demonstration of effects on growth were phenomenological in nature and what was needed to be shown was the mechanism(s) by which neuroendocrine hormones could influence bacterial physiology. Due largely to the availability of an ever growing arsenal of molecular biological techniques, phenomenology was to be eschewed in favor of dissecting molecular mechanisms. While I do not mean to begrudge nor demean the value of mechanistic studies, one may argue that many of the advances in the treatment of disease have been made through the observation of phenomena for which no mechanism at the time of discovery was available. Antibiotic development owes itself largely to the observation of phenomena. While the requirement for molecular analyses currently reigns dominant in the majority of first-tier microbiology journals, the relegation of phenomenological studies to the status of second-class research ignores its historically pivotal role in fueling scientific and medical advances. A number of recent articles examining the failure of genomic-based strategies to lead to the discovery of new antimicrobials that ultimately make the transition from the lab bench to the clinic have addressed this very point (Barrett 2005; Finch 2007).

My reasoning for discussing the relative merits of phenomenology versus molecular analyses is not to point out my own shortcomings in the area of molecular analysis, but to offer a cautionary note to other researchers who may choose to explore microbial endocrinology. Catecholamines, which to date have been the principal neuroendocrine hormones that have been examined in the microbial endocrinology field by virtue of their prominence in the stress response, represent but a tiny sliver of the spectrum of neuroendocrine hormones that can be examined for potential interaction with both pathogenic as well as commensal bacteria. For example, gamma amino butyric acid (GABA), the primary inhibitory neurotransmitter in the mammalian brain, is produced in such large amounts by bacteria in the gut that a role for bacterial-derived GABA has been proposed to account for the altered organ function (encephalopathy) that is part of the pathogenesis of advanced liver disease and sepsis (Minuk 1986; Winder et al. 1988). In fact, GABA produced by bacteria, such as those contaminating a distilled water apparatus, have been found not only to confound neurotransmitter binding studies with mammalian cells (Balcar 1990), but also to possess a high affinity binding protein that resulted in one of the first bioassays for GABA that was entirely bacterial-based (Guthrie and Nicholson-Guthrie 1989; Guthrie et al. 2000). In this book, Chap. 2 by Victoria Roshchina provides an exhaustive review of the wide breadth of neurohormones that are found in prokaryotes that we otherwise only associate with multicellular eukaryotic systems.

By utilizing a microbial endocrinology approach, researchers can further our understanding of how host and bacteria, both commensal and pathogenic, interact in the gut (or at other sites). That approach, in turn, could provide insights into not only homeostasis but also other medical conditions that involve gut pathology that upon verification could enable the design of new, innovative medical interventions. Although researchers realized more than 100 years ago that the mammalian gut is innervated, how this system interacts with the gut microbial flora remains largely a mystery. Further, large amounts of neurochemicals are produced within the gut that find their way into the gut lumen where the possibility of interactions with the gut microflora exist and remain largely unexplored. For example, milligram quantities of serotonin are produced by the gut that can be recovered from the lumen, although the physiological reason for this production is not well understood. Could it be that serotonin produced by the mammalian gut has some hitherto unknown interaction with a specific part of microbial population? Thus, examination of any such sero-tonin–bacterial interaction will depend on both *phenomenology* and molecular analyses to provide as complete a picture as possible of the relevance of microbial endocrinology to both homeostasis and disease.

Further, the bidirectional nature of bacterial neuroendocrine interactions contained within the theory of microbial endocrinology also suggests that bacteria can influence mammalian function. More recent work utilizing metabolomics to compare the blood metabolic profile of conventional-reared and germ-free mice revealed that the gut microbiome contributed to the concentration of neuroactive components in the circulation (Wikoff et al. 2009). That the presence of a microbial community within the gut, and inherent interactions between the host and gut microflora, is crucial to an animal's neurological health was demonstrated in 2004 when Nobuyuki Sudo and colleagues at Kyushu University in Japan examined the role of microbial colonization on the hypothalamic–pituitary–adrenal response to stress in gnotobiotic, germ-free, and conventionally-reared mice (Sudo et al. 2004). Not only did the development of host neural systems that control the physiological response to stress depend on postnatal microbial colonization of the gut, but also reconstitution of gnotobiotic mice with feces from specific pathogen-free mice altered their subsequent neurohormonal stress response. And more recently, Li et al. (2009) at Texas Tech University showed that diet-induced alteration of gut microbial diversity can even affect memory and learning in mice. Thus, we are just beginning to understand the degree to which microbial diversity is crucial not only to the development and regulation of normal gastrointestinal function, but also to how it may interface with the host's neurophysiological system.

And crucially, does this bidirectional nature of bacterial-neuroendocrine interactions contained within microbial endocrinology also imply that gut neuronal activity may as well influence local bacterial ecology and vice versa? In attempting to answer this question, a new hypothesis has recently been proposed based on the ability of bacteria as well as other microbes to both recognize and synthesize neuroendocrine hormones. According to this hypothesis, the microbiota within the intestinal tract comprise a community that interfaces with the mammalian nervous system that innervates the gastrointestinal tract to form a microbial organ which enters into a symbiotic relationship with its mammalian host that is governed by bacterial-neuroendocrine (microbial endocrinology) interactions (Lyte, 2009a). The consideration of a microbial organ within the gut that interfaces with the host through microbial endocrinology-based interactions involving the host's nervous system provides for a new paradigm with which to understand and design new therapeutic approaches for a range of clinical diseases.

1.3   Collaboration and Dissemination

A critical point in the development of microbial endocrinology turned out to be a fortuitous meeting at the 1995 First International Rushmore Conference on Mechanisms in the Pathogenesis of Enteric Diseases, held in Rapid City, South Dakota. Following my presentation, I was approached by a bearded and pony-tailed Richard Haigh, at the time a Ph.D. student at Leicester University in the United Kingdom (and the same Dr. Haigh who is the author of Chap. 16). Richard's interest in my work served as the bridge to Dr. Primrose Freestone who was a postdoctoral fellow in the same lab in the then Department of Microbiology and Immunology within the medical school. Since that time, my association with Primrose Freestone has been, and continues to be, instrumental in furthering the microbial endocrinology field.

During this time, I also consider myself fortunate to have helped interest other investigators to examine bacterial–neuroendocrine interactions. For example, during microbiology meetings both in the United States and in Japan at which I and my colleagues had presentations, I was approached by James Kaper of the University of Maryland and his then postdoctoral fellows Jorge Girón and Vanessa Sperandio. Following in-depth conversations with them regarding my concept of bacteria recognizing hormones and the potential of microbial endocrinology, as well as instructions on how to design experiments utilizing neurochemicals, I sent a shipment from my lab of neurochemicals to help with initial experiments. Nearly 2 years from the time of my first discussion with James Kaper, Vanessa Sperandio and her colleagues published their landmark paper in PNAS demonstrating the ability of catecholamines to influence quorum sensing in *Escherichia coli* O157:H7 (Sperandio et al. 2003).

It is axiomatic that the development of any newly emerging field must of necessity rely on the willingness of researchers to freely exchange ideas. Often, such exchange is either not rewarded with any acknowledgement or, at times, proper citation of papers that were instrumental in the founding of the field. As seems to be increasingly the case in science in general, this may be an unfortunate sign of the current times. Regardless, the further theoretical and experimental development of any field must continue to rely on the free exchange of ideas even though the practical aspects of such scientific exchange often do not match the hoped for ideal embodiment of such ideals, especially the regard for proper citation.

Probably one of the greatest fears when one believes to have discovered a new discipline is whether one is really the first. Nearly 5 years following the 1992 publication of my initial papers that helped lay the foundation for microbial endo-crinology (Lyte 1992; Lyte and Ernst 1992; Lyte 1993) I became aware of a refer-ence to one of the forgotten secrets of microbiologists that one way to get *Clostridium perfringens* to establish in mice, in order to evaluate potential antimi-crobial agents in the treatment of gas gangrene, was to co-inject epinephrine with the *C. perfringens* (Traub et al. 1991). This revelation hit me like a thunderbolt that my fear of not properly citing those before me had been realized. What followed turned out to be a journey into a rich history documenting the ability of neuroendo-crine hormones to influence the pathogenesis of infectious disease and the many attempts that have been made to understand the phenomena (the majority of these studies have been referenced in my 2004 review, (Lyte 2004). In retrospect, given the usage of neuroendocrine hormones, such as the catecholamines, dating as far back as 1930 in the treatment of clinical conditions ranging from urticaria (itching) (Renaud and Miget 1930) to their current prominent role in the maintenance of cardiac and kidney function in critically-ill patients (Singer 2007), it should not come as that much of a surprise that there has been ample opportunity for the obser-vation of neuroendocrine–bacterial interactions in infectious disease. The study of this past literature is highly instructive not simply as only an historical narrative, but also as importantly as a guide to future research design from a translational medicine viewpoint since it provides evidence of the role of microbial endocrinol-ogy both in health and disease in both animals and humans.

1.4   Whither Microbial Endocrinology

Predictions concerning future progress, such as the usual self-serving admonitions about the inevitability of discoveries leading to improvement of the human condi-tion, are fraught with peril. Whatever the direction that microbial endocrinology will take will depend on a myriad of factors.

Among the most crucial of factors will be experimental design. Although I, and my colleague and co-editor Primrose Freestone, have written extensively on this subject before (Freestone and Lyte 2008), there are some points that bear repeating. First and foremost, the choice of environmental conditions, especially those that concern in vitro culture, namely the media with which is chosen to evaluate potential microbiological response(s) to neuroactive agents, will remain the single most important determining criteria to evaluate potential interactions between microbes and neuroactive compounds. The use of rich microbiological media is likely a self-defeating way to approach evaluation of neuroendocrine–microbial interactions. If the microorganism has everything it needs for survival and proliferation already present in the rich media, why would there be any need to actively seek those environmental signals, such as neuroendocrine hormones, in order to establish where they are and then initiate processes needed for sur-vival in nutrient-limited environments that are not akin to in vitro incubation in rich media? The answer is that it is highly unlikely that one would obtain the same response from a microorganism that is exposed to a neurochemical in a rich medium as compared to a nutrient-limited medium.

This above in vitro caveat applies equally to the design of in vivo experiments. As discussed in more detail in the first published review on the field of microbial endocrinology (Lyte 2004), the non-peer reviewed study, which had reported the dramatic ability of norepinephrine to increase the in vitro growth of *Salmonella choleraesuis*, could not be replicated in an in vivo animal model in which the authors had implanted a norepinephrine-containing pellet subcutaneously on the back and weeks later infected the animals via an intraperitoneal route with bacteria (Nietfeld et al. 1999). As discussed in the 2004 review (Lyte 2004), the authors of the 1999 study (Nietfeld et al. 1999) made a number of incorrect experimental design decisions regarding rather standard animal physiology including, but not limited to, assuming that the parenteral administration of a specific hormone will result in a constant defined concentration within any one specific organ or organ system over a period of weeks and further that the body over that time frame will not seek to adapt to elaboration of the hormone by increasing the production of specific degradative enzymes (for that very same neurohormone) whose pathways have been well characterized over the past decades. A simple consideration of the anatomical distance from the peritoneal cavity (into which the bacteria were injected) to the subcutaneous space on the back of the neck (into which the norepinephrine release pellet was inserted), and the body's immediate system-wide response of increased enzymatic degradation of circulating catecholamines, it would be rather problematic to interpret any changes in bacterial growth rates in terms of any neuroendocrine influence. And since the issue of environmental context is pre-eminent in any discussion of microbial endocrinology, would it not have been more physiologically relevant to have evaluated the ability of a stress-related neuroendocrine hormone to affect the growth of *S. choleraesuis* in the gut, which is the natural site of infection with this organism, rather than the peritoneum?

Microbial endocrinology in many ways represents an interdisciplinary discipline that seeks to understand the role of microbes in health and disease that is driven fundamentally by an evolutionary approach (see Chap. 2 for a discussion of the ubiquitous distribution of neuroactive compounds throughout nature). Although much has been made of the use of the term interdisciplinary (and the many variants such as multidisciplinary), the "take home" message may be that in approaching a study of neuroactive compounds in microbiology, microbial endocrinology represents an opportunity for true interaction of microbiologists with individuals from disciplines not readily associated with the study of microbial processes, such as physiologists, endocrinologists, and behaviorists, to name a few. The accumulated evidence which suggests the ability of the gut microbiota to play a role not only in host neural development and function, but in overall behavior as well, has recently been addressed in an excellent review (Forsythe et al. 2010). The utilization of microbial endocrinology-based approaches to understand the etiology of clinical neurological disease that has almost exclusively to date been thought of primarily in neurological terms, has also recently been proposed. For example, the ability of the gut bacterium *Helicobacter pylori* to utilize L-dopa for its own growth has been proposed as a microbial endocrinology-based mechanism to explain the benefit of *H. pylori*-directed antibiotic therapy to improve L-dopa bioavailibility in Parkinson's disease patients and thereby improve drug therapy management (Lyte, 2009b). Additionally, reciprocal gut-brain interactions that occur as a result of postinjury trauma and that are based on evolutionary microbial endocrinology-based constructs has been proposed as a new mechanism by which to both understand, and thereby treat, the severe condition of systemic inflammatory response syndrome that can occur in traumatically-injured patients (Lyte 2009c).

And finally, as the oft-used cliché goes "this is not the end of the story, just the beginning." Recently, I received a call from my colleague in the PNI field, which as described earlier in this chapter, helped shape my ideas concerning microbial endocrinology. In recalling my talks from nearly 18 years ago at which I presented the first evidence for the role of neurohormones in the ability of stress to affect the course of bacterial infection and the somewhat rocky road that I traveled since that time in terms of recognition by the scientific community, he commented that my days of "wandering in the wilderness" appeared to be over. If that is indeed the case, as I do hope the contents of this book will attest to, then I must add to that sentiment that it has only been with the help of my collaborators, students, and technicians over the years, and with a special mention of my longtime collaborator Primrose Freestone in whom one could not ask for a better scientific colleague, that it has been possible to make that journey.

References

Ader, R., and Cohen, N. 1975. Behaviorally conditioned immunosuppression. Psychosom Med 37:333–340.

Ader, R., and Cohen, N. 1993. Psychoneuroimmunology: conditioning and stress. Annu Rev Psychol 44:53–85.

Ader, R., Cohen, N., and Felten, D. 1995. Psychoneuroimmunology: interactions between the nervous system and the immune system. Lancet 345:99–103.

Balcar, V. J. 1990. Presence of a highly efficient "binding" to bacterial contamination can distort data from binding studies. Neurochem Res 15:1237–1238.

Barrett, J. F. 2005. Can biotech deliver new antibiotics? Curr Opin Microbiol 8:498–503.

Felten, D. L., Felten, S. Y., Sladek, J. R., Jr., Notter, M. D., Carlson, S. L., Bellinger, D. L., and Wiegand, S. J. 1990. Fluorescence histochemical techniques for catecholamines as tools in neurobiology. J Microsc 157 (Pt 3):271–283.

Finch, R. 2007. Innovation – drugs and diagnostics. J Antimicrob Chemother 60 Suppl 1:i79–i82.

Fink, G. 2000. Encyclopedia of stress. Academic, San Diego, CA.

Forsythe, P., Sudo, N., Dinan, T., Taylor, V.H., and Bienenstock, J. 2010. Mood and gut feelings. Brain Behav Immun 24:9–16.

Freestone, P. P., and Lyte, M. 2008. Microbial endocrinology: experimental design issues in the study of interkingdom signalling in infectious disease. Adv Appl Microbiol 64:75–105.

Gruchow, H. W. 1979. Catecholamine activity and infectious disease episodes. J Human Stress 5:11–17.

Guthrie, G. D., and Nicholson-Guthrie, C. S. 1989. gamma-Aminobutyric acid uptake by a bacterial system with neurotransmitter binding characteristics. Proc Natl Acad Sci USA 86:7378–7381.

Guthrie, G. D., Nicholson-Guthrie, C. S., and Leary, H. L., Jr. 2000. A bacterial high-affinity GABA binding protein: isolation and characterization. Biochem Biophys Res Commun 268:65–68.

Irwin, M. R. 2008. Human psychoneuroimmunology: 20 years of discovery. Brain Behav Immun 22:129–139.

Li, W., Dowd, S., Scurlock, B., Acosta-Martinez, V., and Lyte, M. 2009. Memory and learning behavior in mice is temporally associated with diet-induced alterations in gut bacteria. Physiol Behav 96:557–567.

Lyte, M. 1992. The role of catecholamines in gram-negative sepsis. Med Hypotheses 37:255–258.

Lyte, M. 1993. The role of microbial endocrinology in infectious disease. J. Endocrinol. 137:343–345.

Lyte, M. 2004. Microbial endocrinology and infectious disease in the 21st century. Trends Microbiol 12:14–20.

Lyte, M., and Ernst, S. 1992. Catecholamine induced growth of gram negative bacteria. Life Sci. 50:203–212.

Lyte, M., and Freestone, P. 2009. Microbial endocrinology comes of age. Microbe 4:169–175.

Lyte, M. 2009a. The microbial organ in the gut as a driver of homeostasis and disease. Med Hypotheses, Nov 8 [Epub ahead of print].

Lyte, M. 2009b. Microbial endocrinology as a basis for improved l-DOPA bioavailability in Parkinson's patients treated for *Helicobacter pylori*. Med Hypotheses, Dec 2 [Epub ahead of print].

Lyte, M. 2009c. Reciprocal gut-brain evolutionary symbiosis provokes and amplifies the postinjury systemic inflammatory response syndrome. Surgery 146:950–954.

Lyte, M., Nelson, S. G., and Baissa, B. 1990a. Examination of the neuroendocrine basis for the social conflict-induced enhancement of immunity in mice. Physiol Behav 48:685–691.

Lyte, M., Nelson, S. G., and Thompson, M. L. 1990b. Innate and adaptive immune responses in a social conflict paradigm. Clin Immunol Immunopathol 57:137–147.

Miczek, K. A., Maxson, S. C., Fish, E. W., and Faccidomo, S. 2001. Aggressive behavioral phenotypes in mice. Behav Brain Res 125:167–181.

Minuk, G. Y. 1986. Gamma-aminobutyric acid (GABA) production by eight common bacterial pathogens. Scand J Infect Dis 18:465–467.

Mullard, A. 2009. Microbiology: Tinker, bacteria, eukaryote, spy. Nature 459:159–161.

Nietfeld, J. C., Yeary, T. J., Basaraba, R. J., and Schauenstein, K. 1999. Norepinephrine stimulates in vitro growth but does not increase pathogenicity of *Salmonella choleraesuis* in an in vivo model. Adv Exp Med Biol 473:249–260.

Parsek, M. R., and Greenberg, E. P. 2005. Sociomicrobiology: the connections between quorum sensing and biofilms. Trends Microbiol 13:27–33.

Peterson, P. K., Chao, C. C., Molitor, T., Murtaugh, M., Strgar, F., and Sharp, B. M. 1991. Stress and the pathogenesis of infectious disease. Rev Infect Dis 13:710–720.

Renaud, M., and Miget, A. 1930. Role favorisant des perturbations locales causees par l' adrenaline sur le developpement des infections microbiennes. C R Seances Soc Biol Fil 103:1052–1054.

Riley, M. A., and Wertz, J. E. 2002. Bacteriocins: evolution, ecology, and application. Annu Rev Microbiol 56:117–137.

Shapiro, J. A. 2007. Bacteria are small but not stupid: cognition, natural genetic engineering and socio-bacteriology. Stud Hist Philos Biol Biomed Sci 38:807–819.

Singer, M. 2007. Catecholamine treatment for shock – equally good or bad? Lancet 370:636–637.

Smith, H. 1996. What happens in vivo to bacterial pathogens? Ann N Y Acad Sci 797:77–92.

Sperandio, V., Torres, A. G., Jarvis, B., Nataro, J. P., and Kaper, J. B. 2003. Bacteria–host communication: the language of hormones. Proc Natl Acad Sci USA 100:8951–8956.

Sudo, N., Chida, Y., Aiba, Y., Sonoda, J., Oyama, N., Yu, X., Kubo, C., and Koga, Y. 2004. Postnatal microbial colonization programs the hypothalamic-pituitary-adrenal system for stress response in mice. J Physiol 558:263–275.

Traub, W. H., Bauer, D., and Wolf, U. Virulence of clinical and fecal isolates of *Clostridium perfringens* Type A for outbred NMRI mice. Chemotherapy 37:426–435.

Webster Marketon, J. I., and Glaser, R. 2008. Stress hormones and immune function. Cell Immunol 252:16–26.

West, S. A., Griffin, A. S., Gardner, A., and Diggle, S. P. 2006. Social evolution theory for micro-organisms. Nat Rev Microbiol 4:597–607.

Wikoff, W., Anfora, A., Liu, J., Schultz, P., Lesley, S., Peters, E., and Siuzdak, G. 2009. Metabolomics analysis reveals large effects of gut microflora on mammalian blood metabolites. Proc Natl Acad Sci USA 106:3698–3703.

Willshaw, G. A., Thirlwell, J., Jones, A. P., Parry, S., Salmon, R. L., and Hickey, M. 1994. Vero cytotoxin-producing *Escherichia coli* O157 in beefburgers linked to an outbreak of diarrhoea, haemorrhagic colitis and Haemolytic uraemic syndrome in Britain. Lett Appl Microbiol 19:304–307.

Winder, T. R., Minuk, G. Y., Sargeant, E. J., and Seland, T. P. 1988. Gamma-aminobutyric acid (GABA) and sepsis-related encephalopathy. Can J Neurol Sci 15:23–25.

Woolf, P. D., McDonald, J. V., Feliciano, D. V., Kelly, M. M., Nichols, D., and Cox, C. 1992. The catecholamine response to multisystem trauma. Arch Surg 127:899–903.

Chapter 2
Evolutionary Considerations of Neurotransmitters in Microbial, Plant, and Animal Cells

Victoria V. Roshchina

2.1 Introduction

The "living" environment of a human includes microorganisms, plants, and animals as well as other human beings. The Relationship between them occurs via what is known as irritation events. The mechanism of irritability appears to have a common base in the form of chemical signals, chemicals which are uniform for every cell. Similar compounds likely to be found in living organisms include acetylcholine, dopamine, norepinephrine, epinephrine, serotonin, and histamine, collectively known as neurotransmitters, and have been found not only in animals (Boron and Boulpaep 2005), but also in plants (Roshchina 1991, 2001a; Murch 2006) and microorganisms (Hsu et al. 1986; Strakhovskaya et al. 1991; Lyte 1992; Oleskin et al. 1998a, b; Tsavkelova et al. 2006; Freestone and Lyte 2008). Thus, the presence of neurotransmitter compounds has been shown in organisms lacking a nervous system and even in unicellular organisms (Roshchina 1991, 2001a). Today, we have more and more evidence that neurotransmitters, which participate in synaptic neurotransmission, are multifunctional substances participating in developmental processes of microorganisms, plants, and animals. Moreover, their universal roles as signal and regulatory compounds are supported by studies that examine their role in and across biological kingdoms (Roshchina 1991, 2001a; Baluska et al. 2005, 2006a, b; Brenner et al. 2006). Any organism may release neurotransmitters, and due to these secretions (Roshchina and Roshchina 1993) the "living environment" influences every other inhabitant of biocenosis, determining relationships between organisms such as microorganism–microorganism, microorganism–plant, microorganism–animal, plant–animal, plant–plant, and animal–animal.

The universal character of their occurrence and similarity of functions at the cellular level should convince scientists to have doubt in the specific name "neurotransmitters" and exchange it, perhaps, for a more wider term such as "biomediators"

V.V. Roshchina (✉)
Institute of Cell Biophysics, Russian Academy of Sciences, Pushchino,
Moscow Region 142290, Russia
e-mail: roshchinavic@mail.ru

M. Lyte and P.P.E. Freestone (eds.), *Microbial Endocrinology*,
Interkingdom Signaling in Infectious Disease and Health,
DOI 10.1007/978-1-4419-5576-0_2, © Springer Science+Business Media, LLC 2010

to make it applicable to any living cell, not only organisms with nervous systems (Roshchina 1991, 2001a). The bioremediator concept permits us to imagine the evolutionary picture, where the neurotransmitter substances were participators of many different cellular processes, including non-synaptic systems of microorganisms and plants.

Non-nervous functions of the neurotransmitters (rather "biomediators" at the cellular level) are analyzed in this chapter, and their respective roles in the different evolutionary kingdoms are compared. This information gathered from species ranging from microorganisms to plants and animals may provide an insight into key problems in cellular endocrinology, and thereby has implications for understanding both health and disease causation. Such analysis can also undoubtedly provide useful perspectives to help guide the further development of the field of microbial endocrinology too.

2.2   Occurrence of Neurotransmitters in Living Organisms

2.2.1   Discoveries

Historical chronologies of the neurotransmitters' discoveries are represented in Table 2.1. The first neurotransmitters were the catecholamines found by the American scientist John Jacob Abel at the end of the nineteenth century in extracts from animal adrenal glands. During the years 1906–1914, the existence of neurotransmitter compounds were identified not only in animals, but also in fungal extracts which were used as medicinal preparations. The twentieth century was the epoch for the discovery of neurotransmitters, mainly by pharmacologists and animal physiologists related to medicine. The roles of the catecholamine compounds in plants and microorganisms became a subject of interest only after 50–70 years of the twentieth century. Pioneering studies included the investigations of Jaffe, Fluck, Riov, Stephenson, Rowatt, Girvin, and Marquardt (see references in monograph of Roshchina 2001a for more detail).

As can be seen from Table 2.2, the concentration range of the neurotransmitter compounds is similar for all three kingdoms of living organisms, although some organs and specialized cells of multicellular organisms may be enriched in these compounds. Over 40 years ago, Soviet physiologist Koshtoyantz (1963) presented a hypothesis that the neurotransmitters are peculiar to all animal cells independently of their position on the evolutionary tree; this view has been confirmed experimentally, and described in several monographs (Buznikov 1967, 1987, 1990) and review (Buznikov et al. 1996). The presence of the neurotransmitters in animals has now been confirmed for all taxa – from Protozoa to Mammalia. As for bacterial cells, no more than 10–12 species have so far been characterized as containing acetylcholine, catecholamines, and serotonin, although histamine has been found in most species of prokaryotes. The issue of mammalian-type hormones in microorganisms has also been considered (Lenard 1992).

Table 2.1 Discovery of neurotransmitters

Neurotransmitter	In microorganisms	In plants	In animals
Acetylcholine	Identified independently by Ewins and Dale (1914) in preparations of ergot spur fungus *Claviceps purpurea* in Great Britain and in bacteria *Pseudomonas fluorescens* (Chet et al. 1973)	In 1947 Emmelin and Feldberg found this substance in stinging trichomes and leaves of common nettle by biological method, based on muscle contraction	In 1921–1926 the presence of acetylcholine has been established in animals by Loewi and Navratil. But earlier, in 1906 student Reid Hunt (worked in USA laboratory of John J. Abel) discovered it in adrenal extracts of animals
Dopamine	Found in infusoria *Tetrahymena pyriformis* by Gundersen and Thompson (1985). Identified in bacterial and fungal microorganisms by Tsavkelova et al. (2000)	In 1944, found in *Hermidium alipes* by Buelow and Gisvold	Discovered in 1950–1952 by pharmacologists Arvid Carlsson, Nils-Åke Hillarp and von Euler in Sweeden
Norepinephrine (noradrenaline)	Identified in microorganisms by Tsavkelova et al. (2000)	In 1956–1958 found in banana fruits in Sweden laboratories organized by Waalkes and Udenfriend	Isolated from adrenal gland extracts of animals in 1897–1898 by John J. Abel
Epinephrine (adrenaline)		In 1972 found in leaves of banana *Musa* by Askar et al. (1972)	Isolated from adrenal gland extracts of animals in 1895 by Polish physiologist Napoleon Cybulski and in 1897 by American John J. Abel
Serotonin	Found in 1986 by Hsu with co-workers in many bacteria	Found in banana fruits (*Musa*) by Bowden et al. (1954)	Discovered by Erspamer in 1940 and Rapport et al. in 1948
Histamine	Found in ergot fungi *Claviceps purpurea* in 1910 by Barger, Dale and Kutscher	Observed in higher plants by Werle and Raub in 1948	In 1919 American John Jacob Abel isolated histamine from pituitary extract of animals

Sources: (Kruk and Pycock 1990; Roshchina 1991, 2001a; Kuklin and Conger 1995; Oleskin 2007; Kulma and Szopa 2007)

Table 2.2 Level of neurotransmitters in living organisms

Neurotransmitter	In microorganisms μg/g of fresh mass or * μmol/L or ** μg/billion of cells	In plants μg/g of fresh mass	In animals μg/g of fresh mass or * nM/L or ** nm/day
Acetylcholine	3.0–6.6	0.1–547	0.326–65200 (0.15–0.2 in brain)
Dopamine	0.45–2.13*	1–4000	<0.888*1214–2425**
Norepinephrine (noradrenaline)	0.21–1.87*	0.1–6,760	0.615–3.23*20–240**
Epinephrine (adrenaline)	No data	0.22–3833	1.9–2.46*30–80**
Serotonin	0.11–50,000**	0.0017–4000	0.21–0.96
Histamine	0.01–3.75	1 (1.34 – pain reaction for human)	0.5–100

Sources: (Fernstrom and Wurtman 1971; Kruk and Pycock 1990; Hsu et al. 1986; Roshchina 1991, 2001a; Oleskin et al. 1998a, b; Tsavkelova et al. 2000)

2.2.1.1 Acetylcholine

In animals, acetylcholine and/or the synthesizing enzyme choline acetyltransferase have been demonstrated in epithelial (airways, alimentary tract, urogenital tract, epidermis), mesothelial (pleura, pericardium), endothelial, muscle and immune cells, mainly in granulocytes, lymphocytes, macrophages, and mast cells (Wessler et al. 2001). Acetylcholine has also been found in Protozoa (Janakidevi et al. 1966a, b). Corrado et al. (2001) showed the synthesis of the molecular acetylcholine during the developmental cycle of *Paramecium primaurelia*. This neurotransmitter has a negative modulating effect on cellular conjugation. But in these unicellular organisms, the presence of functionally related nicotinic and muscarinic receptors and a lytic enzyme acetylcholinesterase has been established. Moreover, the authors could demonstrate (using immunocytochemical and histochemical methods) that the activity of enzyme choline acetyltransferase, which catalyzed acetylcholine synthesis, was located on the surface membrane of mating-competent cells and of mature, but not nonmating-competent *P. primaurelia* cells.

Acetylcholine has been well identified as a component of bacteria (its production was discovered in a strain of *Lactobacillus plantarum*) (Stephenson and Rowatt 1947; Rowatt 1948; Girvin and Stevenson 1954; Marquardt and Falk 1957; Marquardt and Spitznagel 1959). Cell free enzyme(s) participating in the acetylcholine synthesis were also first found in *L. plantarum* (Girvin and Stevenson 1954).

In the plant kingdom, acetylcholine is found in 65 species from 33 different families (Roshchina 1991, 2001a; Wessler et al. 2001; Murch 2006). Acetycholine was synthesized not only free, but also in a conjugated form as well, in particular as conjugates of cholinic esters with plant auxins (Fluck et al. 2000). Acetylcholine is particularly abundant in secretory cells of common nettle stinging hairs, where its concentration reaches 10^{-1} M or 120–180 nmol/g of fresh mass. Together with the histamine contained in the secretion, acetylcholine may provoke a pain response and formation of blisters when the plant comes in contact with human skin.

Kawashima et al. (2007) have attempted to compare the concentration of the neurotransmitter acetylcholine in a wide variety of sources using the same experimental conditions, which involved a radioimmunoassay with high specificity and sensitivity (1 pg/tube). The authors measured the acetylcholine content in samples from the bacteria, archaea, and eucarya domains of the universal phylogenetic tree. The authors compared the concentrations in different groups of bacteria (*Bacillus subtilis*), archaea (*Thermococcus kodakaraensis* KOD1), fungi (shiitake mushroom and yeast), plants (bamboo shoot and fern), and animals (e.g., bloodworm and lugworm). The levels varied considerably, however, with the highest acetylcholine content detected in the top portion of bamboo shoot (2.9 μmol/g), which contained about 80 times of that found in rat brain. Various levels of acetylcholine-synthesizing activity were also detected in extracts from the cells tested, which contained a choline acetyltransferase-like enzyme (sensitive to bromoacetylcholine, a selective inhibitor of choline acetyltransferase). The enzyme activity was found in *T. kodakaraensis* KOD1 (15%), bamboo shoot (91%), shiitake mushroom (51%), bloodworm (91%), and lugworm (81%). Taken together, these findings demonstrate the ubiquitous expression of acetylcholine and acetylcholine-synthesizing activities among life forms without nervous systems, and support the notion that acetylcholine has been expressed and may be active as a local mediator and modulator of physiological functions since the early beginning of life.

2.2.1.2 Catecholamines

In unicellular organisms, biogenic amines are also synthesized. The large amounts of dopamine accumulated by cells of infusoria *Tetrahymena pyriformis* strain NT-1 and secreted into their growth medium were found to depend primarily upon an extracellular, non-enzymatic conversion of tyrosine to L-dihydroxyphenylalanine (Gundersen and Thompson 1985). Recently, the catecholamines norepinephrine and dopamine have been identified in microorganisms by high-performance liquid chromatography by Tsavkelova et al. (2000). Dopamine in concentrations 0.45–2.13 μmol/L was found in the biomass of bacteria *Bacillus cereus*, *B. mycoides*, *B. subtilis*, *Proteus vulgaris*, *Serratia marcescens*, *S. aureus*, and *E. coli*, but was absent in the fungi *Saccharomyces cerevisiae*, *Penicillum chrysogenum*, and *Zoogloea ramigera*. Norepinephrine was found (0.21–1.87 μmol/L) in the bacteria *B. mycoides*, *B. subtilis*, *P. vulgaris*, and *S. marcescens* as well as in fungi such as *S. cerevisiae* (0.21 μmol/L) and *P. chrysogenum* (21.1 μmol/L). It is especially interesting that in many cases, the content of catecholamines in microorganisms is higher than in animals, for example in human, blood norepinephrine is found about

0.04 μmol/L (Kruk and Pycock 1990). Moreover, it was demonstrated that bacteria, in particular *B. subtilis*, may release norepinephrine and dopamine out of the cell and, perhaps, by this way possibly participate in intercellular communication both in microorganism–microorganism and bacteria–host.

In plants, catecholamines have been found in 28 species of 18 plant families (Roshchina 1991, 2001a; Kuklin and Conger 1995; Kulma and Szopa 2007). The amount of dopamine found varies during plant development (Kamo and Mahlberg 1984), and sharply increases during stress (Swiedrych et al. 2004). Of particular note is the finding that increased amounts of dopamine (1–4 mg/g fresh mass) are found in flowers and fruits, in particular in Araceae species (Ponchet et al. 1982). This demonstrates the important role of the catecholamines as neurotransmitters in fertilization as well as in fruit and seed development.

2.2.1.3 Serotonin

Some microorganisms living within parasitic nematodes are also able to synthesize serotonin (Hsu et al. 1986). In the bacterial flora of the ascarid *Ascaris suum*, mainly facultative anaerobes (17 species) produced and excreted serotonin into the culture medium of up to 14.32–500.00 μg/g of fresh mass for *Corynebacterium* sp. (in the tissues of the helminth itself only 0.25 μg serotonin per g fresh mass). The concentration of serotonin, in terms of μg serotonin/10^9 cells for different cultures of microorganisms isolated from helminths is as follows: *Klebsiella pneumoniae* 8.15, *Aeromonas* 26.71, *Citrobacter* 0.58, *Corynebacterium* sp. 14.32–500.00, *Enterobacteria aggiomerans* 2.93, *Shigella* 1.04, *Achromobacter xylosoxidans* 1.66, *Chromobacterium* 3.67, *Achromobacter* 0.15, *Acinetobacter* 11.79, *Streptococcus* 37.52, *Listeria monocytogens* 4.71, and *E. coli* 3.33. Serotonin has also been found in the yeast *Candida guillermondii* and bacterium *Enterococcus faecalis* (Fraikin et al. 1989; Belenikina et al. 1991; Strakhovskaya et al. 1991, 1993). In 1998, Oleskin et al. also established the presence of serotonin in the phototrophic bacterium *Rhodospirillum rubrum* (1 μg/billion of cells ~3–12,500 μg/g of fresh mass) as well as in nonphototrophic bacteria *Streptococcus faecalis* and *E. coli* (50 and 3.3 μg/billion of cells, relatively). The inhibitor of tryptophan hydroxylase, *n*-chlorophenylalanine, affects the growth of the yeast *Candida guillermondii*, but not the development of the bacterium *E. coli*. This suggests that in the latter case, there is an alternative pathway to that found in animals (Oleskin et al. 1998a, b), which is peculiar (Roshchina 1991, 2001a) to plants: tryptophan ⇒ tryptamine ⇒ serotonin.

In plants, serotonin is found in 42 species of 20 plant families (Roshchina 1991, 2001a). Besides free serotonin, conjugated serotonins such as *N*-feruloylserotonin, *N*-(*p*-coumaroyl) serotonin, *N*-(*p*-coumaroyl) serotonin mono-β-D-glucopyranoside have been isolated from safflower *Carthamus tinctorius* L. seed. It should be noted that serotonin in animals (such as rats) may exist in complexes with heparin that prevents the aggregation of thrombocytes (Kondashevskaya et al. 1996).

2.2.1.4   Histamine

Histamine was first found in the ergot fungus *Claviceps purpurea* (Table 2.1), and subsequently in many bacterial and plant cells by Werle and coauthors (1948, 1949). Since then, it has also been observed in many types of foods as the result of microbial activity. Histamine is one of the biogenic amines formed mainly by microbial decarboxylation of amino acids in numerous foods, including fish, cheese, wine, and fermented products. A number of microorganisms can produce histamine. In particular, bacteria such as *Morganella morganii*, *Proteus* sp, and *Klebsiella* sp. are considered strong histamine formers in fish (Ekici and Coskun 2002; Ekici et al. 2006). Fernández et al. (2006) summarized the data on the histamine content as toxicant in food. Histamine poisoning is the most common food borne problem caused by biogenic amines. At non-toxic doses, this histamine can cause intolerance symptoms such as diarrhea, hypotension, headache, pruritus, and flushes. Just 75 mg of histamine, a quantity commonly present in normal meals, can induce symptoms in the majority. One separate problem concerns the histamine formed by microorganisms in animal pathogenesis. Gram-negative bacterial species such as *Branhamella catarrhalis*, *Haemophilus parainfluenzae*, and *Pseudomonas aeruginosa* have been demonstrated to synthesize clinically relevant amounts of histamine in vitro that implicate the bacterial production of histamine in situ as an additional damage factor in acute exacerbations of chronic bronchitis, cystic fibrosis, and pneumonia. Histamine may also increase the virulence of these bacterial species, unlike some Gram-positive species such as *Staphylococcus aureus* and *Streptococcus pneumoniae* (Devalia et al. 1989). Among "non-pathogenic" species, only the *Enterobacteriacae*, as a group, were found to form histamine in significant concentrations.

Significant amounts of histamine have also been observed in higher plants, initially by Werle and Raub in 1948, and subsequently described for 49 plant species belonging to 28 families ranging from basidiomycetes to angiosperms (Roshchina 1991, 2001a). Besides histamine itself, its derivatives *N*-acetylhistamine, *N*, *N*-dimethylhistamine, and feruloylhistamine are also found in plants. Especially high levels are observed in species of the family Urticaceae that could be one of the taxonomic classification signs. The Brazilian stinging shrub *Jatropha urens* (family Euphorbiaceae) contains 1,250 μg histamine per 1,000 hairs. The presence of histamine in stinging hairs is a protective mechanism that serves order to frighten off predatory animals by inducing burns, pain, and allergic reactions. Under stress conditions, a sharp increase of histamine is observed in plants, as in animals. Ekici and Coskun (2002) have determined the histamine content of some commercial vegetable pickles at the range of 16.54 and 74.91 mg/kg (average 30.73 mg/kg). The maximum value (74.91 mg/kg) was obtained from a sample of hot pepper pickles. The amount of histamine varies according to the phase of plant development. For example, in the marine red algae *Furcellaria lumbricalis* (Huds.) Lamour, the occurrence of histamine was from 60 to 500 μg/g fresh mass observed in both non-fertile fronds and sexual-expressed parts, in all regions of the thallus of male, female, and tetrasporophyte (Barwell 1979, 1989). The amount of histamine (in µg/g fresh mass) in the male plant was 90–490 (sometimes up to 1,100), in the female plant 60–120, and in asexual tetra sporophyte 100–500. Especially enriched were the neurotransmitter cells of male plants, as the ramuli were approximately five times higher in histamine than female and asexual plants.

2.2.2   Neurotransmitters as Toxicants

High concentrations of biogenic amines in foodstuffs and beverages can induce a range of toxicological effects (Fernández et al. 2006). Significant attention is needed to control the histamine levels in foods (Bodmer et al. 1999). Histamine poisoning is the most common food borne problem. Besides the compounds naturally found in vegetables and fruits as well as those formed as a result in the result of fermentation of cheese, wine, and sauerkraut, biogenic amines also play an essential role in the metabolism of the histamine-forming bacteria present in foods (Kung et al. 2007). Flushing of the face and neck are symptoms of histamine intoxication, followed by an intense, throbbing headache. Other symptoms include dizziness, itching, faintness, burning of the mouth and throat, and the inability to swallow. Taylor et al. (1978) reported that ingestion of 70–1,000 mg of histamine in a single meal is necessary to elicit any symptoms of toxicity. A level of histamine exceeding 10 mg/100 g of fresh weight is associated with poor product quality indicative of microbial spoilage, with levels of 200 mg histamine per kg of food product accepted as a toxic indicator for fish, and 10 mg/kg for wines, whereas for hot pepper pickles all values are below the level of 1,000 mg/kg. An average food content for histamine of approximately 30 mg/kg can be considered the minimal level for clinical symptoms of toxicity (Ekici and Coskun 2002). These toxicological problems are particularly severe in individuals who, for whatever reason, are deficient in diamine oxidase, the histamine-degrading enzyme. At non-toxic doses, histamine can cause intolerance symptoms such as diarrhea, hypotension, headache, pruritus, and flushes. Just 75 mg of histamine, a quantity commonly present in normal meals, can induce symptoms in the majority of healthy persons with no history of histamine intolerance.

The amount of neurotransmitters in cellular secretions can be increased following unfavorable stimuli, in particular the interactions with other organisms. Large amounts of dopamine are usually secreted by cells of infusoria *Tetrahymena pyriformis* into their growth medium (Gundersen and Thompson 1985). This release of dopamine is especially important during infection, when the animal or plant accumulates some of the neurotransmitters, from one side, and pathogens release neurotransmitters from another side (Romanovskaya and Popenenkova 1971). On northeastern Pacific coasts, the alga *Ulvaria obscura* produces large amounts of dopamine (van Alstyne et al. 2006). This organism, dominant in subtidal "green tide" blooms due to this antiherbivore defense, can be harmful to marine communities, fisheries, and aquaculture facilities because the alga presence is the cause of reduced feeding by echinoderms, mollusks, and arthropods.

Dopamine constituted an average of 4.4% of the alga's dry mass, and was responsible for the decreased feeding by sea urchins (*Strongylocentrotus droebachiensis*). Subsequent experiments demonstrated that dopamine also reduced the feeding rates of snails (*Littorina sitkana*) and isopods (*Idotea wosnesenskii*). This is the first experimental demonstration of a plant (algal) catecholamine functioning as a feeding deterrent.

2.2.3 Components of Cholinergic and Aminergic Systems

In microbial cells, components of cholinergic and aminergic systems similar to those found in mammalian cells, including the complete biosynthetic pathway required for their synthesis (relative synthetases) and their catabolism (cholinesterases, aminooxidases, and others), as well as functional analogs of cholino- and aminoreceptors are shown to be present.

2.2.3.1   Choline Acetyltransferase

The enzymes choline acetyltransferases or choline acetylases (EC 2.3.1.6) participate in the synthesis of acetylcholine from choline and acetic acid (Nachmansohn and Machado 1943). A cell free enzyme with "choline acetylase" activity was present in *Lactobacillus plantarum* (Girvin and Stevenson 1954). This enzyme activity has been also found in many plant species (Roshchina 1991, 2001a).

2.2.3.2   Cholinesterase

Enzymes which degrade acetylcholine to choline and acetic acid are named cholinesterases and were first found in 1937 by Loewi in the hearts of amphibia. The function of acetylcholinesterase at cholinergic synapses of animals is to terminate cholinergic neurotransmission (Augustinsson 1949). However, the enzyme is expressed in tissues that are not directly innervated by cholinergic nerves. Moreover, transient expression in the brain during embryogenesis suggests that acetylcholinesterase may function in the regulation of neurite outgrowth. Overexpression of cholinesterases has also been correlated with tumorigenesis and abnormal megakaryocytopoiesis (Small et al. 1996). Cholinesterase is also found in unicellular animal such as *Paramecium* (Corrado et al. 1999). An immunoblot analysis of the *Paramecium* enzyme revealed that the acetylcholinesterase had a molecular mass from 42 to 133 kDa, as reported for analogous enzyme isolated from higher organisms. Structural homologies between cholinesterases and the adhesion proteins indicate that cholinesterases could also function as cell–cell or cell–substrate adhesion molecules. Abnormal expression of cholinesterases of both types has been detected around the amyloid plaques and neurofibrillary tangles in the brains of patients with Alzheimer's disease (Small et al. 1996).

As for microorganisms, Goldstein and Goldstein (1953) first described the production of cholinesterase by a strain of bacterium *Pseudomonas fluorescens* after the culture was grown with acetylcholine as the sole source of carbon. The *P. fluorescens* enzyme was inducible, mainly, by choline (not as a carbon substrate, but, perhaps, as a source of nitrogen) or by two- to threefold lesser degree by some choline esters: acetylcholine > propionylcholine = benzoylcholine > butyrylcholine > acetyl-β-methylcholine). Addition of glucose completely prevented the induction of *P. fluorescens* enzyme. The pH optimum for growth of the culture and cholinesterase activity was 7.0, although the culture growth was higher in alkaline medium, where spontaneous hydrolysis of acetylcholine is also maximal. The choline oxidase synthesis in the *P. fluorescens* has also been induced by choline. The cholinesterase of the bacterium may hydrolyze acetylcholine or propionylcholine, but to a lesser degree butyrylcholine, benzoylcholine, or acetyl-β-methylcholine. Like cholinesterase in animals, the enzyme activity in *P. fluorescens* was inhibited by neostigmine, with complete inactivation observed at high concentrations (10^{-3}–10^{-2} M) and only partly at the lower levels of 10^{-6} M. These levels of inhibition are similar to that observed in mammalian organ systems. Then, the *P. fluorescens* protein was isolated and characterized (Goldstein 1959; Searle and Goldstein 1957, 1962; Fitch 1963a, b). Moreover, the strains of the *P. fluorescens* tested preferred the acetylcholine for growth promotion over choline, glycerol, glucose, succinate, betaine, and serine (Fitch 1963a). The isolated cholinesterase was inhibited by neostigmine in smaller (1,000 times) concentration, than by physostigmine, but was not depressed by diisopropylfluorophosphate (Fitch 1963b). A bell-shaped substrate saturation curve was observed, and specific activity of the 115-times purified cholinesterase was 10.5 μmol/mg protein/min. The enzyme had the features both of true cholinesterase and acetylcholinesterase (Laing et al. 1967, 1969). Specific activity of the cholinesterase from *P. fluorescens* purified 40-fold by CM-50 Sephadex was up to 70 μmol/mg protein/min. The values of *Km* at pH 7.4 and 37°C were 1.4×10^{-5} M for acetylcholine and 2.0×10^{-5} M for propyonylcholine, respectively, while butyrylcholine and benzoylcholine were not hydrolysable at all. The purified enzyme was inhibited by organophosphorus compounds and neostigmine, but not by physostigmine.

Imshenetskii et al. (1974) showed that a large variety of microorganisms may decompose acetylcholine including 31 strains of bacteria (genera *Arthrobacter* and *Pseudomonas*) and two strains of fungi (from 194 strains studied) that live in soil. Around 100–200 mg of wet biomass of active microbial strains were able to decompose 15–30 μmol of acetylcholine during a 2 h incubation, with the most active strains (50 mg of wet biomass) able to degrade up to 10 μmol/min. This active soil strain was identified as *Arthrobacter simplex* var. *cholinesterasus* var.*nov.* The amount of the decomposed acetylcholine by this microbe was 30 times higher than in other strains (*Pseudomonas fluorescens* – 4 μM/h, *P. aerugenosa* – 1 μM/h), while *Arthrobacter simplex* var. *cholinesterasus*.var *nov.* had an activity of 300 μM/h. Actinomycetes (except two strains) and yeast had no significant cholinesterase activity.

The cholinesterase activity has also been found in lower groups of the plant kingdom: in extracts of Characeae algae *Nitella* by Dettbarn, in 1962 and mycelium of fungi *Physarium polycephalum* by Nakajima and Hatano in 1962, and then a series of classical papers of Jaffe and Fluck with coworkers in 1970–1975 were devoted to the observation of the enzyme in many plant species: ~118 terrestrial species and ten marine algae were identified as having cholinesterase activity (for more details see the relevant references in monographs Roshchina 1991, 2001a). The values of the enzyme activity (the substrate hydrolysis rate) in most higher plants is an average of 1–900 μmol/h/g fresh weight, depending on the plant species. It was specially shown that Bryophytes (mosses, liverworts and hornworts) demonstrate the maximal cholinesterase activity of up to 0.360 μmol/h/g fresh weight (Gupta et al. 2001). Thus, detection of cholinesterase activity could serve as an additional indicator of the acetylcholine presence. Recently, identification, purification, and cloning of maize acetylcholinesterase provided the first direct evidence of the enzyme formation in plants (Sagane et al. 2005). An especially important fact is that the acetylcholinesterase distribution in seedlings is sensitive to gravity, leading to asymmetry of the enzyme distribution (Momonoki 1997).

2.2.3.3 Enzymes of Biogenic Amine Metabolism

The biosynthetic pathway of biogenic amines includes decarboxylation and hydroxylation of corresponding amino acids, in particular phenylalanine for the catecholamines, tryptophan for serotonin, and histidine for histamine (Lawrence 2004). Phenylalanine, precursor of dopamine, norepinephrine and epinephrine, is first hydroxylated, transforming to tyrosine and then to dihydroxyphenylalanine (DOPA). These processes are catalyzed by phenylalanine hydroxylase or phenylalanine monoxidase and tyrosine hydroxylase or tyrosine-3-monoxidase. Dopamine, an immediate precursor of norepinephrine and epinephrine, arises from DOPA through decarboxylation by means of the enzyme decarboxylase dioxyphenylalanine and the decarboxylase of aromatic amino acids (EC 4.1.1.26). Another route of tyrosine transformation is via decarboxylation, when it transforms to tyramine, and then by hydroxylation with the participation of tyramine hydroxylase into dopamine, which is then oxidized to norepinephrine by the copper-containing enzyme β-hydroxylase 3, 4-dioxyphenylethylamine. Then, under the influence of the transmethylase of phenylethanolamines, the formation of epinephrine takes place.

In the catabolism of catecholamines, aminooxidases participate as a whole in oxidative deamination of the catecholamines to metanephrine, normetanephrine, vanillic aldehyde, dehydroxymandelic and vanillic acids. For microorganisms, this metabolism process has not yet been studied. In plants, diamineoxidases play the main role in catecholamine metabolism, unlike animals that use monoaminooxidases for this purpose (Roshchina 2001a). As for catecholamine-*O*-methyltransferases, they are present in all animal tissues, and especially active in nervous cells. In plants, the catecholamine-*O*-methyltransferases pathway is also possible because the last three compounds are ordinary products of plant metabolism (Kuklin and Conger 1995;

Roshchina 2001a; Kulma and Szopa 2007). Besides the above-mentioned ways of metabolism, catecholamines are oxidized by oxygen of air, forming oxidized products – red pigments aminochromes and black-brown pigments melanines which are polymers of indole (found both in plant and animals). The mechanism of oxidation per se is connected with the arising of superoxide radical – active oxygen $\dot{O}_2^-$ Blockade of oxidation of dopamine by superoxide dismutase confirms this possibility. Enzymatic oxidation of catecholamines to melanines by polyphenol oxidase has been also demonstrated (Roshchina 2001a). The above-mentioned enzymes are found only in animals and plants. There is little data for catecholamine oxidation of microorganisms, although monoaminooxidase activity in mycobacteria (Pershin and Nesvadba 1963) and *E. coli* (Takenaka et al. 1997) has been found.

Serotonin is synthesized in plants and animals from tryptophan formed by the shikimate pathway, which has also been proposed for microorganisms (Oleskin et al. 1998a, b; Oleskin 2007). This process proceeds by two pathways: either via 5-hydroxytryptophan or tryptamine formation, or the first step of serotonin biosynthesis via decarboxylation of tryptophan, which then transforms in plants to tryptamine by action of the enzyme tryptophan decarboxylase (EC 4.1.1.27), or by the dacarboxylation of aromatic amino acids (EC 4.1.1.26/27). Then, tryptamine is transformed to serotonin by hydroxylation with participation of the enzymes tryptamine-5-hydroxylase or L-tryptophan-5-hydrolylase (EC 1.14.16.4). Hydroxylation of tryptophan leads to the formation of 5-oxytryptophan in the presence of tryptophan-5-hydroxylase (EC1.14.16.4). At the next stage, 5-oxytryptophan is decarboxylated by the decarboxylase of aromatic acids to yield serotonin. Tryptamine 5-hydroxylase, which converts tryptamine into serotonin and common in animals, was also found as a soluble enzyme that had maximal activity in rice roots (Kang et al. 2007). The tissues of rice seedlings grown in the presence of tryptamine exhibited a dose-dependent increase in serotonin in parallel with enhanced enzyme activity. However, no significant increase in serotonin was observed in rice tissues grown in the presence of tryptophan, suggesting that tryptamine is a bottleneck intermediate substrate for serotonin synthesis. If we compare the enzymes from the different kingdoms, we can see more similarity. In particular, in the plant genus *Arabidopsis*, there is a homolog to part of a DNA binding complex corresponding to the animal tyrosine and tryptophan hydroxylases (Lu et al. 1992). Aminooxidases of biogenic amines may differ in microorganisms in relation to substrate specificity, in particular for the bacterium *Methanosarcina barkeri* and infusoria *Tetrahymena pyriformis* (Yagodina et al. 2000). Both studied enzymes can deaminate serotonin, but not histamine. The existence of one active center for substrate binding is supposed in the aminooxidase of the bacterium, while several centers are thought to exist in the infusoria.

For all living organisms, the biosynthesis pathway of histamine includes histidine decarboxylase which participates in the decarboxylation of histidine (Roshchina 2001a; Boron and Boulpaep 2005; Martín et al. 2005). The gene encoding histidine decarboxylase (*hdcA*) has been identified in different Gram-positive bacteria (Martín et al. 2005). Histidine decarboxylase used to be part of a cluster that included a gene of unknown function (*hdcB*) and a histidine–histamine antiporter gene (*hdcC*) in *Pediococcus parvulus* 276 and *Lactobacillus hilgardii* 321 has been identified (Landete et al. 2005). Catabolism of histamine occurs also via methylation or acetylation in the presence of histamine-*N*-methyltransferase, or histamine-*N*-acetyltransferase, and genes cording of the enzymes have been found in bacteria, plants, and animals (Iyer et al. 2004).

2.2.3.4 Recognition of Neurotransmitters

The presence of neurotransmitters in cells is usually considered in the context of receptors to the compounds, according to concepts of neurotransmitter reception in animals. The main study methodology is based on pharmacological assays, where sensitivity to the neurotransmitter on cellular reaction is analyzed by the use of agonists and antagonists to the neurotransmitter. All perspectives on the issue for non-synaptic systems have a fundamental similarity to studies undertaken for the nerve cell.

For acetylcholine, there are two types of acetylcholine receptor – nicotinic (receptors respond to nicotine) and muscarinic (sensitive to muscarine). Corrado et al. (2001) showed the presence of functionally related nicotinic and muscarinic receptors and its lytic enzyme acetylcholinesterase in the unicellular animal *Paramecium primaurelia*. In plants, the presence of similar receptors has also been shown (Roshchina 2001a). Recently, it was established that muscarinic and nicotinic acetylcholine receptors are involved in the regulation of stomata function – the opening and closing movement – in the plants *Vicia faba* and *Pisum sativum* (Wang et al. 1998; Wang et al. 1999a, 2000). Leng et al. (2000) showed the regulation role of acetylcholine and its antagonists in inward rectified K^+ channels from *Vicia faba* guard cells. Location of the muscarinic receptor was shown in plasmatic membrane and chloroplast membranes (Meng et al. 2001), and cholinesterase activity was found in the cells (Wang et al. 1999b). The germination of plant microspores such as vegetative microspores of horsetail *Equisetum arvense* or pollen (generative microspores) of knight's star *Hippeastrum hybridum* was blocked by the antagonists of acetylcholine, which are linked with nicotinic cholinoreceptors and Na^+/K^+ ion channels (Roshchina and Vikhlyantsev 2009). The nicotinic cholinoreceptors were cyto-chemically identified in the single-cell amoebae *Dictyostelium discoideum*, slugs, and spores, however, the proteins immunologically related to the muscarinic receptors were not present in the spores (Amaroli et al. 2003). Interestingly, the nicotine and acetylcholine as the ligands of human nicotinic cholinoreceptors in culture of epithelial cells HEp-2 may stimulate the growth of *Chlamidia pneumoniae*. (Yamaguchi et al. 2003).

The receptors for biogenic amines, peculiar to highly organized animals, are known as dopamine receptors, adrenoreceptors, and serotonin and histamine receptors. The similar receptors were observed in bacteria (Lyte and Ernst 1993; Freestone et al. 2007) and in plant cells (see monographs Roshchina 1991, 2001a). Alpha and beta adrenergic-like receptors may be involved in catecholamine-induced growth of Gram-negative bacteria (Lyte and Ernst 1993). In particular, Freestone et al. (2007)

showed the blockade of catecholamine-induced growth of *E. coli*, *Salmonella enterica*, and *Yersinia enterocolitica* by adrenergic and dopaminergic receptor antagonists. In plants, adrenoreceptors participate in cytoplasm movement, ion permeability, and membrane potential, in flowering of *Lemna paucicostata*, photophosphorylation, as well as the seed and pollen germination (Roshchina 1991, 2001a; Baburina et al. 2000; Kulma and Szopa 2007). Serotonin- and histamine-sensitive receptors in plants regulate the seed, pollen, and vegetative microspores germination (Roshchina 1991, 2001a, 2004, 2005a). Shmukler et al. (2007) discussed earlier hypotheses of protosynapse for low-organized animals and embryos of high-organized animals, where the distribution of membrane serotonin receptors is restricted to the period of blastomer formation during cleavage and localized in the area of interblastomer contact. The hypothesis was based on their experiments, where the membrane currents of the *Paracentrotus lividus* early embryos have been registered after local application of serotonin drugs with special micropipette. Receptors of neurotransmitters may be linked with ion channels. Moreover, some domains of the ion channels appear to be common with the cytoskeleton, in particular with actin (Cantiello 1997), and so the received chemosignal is likely to spread to the organelles via actomyosin filaments (Roshchina 2005a, 2006a, b).

2.3 Common View on the Neurotransmitter (Biomediator) Functions

The presence of neurotransmitters in any organism leads us to the problem of information transmission within and between the living cells as a whole. Like the genetic code, having a common base in all living organisms in a form of the sequence and combination of several purine and pyrimidine bases, the mechanism of irritability appears to have a common base in the form of chemical signals uniform for every cell. The compounds acetylcholine and biogenic amines named neurotransmitters, besides having specialized mediator function in organisms with nervous systems, also play other roles, not only in animals, but also in microorganisms and plants. From this position, one could call the compounds rather "biomediators," than "neurotransmitters" or "neuromediators" (Roshchina 1989, 1991).

2.3.1 Functions of Neurotransmitters on Different Evolutionary Steps

The function of compounds named neurotransmitters originates from simple chemotaxis and chemosignaling of microbial cells and leads to intercellular communication (Fig. 2.1). The so-called neurotransmitters may regulate (as hormones) growth and development of other unicellular organisms, and be attractants or repellents for them. In higher concentrations the same substances also play a defense role (for saving or

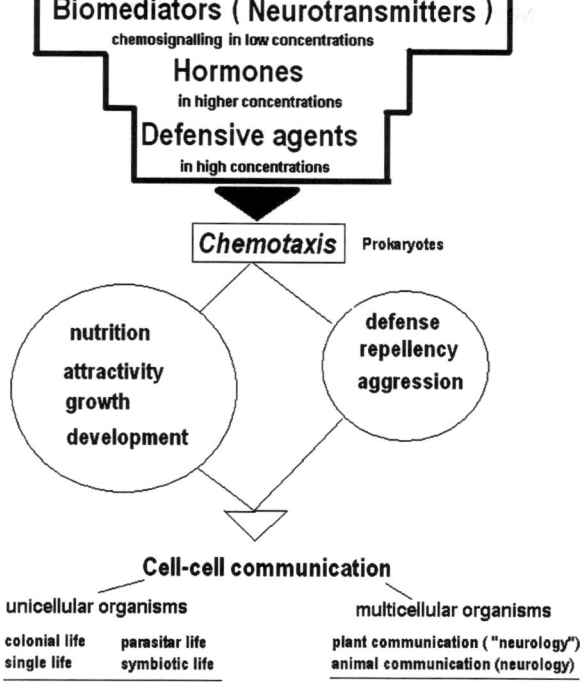

Fig. 2.1 The scheme of the evolution in the neurotransmitter (biomediator) function

aggression) or, in some cases, serve as an origin of cultural food. The following step in evolution includes the development of colonial relations (parasitar or symbiotic) and then the formation of multicellular organisms that forms more specialized function of biomediators in the irritation transfer along the multicellular system. This evolution way leads us to the concept of neurology, not only for animals, but also for plants (Baluska et al. 2005; Brenner et al. 2006; Murch 2006).

In Table 2.3, we compare the main functions of the neurotransmitters in all kingdoms. The realization of the irritation impulse transfer into a cell from the surface or between compartments of the cell occurs with the participation of neurotransmitters, and on cellular level the compounds may induce different reactions. Neurotransmitters are stored in secretory vesicles, and then can be liberated within cell or out. Primary reaction to acetylcholine is often a change in membrane permeability for ions, while other reactions for both the neurotransmitter and biogenic amines are connected with the systems of secondary messengers – cyclic nucleotides, Ca^{2+}, inositol-3-phosphate, etc.

First of all, the neurotransmitter functioning as chemosignal (the neurotransmitter is released from one cell and perceived by another) occurs in certain structural forms – the chink forming between cells or between organelles within cell. We can usually see the chink between plasmic membranes of any contacted cells such as

Table 2.3 The established functions of neurotransmitters in living organisms

Neurotransmitter	Microorganisms	Plants	Animals
Acetylcholine	Regulation of motility	Regulation of membrane permeability and other cellular reactions up to growth and development in many plant species	Regulation of cell proliferation, growth and morphogenesis. The carriage of nerve impulses across the synaptic chink, from one *neuron* to another of impulses across the "motor plate," from a neuron to a muscle cell, where it generates muscle contractions
Dopamine	Stimulation of gram negative and gram positive bacterial growth and virulence	Regulation of many cellular processes from growth and development to defense reactions	Decreases peripheral vascular resistance, increases pulse pressure and mean arterial pressure. The positive chronotropic effect produces a small increase in heart rate as well. Important for forming memories. In embryos of Vertebrata and lower animals may regulate development
Norepinephrine (adrenaline)	Bacterial growth stimulation	Regulation of many cellular processes from growth and development to defense reactions	Increases peripheral vascular resistance, pulse pressure and mean arterial pressure as well as stimulates of the thrombocytes' aggregation
Epinephrine (adrenaline)	Bacterial growth stimulation	Regulation of many cellular processes from ion permeability, growth and development to defense reactions	Induced vasodilation (mainly in skeletal muscle) and vasoconstriction (especially skin and viscera)
Serotonin	Stimulation of growth of culture and cellular aggregation bacteria *Streptococcus faecalis*, yeast *Candida guillermondii*, *E. coli* K-12 and *Rhodospirillum rubrum*. Regulation of membrane potential	Regulation of growth and development of many plant cells	Control of appetite, sleep, memory and learning, temperature regulation, mood, behavior (including sexual and hallucinogenic), vascular function, muscle contraction, endocrine regulation, and depression. In embryos of Vertebrata and lower animals may regulate development

| Histamine | Stimulation of cultural growth and cellular aggregation of *E. coli* K-12 | Regulation of the growth and development at stress | Involves in many allergic reactions and increases permeability of capillaries, arterial pressure is decreased, but increases intracranial pressure that causes headache, smooth musculature of lungs is reduced, causing suffocation, causes the expansion of vessels and the reddening of the skin, the swelling of clothStimulation of the secretion of gastric juice, saliva (digestive hormone) |

Sources: (Anuchin et al. 2007, 2008; Buznikov 1967, 1987, 1990 Faust and Doetsch 1971; Burton et al. 2002; Freestone et al. 2007; Lyte and Ernst 1992, 1993; Lyte et al. 1997; Oleskin et al. 1998a, b; Oleskin 2007; Roshchina 1991, 2001a; Strakhovskaya et al. 1993)

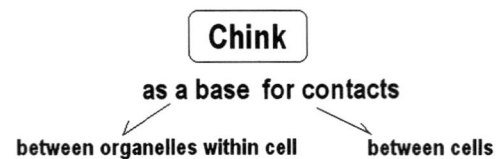

Fig. 2.2 Structure where the neurotransmitters action is possible

the membranes of unicellular contacting organisms and synaptic membranes of cells in organisms with a nervous system (Roshchina 1991, 2001a, b; Buznikov et al. 1996; Shmukler et al. 2007). As seen in the scheme presented in Fig. 2.2, at any membrane contacts, either time-changed or constant, chinks may be formed. There are chinks between endoplasmic reticulum and organelles within cells, or between different cells. Today, constant or temporary chinks between cells or within the cell are considered a necessary structural form for the chemosignal transfer.

As can be seen from Table 2.3, common cellular effects of neurotransmitters in any type of living kingdom cell are the changes in membrane permeability (short-time effects) and the regulation of growth and development (long-time effects). Regulatory function of the neurotransmitters appears to be an ancient function, relating processes occurring both within a cell and the environmental unicellular populations. The secretion that contains neurotransmitters is released out of any cell and may contact with other cells at chemotaxis. The cell, which receives similar chemosignals, responds primarily by the changes in ion permeability and the formation of action potential, and then with various metabolic and growth reactions. Some other aspects and details will be described below.

2.3.1.1   Functions in Microorganisms

The first report, showing acetylcholine production in bacterium strains, was from *L. plantarum* and. *L. odontolyticus* (Stephenson and Rowatt 1947). Approximately 5 μg acetylcholine/mg dry wt. cells/h was formed if the bacteria were grown both in vegetable juice and washed cells. Acetylcholine can be also used as substrate for microorganisms, and regulates their development in special conditions (Imshenetskii et al. 1974). Its regulation of motility peculiar to photosynthesing bacteria *Rhodospirillum rubrum* and *Thiospirillum jenenese* has also been shown (Faust and Doetsch 1971).

Catecholamines can regulate the growth of Gram-negative bacteria, including *E. coli* (where concentration dependent specificity was observed with response to norepinephrine » epinephrine > dopamine), *Y. enterocolitica* and *P.s aeruginosa* (Lyte and Ernst 1992; Freestone et al. 1999). Dopamine also stimulates the cultural growth of *E. coli*, *Y. enterocolica*, *S. enterica*, *S. epidermidis*, etc., and the cellular aggregation and formation of colonies of *E. coli* and *S. epidermidis* (Lyte and Ernst

1993; Neal et al. 2001; Freestone et al. 2007; Anuchin et al. 2007, 2008). Similar effects on Gram-negative bacteria *E. coli, S. enterica* and *Y. enterocolitica* were observed for norepinephrine (Lyte and Ernst 1992, 1993; Lyte et al. 1997; Freestone et al. 1999, 2007; Burton et al. 2002) and on *E .coli* for epinephrine (Anuchin et al. 2007; Freestone et al. 2007). Serotonin stimulated cultural growth and cellular aggregation of bacterial species, including *Streptococcus faecalis*, the yeast *Candida guillermondii*, (Strakhovskaya et al. 1993), *E. coli* K-12 and *Rhodospirillum rubrum* at concentrations of $2 \times 10^{-7} - 2 \times 10^{-5}$ M (Oleskin et al. 1998a, b; Anuchin et al. 2007, 2008). Moreover, histamine showed similar effects on *E. coli* (Anuchin et al. 2007, 2008). Serotonin at $10^{-6} - 10^{-5}$ M concentrations also inhibited light-dependent membrane potential generation in *Rsp. rubrum*, but in the myxobacterium *Polyangium* sp. serotonin stimulates cell aggregation and myxospore formation (Oleskin et al. 1998a, b). At concentrations near 20 μM, serotonin inhibits cell aggregation and microbial culture growth and photo-dependent membrane potential of the bacterium *Rsp. rubrum*. At micromolar amounts, the effects presumably result from the specific action of serotonin as an intercellular communication agent accelerating and possibly synchronizing the development of the microbial cell population. According to Oleskin et al. (1998a, b), the growth stimulation of microorganisms by serotonin over a millimolar to micromolar range has been demonstrated in prokaryotes, both Gram-positive including *Streptococcus faecalis* (Strakhovskaya et al. 1993) and Gram-negative bacteria including *E. coli* and *Rsp. rubrum* (Oleskin et al. 1998a, b). In some cases, such as that for *Bacillus brevis*, the degree of growth stimulation achieves 100% of control. Freestone et al. (2008a, b) showed that catecholamine stress hormones can significantly increase the growth of a wide range of gram negative and gram positive bacteria. Using a novel two-fluorophore chemotaxis assay, it was found that *E. coli* is attracted to epinephrine and norepinephrine (and also increased the bacterial motility and biofilm formation), while it is repelled by indole (Bansal et al. 2007). Moreover, epinephrine/norepinephrine upregulated the expression of genes involved in surface colonization and virulence, while exposure to indole decreased their expression (Bansal et al. 2007). Histamine synthesis by respiratory tract microorganisms: was also observed, and its possible role in pathogenicity considered (Devalia et al. 1989).

Much attention has also been given to the role of colonial organization and intercellular communication in parasite/commensal/symbiont-multicellular host organism systems. Data from the literature on the ability of microorganisms to form plant hormones (biogenic amines) have been reviewed by Tsavkelova et al. (2006), who discuss the *Rhodospirillum rubrum* pathways whereby the biogenic amines are metabolized, and their effects on the development and activity (physiological and biochemical) of the microorganisms are considered. The role as hormones and hormone-like substances is in the formation of association-type (microorganism-host) interactions. The review by Oleskin et al. (2000) suggested that the integrity and coherence of microbial populations (colonies, biofilms, etc.) be viewed as peculiar so-called "super-organisms," which are thought to have become multicellular organisms during the course of evolution. This included such relevant phenomena as apoptosis, bacterial altruism, quorum effects, collective differentiation of microbial cells, and the formation of population-level structures such as an extracellular matrix. Emphasis can also be placed on the channels in colonies and agents of intercellular communication in microbial populations. The involvement of a large number of evolutionarily conserved communicational facilities and patterns of intercellular interactions can therefore be underscored. Moreover, an interesting fact is the 5-hydroxytryptophan conversion to serotonin under UV-irradiation (Fraikin et al. 1989). This neurotransmitter may serve as a protector for microorganisms in similar unfavorable conditions. For example, dinoflagellates (a large group of flagellate protists contained in marine plankton) and green algae *Gonyaulax polyedra* synthesize the protector melatonin, using serotonin as a precursor (Balzer et al. 1993). Circadian rhythms of indoleamines in the dinoflagellate *Gonyaulax polyedra* and persistence of melatonin rhythm in constant darkness, have a relationship to 5-methoxytryptamine.

2.3.1.2 Function in Plants

In plants, neurotransmitters demonstrate a high biological activity, playing a role as chemosignals, regulators of membrane permeability, growth and development regulators, etc. (Roshchina (1991, 2001a). Some examples will be considered below.

A signaling role of acetylcholine is well seen as the participation in plant root-shoot signal transduction (Wang et al. 2003b; Baluska et al. 2004, 2005; Brenner et al. 2006). Acetylcholine causes rooting in leaf explants of in vitro raised tomato (*Lycopersicon esculentum* Miller) seedlings (Bamel et al. 2007). Contractile effects of acetylcholine connected with membrane ion permeability were also observed in the regulation of the stomata function – the opening and closing movement in plants such as *Vicia faba* and *Pisum sativum* (Wang et al. 1998, 1999a, 2000). It was established that muscarinic and nicotinic acetylcholine receptors are involved in the event. A regulatory role for acetylcholine and its antagonists in inward rectified K^+ channels from guard cells protoplasts from leaf stomata of *Vicia faba* was found (Leng et al. 2000). Ca^{2+} and Ca-related systems were found to participate in acetylcholine-regulated signal transduction during stomata opening and closing (Wang et al. 2003a; Meng et al. 2004). A chloride channel in the tonoplast (vacuolar membrane) of *Chara corallina* also responds to acetylcholine (Gong and Bisson 2002). Electric processes participate in the electrical signaling, memory and rapid closure of the carnivorous plant *Dionaea muscipula* Ellis (Venus flytrap), and acetylcholine is thought to include in the phenomenon (Volkov et al. 2009).

Acetylcholine and cholinergic system play essential roles in plant fertilization and breeding. For example, lower activities of acetylcholinesterase and choline acetyltransferase in pistils (Tezuka et al. 2007) or in pollen (Kovaleva and Roshchina 1997) were associated with self-incompatibility. A role for acetylcholine can be proposed as dealt with phytochrome and photoreceptor in the growth regulation as well. Wisniewska and Tretyn 2003). There is a connection between some fungal infections (in particular for the *Fusarium* fungi) and the accumulation of plant growth regulators, gibberellic acid, and auxins. Acetylcholine and antibody against acetylcholinesterase may inhibit biosynthesis of gibberellic acid, one of the main growth hormones (Beri and Gupta 2007). The enzyme may also be included in choline-auxin relations that affected plant growth processes. Direct evidence for the hydrolysis of choline-auxin or indole acetylcholine conjugates by pea cholinesterase has been demonstrated by some authors (Ballal et al. 1993; Bozso et al. 1995; Fluck et al. 2000).

A defense function for catecholamines in the plant cell has also been considered in the literature (Roshchina 1991, 2001a; Szopa et al. 2001; Kulma and Szopa 2007). Increased dopamine content in some algae, in particular *Ulvaria obscura,* has led to the consideration of the neurotransmitter as a feeding deterrent (van Alstyne et al. 2006). This is a novel ecological role for a catecholamine. The confirmation of dopamine production acting as defense mechanism against grazers was done from experiments with isopods, snails, and sea urchin eating the agar-based foods contained exogenous dopamine. Damaged algae were also found to release a water-soluble reddish-black substance (dopachrome) that inhibits the development of brown algal embryos, reduced the rates of macroalgal and epiphyte growth and caused increase mortality in oyster larvae (Nelson et al. 2003). Further, serotonin itself (Roshchina 2001a) and its derivatives, such as melatonin (Posmyk and Janas 2009), may also play a protectory role as antioxidants in various plants.

2.3.1.3 Functions in Animals

Currently, we have information about cellular functions for all animal organisms, including those which lack a nervous system and specialized functions peculiar to multicellular organisms with nervous system. First are related to the growth (similar with microbial and plant systems) and morphogenetic reactions. According to modern concepts, acetylcholine and serotonin may play a morphogenetic role in animals – from lower to higher ones (Buznikov 1990; Buznikov et al. 1996; Lauder and Schambra 1999).

The specialized function of neurochemical compounds concerned with the transmission of signals from one neuron to the next across synapses has been considered almost exclusively for neuronal systems as described in classical animal physiology. Neurotransmitters are also found at the axon endings of motor neurons, where they stimulate the muscle fibers to contract. The first of the neurotransmitters to be studied, acetylcholine, transfers nerve impulses from one neuron to another, where it propagates nerve impulses in the receiving neuron, or from a neuron to a muscle cell, where it generates muscle contractions. Moreover, genetic defects of acetylcholine signaling promote protein degradation in muscle cells (Szewczyk et al. 2000). It is obviously important to have proper nervous system and muscle functioning. In the adult nervous system, neurotransmitters mediate cellular communication within neuronal circuits. In developing tissues and primitive organisms, neurotransmitters subserve growth regulatory and morphogenetic functions as regulators of embryogenesis (Buznikov 2007). They regulate growth, differentiation, and plasticity of developing central nervous system neurons. Cellular effects of acetylcholine in animals may also be related to pathogenesis of diseases such as acute and chronic inflammation, local and systemic infection, dementia, atherosclerosis, and finally cancer (Wessler et al. 2001).

2.3.1.4 Possible Evolution of Neurotransmitter Reception

Since neurotransmitters are found in all living organisms – from unicellular to multicellular ones, Christophersen (1991) has described their possible evolution in terms of the molecular structure of neurotransmitters and adaptive variance in their metabolism, like that known for hormone receptors (Csaba 1980). The similarity of domains in signal receptors (Berman et al. 1991) was seen to compare with the physicochemical properties of signal receptor domains as the basis for sequence comparison. Christophersen (1991) advanced the hypothesis that all metabolites, even minor ones, are expressed as a result of stimuli and are directed against or support actions of receptor-based systems that reflect the evolution of receptors. For example, there is a similarity in some domains of rhodopsin, bacteriorhodopsin, and neurotransmitter receptors (Pertseva 1989, 1990a, b; Fryxell and Meyerowitz 1991). Recently, transgenic technique has permitted the expression of the human dopamine receptor in the potato *Solanum tuberosum* (Skirycz et al. 2005). A blockade of catecholamine-induced growth by adrenergic and dopaminergic receptor antagonists has been also observed for *E. coli* O157:H7, *S. enterica* and *Y. enterocolitica* (Freestone et al. 2007) The similarity and universality of basic endocrine mechanisms of the living world are shown in the examples of the development of receptor-based mechanisms of protozoa and invertebrates (Csaba and Muller 1996). First of all, there are conservative parts or domains in modern cholino- or aminoreceptor, which are also found in prokaryotes and had no changed in the evolution (Pertseva 1989, 1990a, b). Homology of some bacterial proteins (from *Mycobacterium smegmatis*, *Corynebacterium glutamicum*, and *Halobacterium salinarum*) to mammalian neurotransmitter transporters (for example vesicular monoamine transporter) was observed as well (Vardy et al. 2005). Today, molecular evolution of the nicotinic acetylcholine receptor has also been confirmed by the multigene family in excitable cells of highly organized animals (Le Novere and Changeux 1995).

2.3.2 Participation of Neurotransmitters in Chemical Relations Between Organisms

2.3.2.1 Microorganism–Microorganism Relations

Communication between microorganisms through their secretions (extracellular products released) enriched in hormones or neuromediators is proposed in many reports (Kaprelyants and Kell 1996; Kaprelyants et al. 1999; Oleskin et al. 2000;

Kagarlitskii et al. 2003; Oleskin and Kirovskaya 2006; Oleskin 2007). Neurotransmitters participate in the communication with each other for growth, in particular serotonin as an intercellular communication agent accelerating and possibly synchronizing development of the microbial cells. Exogenous serotonin stimulates the growth of yeast *Candida guillermondii*, and the Gram-positive bacterium *Streptococcus faecalis* at low concentration near 10^{-7} M added with a periodicity of 2 h (Strakhovskaya et al. 1993). Photoactivation of the synthesis of endogenous serotonin in cells exposed to UV light at 280–360 nm led to the photostimulation of the same cultivated cells in lag-phase (Strakhovskaya et al. 1991; Belenikina et al. 1991). Exogenous serotonin at 2×10^{-7}–10^{-5} M also accelerates culture growth and induces cell aggregation in *E. coli* and *R. rubrum* (Oleskin et al. 1998a, b). Moreover, dopamine and norepinephrine stimulate the growth of *E. coli*, *S. enterica*, *Y. enterocolitica*, and the staphylococci as well as the yeast *Saccharomyces cerevisiae* (Neal et al. 2001; Kagarlitskii et al. 2003; Oleskin and Kirovskaya 2006; Freestone et al. 2007).

2.3.2.2 Microorganism–Plant Relations

The communications of plant–microorganisms or plant–fungi via neurotransmitters is still a relatively unexplored field. Although the presence of the compounds is documented for some fungi and rhizobial bacteria (Roshchina 2001a), we can only speculate that there is a role for microbial-produced hormones in plant physiology. However, recently Ishihara et al. (2008) found that the rice pathogenic infection by fungi *Bipolaris oryzae* (the formation of brown spots on the leaves) leads to the enhanced serotonin production as a defensive response. In the defensive mechanism, the tryptophan pathway is involved as well. The pathway enzymes of rice have been characterized (Kang et al. 2007).

2.3.2.3 Microorganism–Animal Relations

Microorganisms may live within an animal organism and have simple symbiotic or parasitic relationships with their host. The example of non-parasitic cooperation can be found in the marine sponge that used acetylcholine and its hydrolyzing enzyme acetylcholinesterase of the associated bacterium *Arthrobacter ilicis* (Mohapatra and Bapujr 1998). The clinical aspect suggested by microorganism–animal interactions based on hormones is understandably of special interest. Evans with co-workers first reported in 1948 that catecholamines such as epinephrine were able to enhance bacterial infections. Presently, we know that they may stimulate the growth of Gram-negative bacteria (Lyte and Ernst 1992, 1993). The concept of "microbial endocrinology", in which pathogens are considered to exploit the host effector's molecules as environmental signals promoting growth and virulence factor deployment, has been proposed (Lyte 1992; Lyte and Ernst 1993, Freestone et al. 2008a, b). Cells of bacteria and fungi release neurotransmitters (for example norepinephrine and dopamine) out into the matrix of cellular cover as shown, in particular the bacterium *Bacillus subtilis*, and with the compounds participate in intercellular communication (Oleskin et al. 2000). Matrix contained biopolymers permit low-molecular neurotransmitters to diffuse among the colonial population. In this case, the compounds serve as chemosignals or information agents of short-radius activity. The formulation of the hypothesis regarding the microbial recognition of catecholamines produced during periods of stress as a potential mechanism by which bacteria can utilize the host's environment to initiate pathogenic process was formulated in 1992 (Lyte 1992; Lyte and Ernst 1993) and developed (Lyte et al. 1996; Lyte and Bailey 1997; Freestone et al. 2008 a,b) showed that norepinephrine stimulates the growth of low inocula of commensal and pathogenic *E. coli* in a minimal medium supplemented with serum. Norepinephrine also forms a complex with transferrin-bound iron in blood or serum, and Freestone et al. (1999, 2000) demonstrated that norepinephrine supplies iron for bacterial growth in the presence of transferrin or lactoferrin. Utilization of iron-catecholamine complexes involving ferric reductase activity has also been found for *Listeria monocytogenes* (Coulanges et al. 1997).

Other examples are changes in the blood and tissue histamine content in rabbits when sensitized with streptococci combined with heart muscle extract (Kozlov 1972) or as well in those of serotonin and histamine in the organs infected with bacterium *Bacterium prodigious*. The review of Freestone et al. (2008a) reveals that responsiveness to human stress neurohormones is widespread in the microbial world and relates to the new concept of microbial endocrinology.

2.3.2.4 Plant–Plant Relations

In the relationships between different plant species, neurotransmitters may play a role of attractant or repellent for normal coexistence (Roshchina 1991, 2001a). Plant microspores such as vegetative microspores of horse-tail *Equisetum arvense* from Cryptogam (spore-bearing) plants or various generative microspores (pollens) from Phanerogams (seed-bearing) plants are unicellular structures containing acetylcholine, catecholamines and histamine (Roshchina 2001a) and are the specific objects of microbiology having medicinal areas of the interests (Roshchina 2006b), acting as drugs or allergenous agents. An especially significant role of neurotransmitter compounds is seen in the pollen–pollen interaction named pollen allelopathy and pollen–pistil relations during pollination that regulate fertilization of certain plant species (Roshchina 2001b, 2007, 2008). Catecholamines stimulate the microspores germination (Roshchina 2001a, 2004, 2009). Fungi and other microorganisms living within many plant cells also appear to release neurotransmitters that act as plant growth regulators. We may only speculate on the biological significance of these observations as yet.

2.3.2.5 Plant–Animal Relations

Participation of neurotransmitters in plant–animal relations has been evidently shown for dopamine (van Alstyne et al. 2006). This is dangerous for marine communities, fisheries, and aquaculture facilities due to similar antiherbivore defense that is the cause of reduced feeding by echinoderms, molluscs, and arthropods (see above in Sect. 2.1.2). Role of neurotransmitters excreted in the plant–animal relations is "terra incognita" as yet.

2.3.2.6   Animal–Animal Relations

Neurosecretion utilizes mechanisms common to all eukaryotic membrane transport, and the process should be a model of the secretion as a whole (Bajjalieh and Scheller 1995). The role of neurotransmitters in the contacts with other organisms may be seen from the effects of the secretions released. For instance, large amounts of dopamine are secreted by cells of infusoria *Tetrahymena pyriformis* into their growth medium (Gundersen and Thompson 1985). Secretions from cones of *Drosophila* contain acetylcholine (Yao et al. 2000), and dopamine and norepinephrine are found in the salivary glands and brain of the tick *Boophilus microplus* (Megaw and Robertson 1974). Exogenous neurotransmitters such as dopamine and serotonin may act as both growth stimulators or as defense agents (Boucek and Alvarez 1970; Yamamoto et al. 1999).

2.3.2.7   Biomediator Role of Neurotransmitters

Non-neurotransmitter functions of the compounds known as neurotransmitters are especially important in the relationships of bacteria and fungi with plants and animals. It appears to be a significant factor in nature. Based on our present knowledge, one could imagine neurotransmitters rather as biomediators that via cellular secretions participate in cell–cell communications in biocenosis, i.e., a group of interacting organisms that live in a particular habitat and form a self-regulating ecological community (Fig. 2.3). We think that information about intracellular location of the compounds and their release within any cell as well as

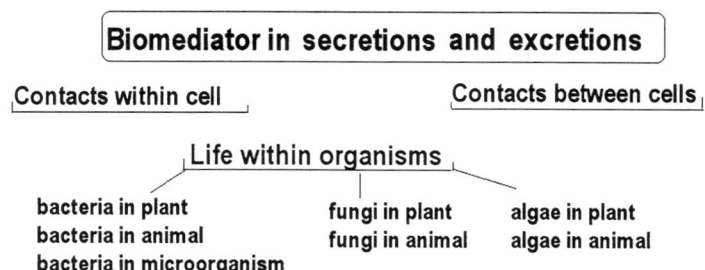

Fig. 2.3 Possible relationships with participation of biomediators

their effects on the cellular organelles could be useful in the study of cell endocrinology. The role of acetylcholine and biogenic amines as intracellular regulators has been confirmed for sea animals (Buznikov 1990; Buznikov et al. 1996) and some plants (Roshchina 1989, 1990a, b, 1991, 2006a, b). A special case is related to the life of microorganisms within host cell of animal or plant. The release of neurotransmitters should occur within and out the guest cell (independently parasitar or not). A universal (biomediator) role of neurotransmitters may be a subject of future investigations.

2.4 Use of Microorganisms and Medicinal Plants Enriched in Neurotransmitters

2.4.1 Microbial Neurotransmitters

Stress stimulates the formation and releasing of biogenic amines, in particular epinephrine, a hormone produced during stress that affects heart rate, blood circulation and other functions of the body. Microorganisms possess the ability to recognize hormones within the host and utilize them to adapt to their surroundings. Norepinephrine and epinephrine, which are released during human stress responses, may act as environmental cues to alter the growth of individual microbes. The growth stimulation of 43 oral bacteria by norepinephrine and epinephrine was found (Roberts et al. 2002), especially for *Actinomyces naeslundii*, *A.s gerenscseriae*, *Eikenella corrodens*, and *Campylobacter gracilis*, and suggest that stress that induces changes in local catecholamine levels in the mouth may play a significant role in the etiology and pathogenesis of periodontal disease. The bacteria-induced enhanced level of the compounds may be recommended as a valuable diagnostic test in medical practice.

Chemical reactions performed by microorganisms have been used as a modern tool in chemistry. In work of Boaventura et al. (2004), the ability of the fungi *Beauveria bassiana* and *Aspergillus niger* to modify the chemical structure of indole compounds was studied. *B. bassiana* was able to transform 3-indolylacetonitrile into 3-methylindole, while *A. niger* transformed tryptamine into 5-hydroxyindole-3-acetamide. These fungi were able to perform both reduction and oxidation of the indole compounds fed, the oxidation occurring with improved levels of oxygen uptake. The synthetic use of microorganisms to perform reactions in the indole nucleus of serotonin is of industrial interest as a way to synthesize active indole derivatives and this area has attracted great attention. According to Heller et al. (2004), serotonin also enhances the activity of membrane-sensitive drug amphotericin B against *A. fumigatus* in vitro. Thus, the combination of known drugs with biogenic amines may lead to the promotion of medicinal effects.

2.4.2 Plant Neurotransmitters

Plants may be suitable for medicine as a source of neurotransmitters and antineurotransmitter drugs or as a polygon for testing of the neurotransmitters and antineurotransmitter compounds as well as model system to study in cell endocrinology.

Neurotransmitters themselves can serve as an active matter of pharmacologically valuable plants. Examples of such plants are published earlier (Roshchina 2001a). Among them are recommendations for practical medicine regarding acetylcholine-enriched food and medicinal species, for example *Digitalis ferruginea* and *Urtica dioica*, catecholamine-enriched *Musa* sp. (Roshchina 2001a), serotonin-enriched *Hippophae rhamnoides*, *Juglans nigra*, and *J. regia* (Bell and Jansen 1971; Badria 2002). Useful features of medicinal plants enriched in neurotransmitters are likely connected with the formation of non-hazard complexes with neurotransmitters such as conjugates of auxins (Ballal et al. 1993; Bozso et al. 1995; Fluck et al. 2000) or phenol–histamine (Hikino et al. 1983). Possession of acetylcholine receptor binding activity is peculiar to many medicinal plants used to improve failing human memory (Wake et al. 2000; Luedtke et al. 2003). Antineurotransmitter natural compounds may also be used in medicine. Agonists or antagonists of neurotransmitters as well as anticholinesterase compounds, mainly alkaloids (Schmeller et al. 1997) or terpenoids (Atta-ur-Rahman et al. 2001), occur in pharmacologically valuable plant material and are effective against diseases from ancient time (Roshchina 2001a). For example, the alkaloids berberine, palmatine, and sanguinarine (inhibitors of cholinesterases, choline acetyltransferase or some receptors) are toxic to insects and vertebrates and inhibit the multiplication of bacteria, fungi, and viruses (Schmeller et al. 1997).

Model systems of plants may also be used for a drug testing. New approaches to the testing of neurotransmitter and antineurotransmitter compounds may be the plant biosensors found among sensitive microobjects, in particular plant microspores such as vegetative horse tail microspores or the generative male microspores named pollen (Roshchina 2004, 2006a, b, 2007). Biosensors are analytical systems, which contain sensitive biological elements and detectors. Intact plant cells are a possible biosensor, having a natural structure that determines their high activity and stability (Roshchina 2006a, b; Budantsev and Roshchina 2004, 2007). Changes in the germination and autofluorescence of unicellular microspores of plants as well as their cholinesterase activity were considered possible biosensor reactions (Roshchina 2005a, b). They could serve as biosensors for medicinal drugs such as known agonists and antagonists of neurotransmitters, instead of animals with the necessity of their vivisection.

Plants appear to be model systems for cell endocrinology. Unicellular plant systems such as the above-mentioned microspores are also suitable models for cellular endocrinology considered in any cell of both unicellular and multicellular organisms. It is a way to an understanding of neurotransmitter occurrence in organelles and different compartments that is based on the concept of universal mechanisms in intracellular chemical signaling from plasmalemma to organelles (Buznikov 1990; Roshchina 1989, 1990a).

The study of neurotransmitter function and location within a cell could be done with fluorescent compounds from microbial and plant cells that bind with the receptors or enzyme of neurotransmitter metabolism (Roshchina 2008). For example, the fluorescent antineurotransmitters *d*-tubocurarine, muscarine (Roshchina 2005a, b, 2008), and some Bodipy derivatives of neurotransmitters (Roshchina et al. 2003; Roshchina 2008) are used as fluorescent natural dyes and markers because they bind with cellular receptors.

2.5 Conclusion

Examination of the compounds known as neurotransmitters or biomediators reveals a similarity in their main functions at the cellular level for all living organisms. These compounds change the membrane ion permeability, electrical characteristics of the cells and in final we see the integral response of the cell or organism as a whole – stimulation or inhibition of growth and development. Neurotransmitters regulate their own metabolic processes within a cell and the relationships (allelopathy) between neighbors with biocenosis, may serve as attractants or repellents as well as oxidative agents (biogenic amines). Interactions between microorganism–microorganism, microorganism–animal (human) and microorganism–plant play essential roles in the environment, and thus neurotransmitter compounds should be considered universal agents of irritation in this living relationship. We should know that our understanding of these relationships is small and that we are just at the beginning. However, the recognition of such relationships is increasingly changing the medical and pharmacological perspectives regarding the non-nervous functions of the neurotransmitters.

References

Amaroli A, Gallus L., Passalacqua M, Falugi C., Viarengo A., and Corrado M.U.D. 2003. Detection of cholinesterase activities and acetylcholine receptors during the developmental cycle of *Dictyostelium discoideum*. Eur. J. Protistol. 39: 213–222.

Anuchin, A.M., Chuvelev, D.I., Kirovskaya, T.A., and Oleskin, A.V. 2007. The effect of monoamine neuromediators on the growth of culture and cellular aggregation of *Escherichia coli* K-12. In Reception and Intracellular Signaling. Proceedings of International conference, 5–7 June 2007, Pushchino, Ed V.P. Zinchenko. pp. 241–243. Pushchino: Institute of Cell Biophysics RAS.

Anuchin, A.M., Chuvelev, D.I., Kirovskaya, T.A., and Oleskin, A.V. 2008. Effects of monoamine neuromediators on the growth-related variables of *Escherichia coli* K-12. Microbiology 77(6): 674–680.

Askar, A., Rubach K. and Schormüller, J. 1972. Dünnschichtchromatographische Trennung der in Bananen vorkommenden Amin-Fraktion. Chem. Microbiol. Technol. Lebensm. 1: 187–190.

Atta-ur-Rahman, Parveen, S., Khalid, A., Farooq, A., and Chouldhary, M.I. 2001. Acetyl and butyryl cholinesterase-inhibiting triterpenoid alkaloids from *Buxus papillosa*. Phytochemistry 58: 963–968.

Augustinsson, K.B. 1949. Substrate concentration and specificity of choline ester-splitting enzymes. Arch. Biochem. 23: 111.

Baburina O., Shabala S., and Newman I. 2000. Verapamil-induced kinetics of ion flux in oat seedlings. Aust. J. Plant Physiol. 27: 1031–1040.

Badria, F.A. 2002. Melatonin, serotonin, and tryptamine in some Egyptian food and medicinal plants. J. Med. Food 5: 153–157

Bajjalieh, S.M. and Scheller R.H. 1995. The biochemistry of neurotransmitter secretion. J. Biol. Chem. 270 (5): 1971–1974.

Ballal, S., Ellias, R., Fluck, R., Jameton, R, Leber, P., Liri, R. and Salama, D. 1993. The synthesis and bioassay of indole-3-acetylcholine. Plant Physiol. and Biochem. 31: 249–255.

Baluska, F., Mancuso, S., Volkmann, D., and Barlow, P. 2004. Root apices as plant command centers: the unique "brain-like" status of the root apex transition zone. Biologia (Bratislava) 59 Suppl. 13: 7–19.

Baluska, F., Volkmann, D., and Menzel, D. 2005. Plant synapses: actin-based domains for cell-to-cell communication. Trends Plant Sci. 10: 106–111.

Baluska, F., Hlavacka, A., Mancuso, S., and Barlow, P.W. 2006a. Neurobiological view of plants and their body plan. In Communication in Plants – Neuronal Aspects of Plant Life, F. Baluska, S. Mancuso, and D. Volkmann, eds. pp. 19–35. Berlin: Springer.

Baluska F, Mancuso S, and Volkmann D, eds. 2006b. Communication in Plants – Neuronal Aspects of Plant Life. Berlin: Springer.

Balzer, I., Poeggeler, B., and Hardeland, R. 1993. Circadian rhythms of indoleamines in a dinoflagellate, *Gonyaulax polyedra*: persistence of melatonin rhythm in constant darkness and relationship to 5-methoxytryptamine. In Melatonin and the Pineal Gland: From Basic Science to Clinical Applications, Y. Touitou, J. Arendt, and P. Pevet, eds. pp. 83–186. Amsterdam: Excerpta Medica.

Bamel K, Gupta S.C., and Gupta, R (2007) Acetylcholine causes rooting in leaf explants of in vitro raised tomato (*Lycopersicon esculentum* Miller) seedlings. Life Sci. 80(24–25): 2393–2396.

Bansal, T, Englert, D., Lee, J., Hegde, M., Wood, T.K. and Jayaraman, A. 2007. Differential effects of epinephrine, norepinephrine, and indole on *Escherichia coli* O157:H7 chemotaxis, colonization, and gene expression. Infect. Immun. 75(9): 4597–4607.

Barwell, C.J. 1979. The occurrence of histamine in the red alga of *Furcellaria lumbricalis* Lamour. Bot. Mar. 22: 399–401

Barwell, C.J. 1989. Distribution of histamine in the thallus of *Furcellaria lumbricalis*. J. Appl. Phycol. 1: 341–344.

Belenikina, N.S., Strakhovskaya, M.G., and Fraikin, G.Ya. 1991. Near-UV activation of yeast growth. J. Photochem. Photobiol. B 10: 51–55

Bell, E.A. and Jansen, D.H. 1971. Medical and ecological considerations of L-dopa and 5-HTP in seeds. Nature 229: 136–137.

Beri, V., and Gupta, R 2007. Acetylcholinesterase inhibitors neostigmine and physostigmine inhibit induction of alpha-amylase activity during seed germination in barley, *Hordeum vulgare* var. Jyoti. Life Sci. 80: 2386–2388.

Berman, A.L., Dityatev, A.E., and Frishman, D.I. 1991. Physicochemical properties of signal receptor domains as the basis for sequence comparison. Comp. Biochem. Physiol. B 98: 445–449.

Boaventura, M.A.D., Lopes, R.F.A., and Takahashi, J.A. 2004. Microorganisms as tools in modern chemistry: the biotransformation of 3-indolylacetonitrile and tryptamine by fungi. Braz. J. Microbiol. 35: 345–347.

Bodmer, S., Imark, C., and Kneubühl, M. 1999. Biogenic amines in foods: histamine and food processing. Inflamm. Res. 48: 296–300.

Boron, W.F. and Boulpaep, E.L. 2005. Medical Physiology: A Cellular and Molecular Approach. Philadelphia, PA: Elsevier/Saunders.

Boucek, R.J. and Alvarez, T.R. 1970. 5-Hydroxytryptamine: a cytospecific growth stimulator of cultured fibroblasts. Science 167: 898–899.

Bozso, B.A., Fluck, R.A., Jameton, R.A., Leber, P.A., and Varnes, J.G. 1995. A versatile and efficient methodology for the preparation of choline ester auxin conjugates. Phytochemistry 40: 1027–1031.

Brenner, E.D., Stahlberg, R., Mancuso, S., Vivanco, J.M., Baluska, F., and van Volkenburgh, E. 2006. Plant neurobiology: an integrated view of plant signaling. Trends Plant Sci. 11: 413–419.

Budantsev, A.Yu. and Roshchina, V.V. 2007. Cholinesterase activity as a biosensor reaction for natural allelochemicals: pesticides and pharmaceuticals. In Cell Diagnostics. Images, Biophysical and Biochemical Processes in Allelopathy, V.V. Roshchina and S.S. Narwal, eds. pp. 127–146. Plymouth: Science Publisher.

Budantsev, A.Yu. and Roshchina, V.V. 2004. Testing alkaloids as acetylcholinesterase activity inhibitors. Farmatsiya (Moscow) 5: 37–39.

Burton, C.L., Chhabra, S.R., Swift S., Baldwin, T.J, Withers, H., Hill, St. J., and Williams, P. 2002. The Growth response of *Escherichia coli* to neurotransmitters and related catecholamine drugs requires a functional enterobactin biosynthesis and uptake system. Infect Immun 70: 5913–5923.

Buznikov, G.A. 1967. Low Molecular Weight Regulators in Embryonic Development. Moscow: Nauka, 265 pp.

Buznikov, G.A. 1987. Neurotransmitters in Embryogenesis. Moscow: Nauka, 232 pp.

Buznikov, G.A. 1990. Neurotransmitters in Embryogenesis. Chur, Switzerland: Harwood Academic Press, 526 pp.

Buznikov, G.A. 2007. Preneuronal transmitters as regulators of embryogenesis. Current state of problem. Russ. J. Develop. Biol. (Ontogenesis) 38: 262–270.

Buznikov. G.A., Shmukler, Y.B., and Lauder, J.M. 1996. From oocyte to neuron: do neurotransmitters function in the same way throughout development? Cell Mol. Neurobiol. 16: 537–559

Cantiello, H.F. 1997. Role of actin filaments organization in cell volume and ion channel regulation. J. Exp. Zool. 279: 425–435.

Christophersen, C. 1991. Evolution in molecular structure and adaptive variance in metabolism. Comp. Biochem. Physiol. B 98: 427–443.

Corrado, D.M.U., Politi, H., Trielli, F., Angelini, C., and Falugi, C.1999. Evidence for the presence of a mammalian-like cholinesterase in *Paramecium primaurelia* (Protista, Ciliophora) developmental cycle. J. Exp. Zool. 283: 102–105.

Corrado, D.M.U., Ballarini, P., and Falugi, C. 2001 Synthesis of the molecular acetylcholine during the developmental cycle of *Paramecium primaurelia* (Protista, Ciliophora) and its possible function in conjugation. J. Exp. Biol. 204: 1901–1907.

Coulanges, V., Andre, P., Ziegler, O., Buchheit, L., and Vidon, J.-M. 1997. Utilization of iron-catecholamine complexes involving ferric reductase activity in *Listeria monocytogenes*. Infect. Immun. 65: 2778–2785.

Csaba, G. 1980. Phylogeny and ontogeny of hormone receptors: the selection theory of receptor formation and hormonal imprinting. Biol. Rev. 55: 47–63.

Csaba, G. and Muller, W.E.G. 1996. Signalling Mechanisms in Protozoa and Invertebrates. Berlin: Springer.

Devalia, J.L., Grady, D., Harmanyeri, Y., Tabaqchali, S., and. Davies, R. J. 1989. Histamine synthesis by respiratory tract micro-organisms: possible role in pathogenicity. J. Clin. Pathol. 42: 516–522.

Ekici, K. and Coskun, H. 2002. Histamine content of some commercial vegetable pickles. Proceedings of ICNP-2002 – Trabzon, Turkye. pp. 162–164.

Ekici, K., Coskun, H., Tarakci, Z., Ondul, E., and Sekeroglu, R. 2006. The contribution of herbs to the accumulation of histamine in "otlu" cheese. J. Food Biochem. 30: 362–371.

Evans, D.G., Miles, A.A., and. Niven, J.S.F. 1948. The enhancement of bacterial infections by adrenaline. Br. J. Exp. Pathol. 29: 20–39.

Faust, M.A. and Doetsch, R.N. 1971. Effect of drugs that alter excitable membranes on the motility of *Rhodospirillum rubrum* and *Thiospirillum rubrum* and *Thiospirillum jenense*. Can. J. Microbiol. 17: 191–196.

Fernández, M., del Río, B., Linares, D.M., Martín M.C., and Alvarez M.A. 2006. Real-time polymerase chain reaction for quantitative detection of histamine-producing bacteria: use in cheese production. J. Dairy Sci. 89: 3763–3769.

Fernstrom, J.D. and Wurtman, R.J. 1971. Brain serotonin content: physiological dependence on plasma tryptophan levels. Science 173: 149–152.

Fitch, W.M. 1963a Studies on a cholinesterase of *Pseudomonas fluorescens*. I. Enzyme induction and the metabolism of acetylcholine. Biochemistry 2: 1217–1221.

Fitch. W.M. 1963b. Studies on a cholinesterase of *Pseudomonas fluorescens*. II. Purification and properties. Biochemistry 2(6): 1221–1227.

Fluck, R.A., Leber, P.A., Lieser, J.D., Szczerbicki, S.K., Varnes, J.G., Vitale, M.A., and Wolfe, E.E. 2000. Choline conjugates of auxins. I. Direct evidence for the hydrolysis of choline-auxin conjugates by pea cholinesterase. Plant Physiol. Biochem. 38: 301–308.

Fraikin, G.Ya., Strakhovskaya, M.G., Ivanova, E.V., and Rubin, A.B. 1989. Near-UV activation of enzymatic conversion of 5-hydroxytryphophan to serotonin. Photochem. Photobiol. 49: 475–477.

Freestone, P.P.E. and Lyte, M. 2008. Microbial endocrinology: experimental design issues in the study of interkingdom signalling in infectious disease .Adv. Appl. Microbiol. 64: 75–105.

Freestone, P. E., Haigh, R. D. Williams, P. H., and M. Lyte. 1999. Stimulation of bacterial growth by heat-stable, norepinephrine-induced autoinducers. FEMS Microbiol. Lett. 172: 53–60.

Freestone, P.E., Lyte, M., Neal, C.P., Maggs, A.F., Haigh, R.D., and Williams, P.H. 2000. The mammalian neuroendocrine hormone norepinephrine supplies iron for bacterial growth in the presence of transferrin or lactoferrin. J. Bacteriol. 182: 6091–6098.

Freestone, P.E., Haigh, R.D., and Lyte, M. 2007. Blockade of catecholamine-induced growth by adrenergic and dopaminergic receptor antagonists in *Escherichia coli* O157: H7, *Salmonella enterica* and *Yersinia enterocolitica*. BMC Microbiol. 7: 8–11.

Freestone, .P.P.E, Sandrini, S.M., Haigh, R.D., and Lyte, M. 2008a. Microbial endocrinology: how stress influences susceptibility to infection. Trends Microbiol. 16: 55–64.

Freestone, P.P.E., Haigh, R.D., and Lyte, M. 2008b. Catecholamine inotrope resuscitation of antibiotic-damaged staphylococci and its blockade by specific receptor antagonists. J. Infect. Dis. 197: 1044–1052.

Fryxell, K.J. and Meyerowitz, E.M. 1991. The evolution of rhodopsins and neurotransmitter receptors. J. Mol. Evol. 33: 367–378.

Girvin, G.T. and Stevenson, J.W. 1954. Cell free "choline acetylase" from *Lactobacillus plantarum*. Can. J. Biochem. Physiol. 32: 131–146.

Goldstein, D.B. 1959. Induction of cholinesterase biosynthesis in *Pseudomonas fluorescens*. J. Bacteriol. 78: 695–702.

Goldstein, D.B. and Goldstein, A. 1953. An adaptive bacterial cholinesterase from *Pseudomonas* species. J. Gen. Microbiol. 8: 8–17.

Gong, X.O. and Bisson, M.A. 2002. Acetylcholine-activated Cl⁻ channel in the *Chara* tonoplast. J. Membr. Biol. 188: 107–113.

Gundersen, R.E. and Thompson, G.A. Jr. 1985. Further studies of dopamine metabolism and function in *Tetrahymena*. J. Eukaryot. Microbiol. 32: 25–31.

Gupta, A., Thakur, S.S., Uniyal, P.L., and Gupta, R. 2001. A survey of Bryophytes for presence of cholinesterase activity. Am. J. Bot. 88: 2133–2135.

Heller, I., Leitner, S., Dierich, M. P., and Lass-Flörl, C. 2004. Serotonin (5-HT) enhances the activity of amphotericin B against *Aspergillus fumigatus in vitro*. Int. J. Antimicrob. Agents 24: 401–404.

Hikino, H., Ogata, M., and Konno, C. 1983. Structure of feruloylhistamine, a hypotensive principle of *Ephedra* roots. Planta Med. 48: 108–109.

Hsu, S.C., Johansson, K.R., and Donahue, M.J. 1986. The bacterial flora of the intestine of *Ascaris suum* and 5-hydroxytryptamine production. J. Parasitol. 72: 545–549.

Imshenetskii, A.A., Popova, L.S., and Kirilova, N.F. 1974. Microorganisms decomposing acetylcholine. Microbiology (in Russian) 43(6): 986–991.

Ishihara, A., Hashimoto, Y., Tanaka, C., Dubouzet, J. G., Nakao, T., Matsuda, F., Nishioka, T., Miyagawa, H., and Wakasa, K. 2008. The tryptophan pathway is involved in the defense responses of rice against pathogenic infection via serotonin production. Plant J. 54(3): 481–495.

Iyer, L.M., Aravind, L., Coon, S.L., Klein, D.C., and Koonin, E.V. 2004. Evolution of cell–cell signaling in animals: did late horizontal gene transfer from bacteria have a role? Trends Genet. 20(7): 292–299.

Janakidevi, K., Dewey, V.C., and Kidder, G.W. 1966a. Serotonin in protozoa. Arch. Biochem. Biophys. 113: 758–759.

Janakidevi, K., Dewey, V.C., and Kidder, G.W. 1966b. The biosynthesis of catecholamines in two genera of Protozoa. J. Biol. Chem. 241: 2576–2578.

Kagarlitskii, G.O, Kirovskaya, T.A., and Oleskin, A.V. 2003. The effects of neuromediator amines on the growth and respiration of microorganisms. In Biopolytics, Seminar of Biological Faculty of MGU, pp.13–17.

Kamo, K.K. and Mahlberg, P.G. 1984. Dopamine biosynthesis at different stages of plant development in *Papaver somniferum*. J. Nat. Prod. 47: 682–686.

Kang, S., Kang, K., Lee, K., and Back, K. 2007. Characterization of tryptamine 5-hydroxylase and serotonin synthesis in rice plants. Plant Cell Rep. 26: 2009–2015.

Kaprelyants, A.S., and Kell, D.B. 1996. Do bacteria need to communicate with each other for growth? Trends Microbiol. 4: 237–242.

Kaprelyants, A.S., Mukamolova, G.V., Kormer, S.S., Weichart, D.H., Young, M., and Kell, D.B. 1999. Intercellular signaling and the multiplication of prokaryotes. In Microbial Signaling and Communication, R. England, et al., eds. Society for General Microbiology Symposium 57, pp. 33–69. Cambridge: Cambridge University Press.

Kawashima, K., Misawa, H., Moriwaki, Y., Fujii, Y.X., Fujii, T., Horiuchi, Y., Yamada, T., Imanaka, T., and Kamekura, M. 2007. Ubiquitous expression of acetylcholine and its biological functions in life forms without nervous systems. Life Sci. 80: 2206–2209.

Kondashevskaya, M.V., Lyapina, L.A., and Smolina, T.Yu. 1996. Complexes of high- and low-molecular heparin with serotonin and their physiological features. Vestn. MGU Ser. Biol. 16: 17–20.

Koshtoyantz, Ch.S. 1963. Problems of enzymochemistry of the processes of excitation and depression and evolution of the function of nervous system. 17th Bakh Lection (USSR), AN SSSR Publ. House, Moscow, 31 pp.

Kovaleva, L.V. and Roshchina, V.V. 1997. Does cholinesterase participate in the intercellular interaction in pollen-pistil system? Biol. Plant. 39(2): 207–213

Kozlov, G.S. 1972. Changes in the blood and tissue histamine content in rabbits when sensitized with streptococci combined with heart muscle extract. Bull. Exp. Biol. Med. 74: 1028–1029.

Kruk, Z.L. and Pycock, C.J. 1990. Neurotransmitters and Drugs. New York: Chapman and Hall.

Kuklin, A.I. and Conger, B.V. 1995. Catecholamines in plants. J.Plant Growth Regul. 14: 91–97.

Kulma, A. and Szopa, J. 2007. Catecholamines are active compounds in plant. Plant Sci. 172: 433–440.

Kung, H.F., Tsai, Y.H., and Wei, C.I. 2007. Histamine and other biogenic amines and histamine-forming bacteria in miso products. Food Chem. 101: 351–356.

Laing, A.C., Miller, H.R., and Bricknell, K.S. 1967. Purification and properties of the inducible cholinesterase of *Pseudomonas fluorescence* (Goldstein). Can. J. Biochem. 45: 1711–1724.

Laing, A.C., Miller, H.R., and Patterson, K.M. 1969. Purification of bacterial cholinesterase. Can. J. Biochem. 47: 219–220.

Landete, J. M., Ferrer, S., and Pardo. I. 2005. Which lactic acid bacteria are responsible for histamine production in wine? J. Appl. Microbiol. 99:580–586.

Lauder, J.M. and Schambra, U.B. 1999. Morphogenetic roles of acetylcholine. Environmental health perspectives. Rev. Environ. Health 107(Suppl. 1): 65–69.

Lawrence, S.A. 2004. Amines: Synthesis, Properties and Applications. Cambridge: Cambridge University Press.

Le Novere, N. and Changeux, J.P. 1995. Molecular evolution of the nicotinic acetylcholine receptor: an example of multigene family in excitable cells. J. Mol. Evol. 40: 155–172.

Lenard, J. 1992. Mammalian hormones in microbial cells. Trends Biochem. Sci. 17: 147–150.

Leng, Q., Hua, B., Guo, Y., and Lou, C. 2000. Regulating role of acetylcholine and its antagonists in inward rectified K$^+$ channels from guard cells protoplasts of *Vicia faba* Science in China. Ser. C. 43(2): 217–224.

Lu, G., DeLisle, A.J., de Vetten, N.C., and Ferl, R.J. 1992. Brain proteins in plants: an *Arabidopsis* homolog to neurotransmitter pathway activators is part of a DNA binding complex. Proc. Natl. Acad. Sci. USA 89: 11490–11494.

Luedtke, R.R., Freeman, R.A., Volk, M., Arfan, M., and Reinecke, M.G. 2003. Pharmacological survey of medicinal plants for activity at dopamine receptor subtypes. II. Screen for binding activity at the D1 and D2 dopamine receptor subtypes. Pharm. Biol. 41: 45–58.

Lyte, M. 1992. The role of microbial endocrinology in infection disease. J. Endocrinol. 137: 343–345.

Lyte, M. and Bailey, M.T. 1997. Neuroendocrine-bacterial interactions in neurotoxin-induced model of trauma. J. Surg. Res. 70: 195–201.

Lyte, M. and Ernst, S. 1992. Catecholamine induced growth of gram negative bacteria. Life Sci. 50: 203–212.

Lyte, M. and Ernst, S. 1993. Alpha and beta adrenergic receptor involvement in catecholamine-induced growth of gram-negative bacteria. Biochem. Biophys. Res. Commun. 190: 447–452.

Lyte, M., Frank, C.D., and Green, B.T. 1996. Production of an autoinducer of growth by norepinephrine cultured *Escherichia coli* O157: H7. FEMS Microbiol. Lett. 139: 155–159.

Lyte, M., Arulanandam, B., Nguyen, K., Frank, C., Erickson, A., and Francis, D. 1997. Norepinephrine induced growth and expression of virulence associated factors in enterotoxigenic and enterohemorrhagic strains of *Escherichia coli*. Adv. Exp. Med. Biol. 412: 331–339.

Marquardt, P. and Falk, H. 1957. Vorkommen und Syntheses von Acetylcholin in Pflanzen and Bakterien. Arzneimittelforschung 7: 203–211.

Marquardt, P. and Spitznagel, G. 1959. Bakterielle Acetylcholine Bildung in Kunstlichen Nahrboden. Arzneimittelforschung 9: 456–465.

Martín, M.C., Fernández, M., Linares, D.M., and Alvarez, M.A. 2005. Sequencing, characterization and transcriptional analysis of the histidine decarboxylase operon of *Lactobacillus buchneri*. Microbiology 151: 1219–1228.

Megaw, M.W.J. and Robertson, H.A. 1974. Dopamine and noradrenaline in the salivary glands and brain of the tick, *Boophilus microplus*: effect of reserpine. Cell. Mol. Life Sci. 30: 1261–1262.

Meng, F., Liu, X., Zhang, S., and Lou, C. 2001. Localization of muscarinic acetylcholine receptor in plant guard cells. Chin. Sci. Bull. 46: 586–589.

Meng, F., Miao, L., Zhang, S., and Lou, C. 2004. Ca^{2+} is involved in muscarine acetylcholine receptor –mediated acetylcholine signal transduction in guard cells of *Vicia faba*. Chin. Sci. Bull. 49(5): 471–475.

Mohapatra, B.R. and Bapujr, M. 1998. Characterization of acetylcholinesterase from *Arthrobacter ilicis* associated with the marine sponge. J. Appl. Microbiol. 84(3): 393–398.

Momonoki, Y.S. 1997. Asymmetric distribution of acetylcholinesterase in gravistimulated maize seedlings. Plant Physiol. 114: 47–53.

Murch, S.J. 2006. Neurotransmitters, neuroregulators and neurotoxins in plants. In Communication in Plants – Neuronal Aspects of Plant Life, F. Baluska, S. Mancuso, and D. Volkmann, eds. pp. 137–151. Berlin: Springer.

Nachmansohn, D. and Machado, A.L. 1943. The formation of acetylcholine. A new enzyme "choline acetylase". J. Neurophysiol. 6: 397–403.

Neal, C.P., Freestone, P.P.E., Maggs, A.F., Haigh, R.D., Williams, P.H., and Lyte, M. 2001. Catecholamine inotropes as growth factors for *Staphylococcus epidermidis* and other coagulase-negative staphylococci. FEMS Microbiol. Lett. 194: 163–169.

Nelson T., Lee, D., and Smith, B. 2003. Are 'green tides' harmful algal blooms? Toxic properties of water-soluble extracts from two bloom-forming macroalgae, *Ulva fenestrate* and *Ulvaria obscura* (Ulvophyceae). J. Phycol. 39: 874–879.

Oleskin, A.V. 2007. Biopolitics. Moscow: Nauchnii Mir., 508 pp.

Oleskin, A.V. and Kirovskaya, T.A., 2006. Research on population organization and communication in microorganisms. Microbiology (Russia) 75: 440–445.

Oleskin, A.V., Kirovskaya, T.A., Botvinko, I.V., and Lysak, L.V., 1998a. Effects of serotonin (5-hydroxytryptamine) on the growth and differentiation of microorganisms. Microbiology (Russia) 67: 305–312.

Oleskin, A.V., Botvinko, I.V., and Kirovskaya, T.A. 1998b. Microbial endocrinology and biopolytics. Vestn. Mosc. Univ. Ser. Biol. 4: 3–10.

Oleskin, A.V., Botvinko, I.V., and Tsavkelova, E.A. 2000. Colonial organization and intercellular communication of microorganisms. Microbiology (Russia) 69: 309–327.

Pershin, G.N., and Nesvadba, V.V. 1963. A study of monoamino oxidase activity in mycobacteria. Bull. Exp. Biol. Med. 56: 81–84.

Pertseva, M.N. 1989. Molecular Base of the Development of the Hormone-Competency. Leningrad: Nauka, 310 pp.

Pertseva, M.N. 1990a. The path of the evolution of the hormonal signal realization system. Sechenov Physiol. J. USSR 76: 1126–1137.

Pertseva, M.N. 1990b. Is the evolution similarity between chemosignalling systems of eukaryotes and prokaryotes? J. Evol. Biochem. Physiol. (Russia) 26: 505–513.

Ponchet, M., Martin-Tanguy, J., Marais, A., and Martin, C. 1982. Hydroxycinnamoyl acid amides and aromatic amines in the influorescences of some Araceae species. Phytochemistry 21: 2865–2869.

Posmyk, M.M. and Janas, K.M. 2009. Melatonin in plants. Acta Physiol. Plant 31: 1–11.

Roberts, A., Matthews, J.B., Socransky, S.S., Freestone, P.P.E, Williams, P.H., and Chapple, I.L.C. 2002. Stress and the periodontal diseases: effects of catecholamines on the growth of periodontal bacteria in vitro. Oral Microbiol. Immunol. 17: 296–303.

Romanovskaya, M.G. and Popenenkova, Z.A. 1971. Effect of vetrazine, chloracizine, and chlorpromazine on histamine and serotonin content in organs of rabbits with *Bacterium prodigiosum* bacteriemia. Bull Exp. Biol. Med. 520–522.

Roshchina, V.V. 1989. Biomediators in chloroplasts of higher plants. I. The interaction with photosynthetic membranes. Photosynthetica 23: 197–206.

Roshchina,V.V. 1990a. Biomediators in chloroplasts of higher plants. 3. Effect of dopamine on photochemical activity. Photosynthetica 24: 117–121.

Roshchina,V.V. 1990b. Biomediators in chloroplasts of higher plants. 4. Reception by photosynthetic membranes. Photosynthetica 24: 539–549.

Roshchina, V.V. 1991. Biomediators in plants. Acetylcholine and biogenic amines. Pushchino: Biological Center of USSR Academy of Sciences, 192 pp.

Roshchina,V.V. 2001a. Neurotransmitters in plant life. Plymouth: Science Publ., 283 pp.

Roshchina, V.V. 2001b. Molecular-cellular mechanisms in pollen allelopathy. Allellopathy J. 8: 3–25.

Roshchina,V.V. 2004. Cellular models to study the allelopathic mechanisms. Allelopathy J. 13: 3–16.

Roshchina,V.V. 2005a. Contractile proteins in chemical signal transduction in plant microspores. Biol. Bull. 32: 281–286.

Roshchina, V.V. 2005b. Allelochemicals as fluorescent markers, dyes and probes. Allelopathy J. 16: 31–46.

Roshchina, V.V. 2006a. Chemosignaling in plant microspore cells. Biol. Bull. 33: 414–420.

Roshchina, V.V. 2006b. Plant microspores as biosensors. Trends Mod. Biol. (Russia) 126: 262–274.

Roshchina, V.V. 2007. Cellular models as biosensors. In Cell Diagnostics: Images, Biophysical and Biochemical Processes in Allelopathy, V.V. Roshchina and S.S. Narwal, eds. pp. 5–22. Plymouth: Science Publisher.

Roshchina, V.V. 2008. Fluorescing World of Plant Secreting Cells. Plymouth: Science Publishers, 338 pp.

Roshchina, V.V. 2009. Effects of proteins, oxidants and antioxidants on germination of plant microspores. Allelopathy J. 23(1): 37–50.

Roshchina, V.V. and Roshchina, V.D. 1993. The Excretory Function of Higher Plants. Berlin: Springer, 314 pp.

Roshchina, V.V. and Vikhlyantsev, I.M. 2009. Mechanisms of chemosignalling in allelopathy: role of Ion channels and cytoskeleton in development of plant microspores. Allelopathy J. 23(1): 25–36.

Roshchina, V.V., Bezuglov, V.V., Markova, L.N., Sakharova, N.Yu., Buznikov, G.A., Karnaukhov, V.N., and Chailakhyan, L.M. 2003. Interaction of living cells with fluorescent derivatives of biogenic amines. Dokl. Russ. Acad. Sci. 393: 832–835.

Rowatt, E. 1948. The relation of pantothenic acid to acetylcholine formation by a strain of *Lactobacillus plantarum*. J. Gen. Microbiol. 2: 25–30.

Sagane, Y., Nakagawa, T., Yamamoto, K., Michikawa, S., Oguri, S., and Momonoki, Y.S. 2005. Molecular characterization of maize acetylcholinesterase. A novel enzyme family in the plant kingdom. Plant Physiol. 138: 1359–1371.

Schmeller, T., Latz-Brüning, B., and Wink, M. 1997. Biochemical activities of berberine, palmatine and sanguinarine mediating chemical defence against microorganisms and herbivores. Phytochemistry 44: 257–266.

Searle, B.W. and Goldstein, A. 1957. Neostigmine resistance in a cholinesterase-containing *Pseudomonas*: a model for the study of acquired drug resistance. J. Pharmacol. Exp. Ther. 119: 182.

Searle, B.W. and Goldstein, A. 1962. Mutation to neostigmine resistance in a cholinesterase-containing *Pseudomonas*. J. Bacteriol. 83: 789–796.

Shmukler, Yu.B., Tosti, E., and Silvestre, F. 2007. Effect of local microapplication of serotoninergic drugs on membrane currents of *Paracentrotus lividus* early embryos. Russ. J. Develop. Biol. (Ontogenesis) 38: 254–261.

Skirycz, A., Swiedrych, A., and Szopa, J. 2005. Expression of human dopamine receptor in potato (*Solanum tuberosum*) results in altered tuber carbon metabolism. BMC Plant Biol. 5: 1–15.

Small, D.H., Michaelson, S., and Sberna, G. 1996. Non-classical actions of cholinesterases: role in cellular differentiation, tumorigenesis and Alzheimer's disease. Neurochem. Int. 28(5): 453–483.

Stephenson, M. and Rowatt, E. with participation of Harrison K. in addendum. 1947. The production of acetylcholine by a strain of *Lactobacillus plantarum*. J. Gen. Microbiol. 1: 279–298.

Strakhovskaya, M.G., Belenikina, N.S., and Fraikin, G.Ya. 1991. Yeast growth activation by UV light in the range of 280–380 nm. Microbiology (Russia) 60: 292–297.

Strakhovskaya, M.G., Ivanova, E.V., and Fraikin, G.Ya. 1993. Stimulatory effect of serotonin on the growth of the yeast *Candida guillermondii* and the bacterium *Streptococcus faecalis*. Microbiology (Russia) 62: 46–49.

Swiedrych, A., Kukuła, K.L., Skirycz,A., and Szopa, J. 2004. The catecholamine biosynthesis route in potato is affected by stress. Plant Physiol. Biochem. 42: 593–600.

Szewczyk, N.J., Hartman, J.J., Barmada, S.J., and Jacobso, L.A. 2000. Genetic defects in acetylcholine signalling promote protein degradation in muscle cells of *Caenorhabditis elegans*. J. Cell Sci. 133:2003–2010.

Szopa, J., Wilczynski, G., Fiehn, O., Wenczel, A., and Willmitzer, L. 2001. Identification and quantification of catecholamines in potato plants (*Solanum tuberosum*) by GC-MS. Phytochemistry 58: 315–320.

Takenaka, Y., Roh, J.H., Suzuki, H., Yamamoto, K., and Kumaga, H. 1997. Metal ionic induction: expression of monoamine oxidase gene of *Escherichia coli* is induced by copper ion. J. Ferment. Bioeng. 83: 194–196.

Taylor, S.L., Leatherwood, M., and Cieber, E.R. 1978. Histamine in sauerkraut. J. Food Sci. 43: 1030–1032.

Tezuka, T., Akita, I., Yoshino N., and Suzuki, Y. 2007. Regulation of self-incompatibility by acetylcholine and cAMP in *Lilium longiflorum*. J. Plant Physiol. 164: 878–885.

Tsavkelova, E.A., Botvinko, I.V., Kudrin, V.S., and Oleskin, A.V. 2000. Detection of neurotransmitter amines in microorganisms using of high performance liquid chromatography. Dokl Biochem. 372: 115–117 (in Russian issue 840–842).

Tsavkelova, E.A.,Klimova, S.Yu., Cherdyntseva, T.A., and Netrusov, A.I. 2006. Hormones and hormone-like substances of microorganisms: a review. Appl. Biochem. Microbiol. (Russia) 42: 229–235.

van Alstyne, K.L., Nelson, A.V., Vyvyan J.R., and Cancilla D.A. 2006. Dopamine functions as an antiherbivore defense in the temperate green alga *Ulvaria obscura*. Oecologia 148: 304–311.

Vardy, E., Steiner-Mordoch, S., and Schuldiner, S. 2005. Characterization of bacterial drug anti-porters homologous to mammalian neurotransmitter transporters. J. Bacteriol. 187: 7518–7525.

Volkov, A.G., Carrell, H., Baldwin, A., and Markin V.S. 2009. Electrical memory in Venus flytrap. Bioelectrochemistry 74(1): 23–28.

Wake, G., Court, J., Pickering, A., Lewis, R., Wilkins. R., and Perry, E. 2000. CNS acetylcholine receptor activity in European medicinal plants traditionally used to improve failing memory. J. Ethnopharmacol. 69: 105–114.

Wang, H., Wang, X, Zhang, S. and Lou, C. 1998. Nicotinic acetylcholine receptor is involved in acetylcholine regulating of stomatal movement. Science in China, ser.C 41: 650–656.

Wang, H., Wang, X., and Lou, C. 1999a. Relationship between acetylcholine and stomatal movement. Acta Botanica Sinica 41: 171–175.

Wang, H., Wang, X., and Zhang, S. 1999b. Extensive distribution of acetylcholinesterase in guard cells of Vicia faba. Acta Botanica Sinica 41: 364–367.

Wang, H., Wan, X., Zhang, S., and Lou, C. 2000. Muscarinic acetylcholine receptor involved in acetylcholine regulating of stomatal function. Chin. Sci. Bull. 45: 250–252.

Wang, H., Zhang, S., Wang, X., and Lou, C. 2003a. Involvement of Ca^{2+}/CaM in the signal trans-duction of acetylcholine-regulating stomatal movement. Chin. Sci. Bull. 48: 351–354.

Wang, H., Zhang, S., Wang, X., and Lou, C. 2003b. Role of acetylcholine on plant root-shoot signal transduction. Chin. Sci. Bull. 48: 570–573.

Werle, E. and Pechmann, E. 1949. Über die Diamin-oxydase der Pflanzen und ihre adaptative Bildung durch Bakterien. Liebig Ann. Chem. 562: 44–60.

Werle, E. and Raub, A. 1948. Über Vorkommen, Bildung und Abbau biogener Amine bei Pflanzen unter besonderer Beruck-sichtigung des Histamins. Biochem. Z. 318: 538–553.

Wessler, I., Kilbinger, H., Bittinger, F., and Kirkpatrick, C.J. 2001. The non-neuronal cholinergic system: the biological role of non-neuronal acetylcholine in plants and humans. Jpn. J. Pharmacol. 85: 2–10.

Wisniewska, J. and Tretyn, A. 2003. Acetylcholinesterase activity in Lycopersicon esculentum and its phytochrome mutants. Plant Physiol. Biochem. 41: 711–717.

Yagodina, O.V., Nikol'skaya, E.B., Shemarova, I.Y., and Khovanskikh, A.E. 2000. Amine oxidase in unicellular microorganisms Methanosarcina barkeri and Tetrahymena pyriformis. J. Evol. Biochem. Physiol. (Russia) 36: 244–248.

Yamaguchi, H., Friedman, H., and Yamamoto, Y. 2003. Involvement of nicotinic acetylcholine receptors in controlling Chlamidia pneumoniae growth in epithelial HEp-2 Cells. Infect. Immun. 71(6): 3645–3647.

Yamamoto, H., Shimizu, K., Tachibana, A., and Fusetani, N. 1999. Roles of dopamine and sero-tonin in larval attachment of the barnacle, Balanus amphitrite. J. Exp. Zool. 284: 746–758.

Yao, W.D., Rusch, J., Poo, M., and Wu, C.F. 2000. Spontaneous acetylcholine secretion from developing growth cones of Drosophila central neurons in culture: effects of cAMP-pathway mutations. J. Neurosci. 20: 2626–2637.

Chapter 3
Mechanisms by Which Catecholamines Induce Growth in Gram-Negative and Gram-Positive Human Pathogens

Primrose P.E. Freestone and Sara Sandrini

3.1 Introduction: The Importance of Iron in Bacterial Growth

Iron is essential for the growth of most bacterial pathogens (Ratledge and Dover 2000). The iron restriction imposed by mammalian tissues therefore represents an important innate immune defence against the growth of infectious agents. In humans and all other animals, iron assimilated from the diet enters into the blood and is transported by forming a complex with the high affinity iron binding protein transferrin (Tf). Several classes of Tf exist: serum Tf, lactoferrin (Lf) which is found in extra-cellular fluids such as mucosal secretions; and ovo-transferrin which is found in the albumin of eggs. Tf and Lf have similar structures consisting of single-chain glycoprotein with a molecular weight of 80,000, which contains two similar, but not identical, binding sites for iron (designated N-terminal and C-terminal) (Lambert et al. 2005). Both the N-and C-terminal Tf and Lf iron binding sites have binding constants for iron in the region of 10^{20}M, though the affinity varies a little between the two sites. Tf and Lf normally exist in a partially iron saturated state, which ensures that the free iron concentration in host tissues is less than 10^{-16}M, a concentration several log orders lower than that required for the growth of most bacterial pathogens.

Not unexpectedly, in response to the extremely limited iron availability of host tissues, infectious bacteria have evolved an array of mechanisms by which to steal this essential nutrient (Ratledge and Dover 2000). These include Tf-binding proteins (Tpbs), highly conserved bacterial proteins that bind specifically to the Tf of their particular host species. (Ratledge and Dover 2000). For several species, inactivation of Tbps can result in significantly reduced virulence, thereby showing the importance to the infectious disease process of bacterial accession of Tf (and Lf)-sequestered iron in vivo. Another strategy that pathogenic bacteria often employ to

P.P.E. Freestone (✉) and S. Sandrini
Department of Infection, Immunity and Inflammation, School of Medicine,
University of Leicester, University Road, Leicester, LE1 9HN, UK
e-mail: ppef1@le.ac.uk

M. Lyte and P.P.E. Freestone (eds.), *Microbial Endocrinology*,
Interkingdom Signaling in Infectious Disease and Health,
DOI 10.1007/978-1-4419-5576-0_3, © Springer Science+Business Media, LLC 2010

scavenge nutritionally essential iron is the production and utilisation of siderophores, low molecular weight secreted catecholate or hydroxamate molecules that possess very high affinity for ferric iron (Ratledge and Dover 2000). However, siderophores are not always able to directly acquire iron from high affinity ferric iron binding proteins such as Tf and Lf, but as will shown later in this chapter, siderophores are, for Gram-negative bacteria important elements in the mechanism by which catecholamines can induce bacterial growth.

3.2   The Spectrum of Bacterial Catecholamine Growth Induction

In serum- or blood-supplemented media, the magnitude of bacterial growth stimulation possible with catecholamine stress hormones can be in 20 h or less up to a 5 logs higher than un-supplemented control cultures (reviewed in Freestone et al. 2008a). Figure 3.1 shows the wide variety of bacterial species that have been reported to be catecholamine-responsive. Although the spectrum of bacteria

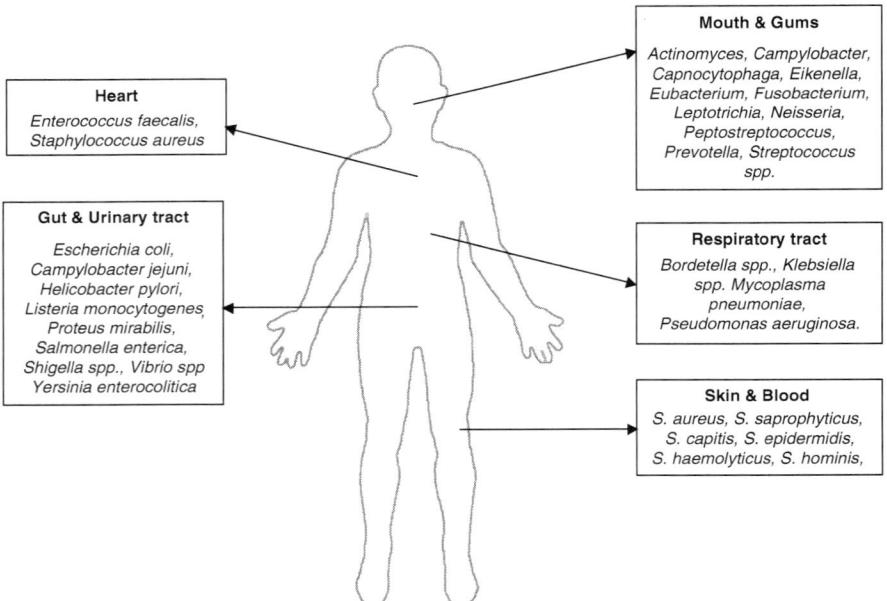

Fig. 3.1 The spectrum of stress-hormone responsive bacteria. Over 50 different types of bacteria occupying all the major niches of the mammalian body have now been shown to be stress hormone responsive in terms of growth induction or enhancement of virulence. Freestone et al. (2008a) gives a detailed explanation of how stress hormones affect the physiology and infectivity of the species shown

shown is not surprisingly weighted towards species inhabiting the highly innervated gastrointestinal tract, catecholamines have a ubiquitous distribution throughout the mammalian body, and it is now clear that bacteria occupying a wide variety of locations might have similarly evolved the ability to sense changes in the stress hormone levels of their host. For example, in humans, exposure to chronic emotional stress is a significant risk factor in the development of periodontal disease, an inflammatory gum condition for which there is strong evidence of bacterial involvement. Stress hormones such as cortisol, epinephrine and norepinephrine have all been isolated from saliva and the fluid occupying the gingival crevicula, and are known to increase during stress. Using a serum-based medium and anaerobic culture conditions, Roberts and co-workers showed that norepinephrine and epinephrine were recognised as stimulatory agents by oral bacteria implicated in causing periodontal disease (Roberts et al. 2002, 2005). Out of the 43 species tested, more than half showed significant catecholamine-growth enhancement, thereby providing additional insight into why stress can initiate or exacerbate gum disease (for a more in depth description of this investigation see Chap. 8). Several reports have also shown neuroendocrine hormones to be potent growth factors for respiratory pathogens. These include *Klebsiella pneumoniae*, (Freestone et al. 1999) as well as various *Bordetella* species (Anderson and Armstrong (2006), such as *B. bronchiseptica* and *B. pertussis*. Additionally, Anderson and Armstrong (2006) showed that in serum-based media, norepinephrine stimulated growth of *B. bronchiseptica* as well as inducing expression of BfeA, a siderophore receptor is important for growth in vivo.

Stress neurohormone enhancement of growth and virulence is not restricted to pathogens of mammalian or avian hosts, as it has been shown that norepinephrine can increase the growth or infectivity of bacteria causing disease in frogs (*Aeromonas hydrophila*) (Kinney et al. 1999) and even non-vertebrates such as oysters (*Vibrio* species) (Lacoste et al. 2001).

How do stress hormones stimulate bacterial growth? Insight into the underlying mechanism may be provided by understanding that the iron-binding catechol moiety found in siderophores, such as enterobactin, is also present in the neuroendocrine catecholamine family of stress hormones, and related inotropic agents (Freestone et al. 2002) (Fig. 3.2). Studies from our laboratories and those of others have shown that norepinephrine, epinephrine and dopamine, the inotropes isoprenaline, dobutamine and certain of their metabolites (such as dihydroxymandelic acid and dihydroxyphenylglycol) as well as plant extracts containing catechol compounds (such as catechin, caffeic, chlorogenic and tannic acids) (see Chap. 4) all share the ability to induce bacterial growth (Anderson and Armstrong 2006, 2008; Coulanges et al. 1998; Freestone et al. 2000, 2002, 2003, 2007a,b, 2008a,b; Roberts et al. 2002, 2005; Williams et al. 2006). The underlying mechanism of this growth enhancement has been shown to involve the catechol-containing compounds enabling bacterial acquisition of normally inaccessible Tf or Lf-complexed iron, or in the case of Gram-negative species, induction of a novel autoinducing growth stimulator (Lyte et al. 1996; Freestone et al. 1999). Both mechanisms will be discussed in detail.

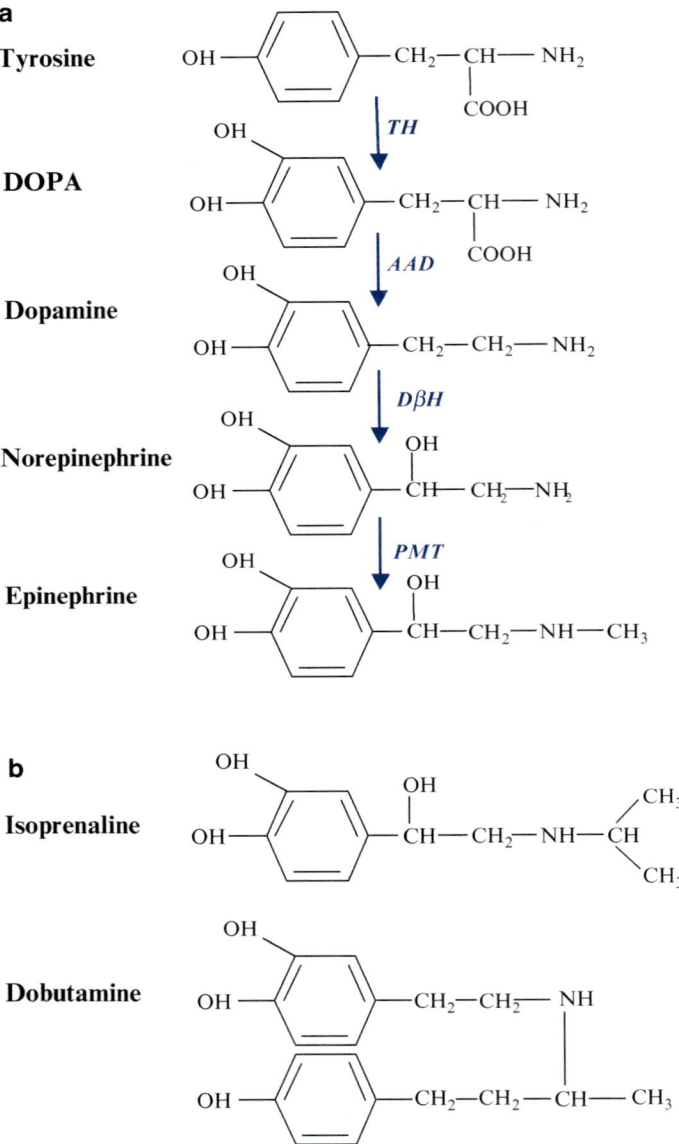

Fig. 3.2 Catecholamine structures. Catecholamines are synthesised from L-DOPA, obtained from dietary sources (principally the amino acids tyrosine and phenylalanine). (**a**) The catecholamine biosynthetic pathway: *TH* tyrosine hydroxylase, *AAD* aromatic L-amino decarboxylase, *DβH* dopamine β-hydroxylase, *PMT* phenylethanolamine N-methyltransferase. Note that *PMT* is not expressed in cells of the enteric nervous system. (**b**) structures of the principal catecholamine inotropes dobutamine and isoprenaline

3.3 Catecholamine-Transferrin and Lactoferrin Interactions

Serum-containing media has been used extensively in the in vitro analyses of stress hormone effects on the growth of bacteria (Lyte 2004, Freestone et al. 2008a). Most bacteria grow poorly in such media largely because of the iron restriction imposed by Tf. To work out the mechanism of how catecholamines were inducing growth in such media, we carried out a series of biochemical fractionations and radio-labelling of mammalian serum with ^{55}Fe (Freestone et al. 2000). Combination of the serum fractions with norepinephrine showed that the stress hormone was apparently interacting with serum-Tf. Tf binds 2 ferric iron atoms and therefore exists in up to four iron binding states, iron-free (apo), iron-replete (diferric, holo) and mono-ferric, with the single iron bound to either the N-or C-terminal binding sites (Lambert et al. 2005). These iron-binding forms of Tf can conveniently be separated using denaturing urea acrylamide gel electrophoresis (Freestone et al. 2000). Initially, using norepinephrine (Freestone et al. 2000) and dopamine (Neal et al. 2001) as our test compounds, we discovered that incubation of purified iron replete holo-Tf with the catecholamines resulted in a marked loss of Tf bound iron (Fig. 3.3a). We found that catecholamine-mediated iron release from Tf was concentration-dependent, and that the concentration of catecholamine required for Tf iron removal correlated closely with that required to induce bacterial growth (Fig. 3.3b); (Freestone et al. 2000, 2002, 2003, 2007a, b; Neal et al. 2001).

To determine whether catecholamines were delivering Tf-derived iron directly to bacteria, we attempted to prepare ^{55}Fe-complexed norepinephrine by incubating the catecholamine with radiolabelled ^{55}Fe-Tf and separating the mixture on a sephadex gel filtration column (Freestone et al. 2000). Interestingly, identical elution profiles for ^{55}Fe-Tf were obtained in the absence and presence of norepinephrine, and no low molecular weight peak of radioactivity corresponding to ^{55}Fe-norepinephrine was observed in the non-denaturing conditions of the Sephadex column, although urea-PAGE clearly indicated norepinephrine-dependent loss of iron from ^{55}Fe-Tf. This indicated that formation of a relatively stable complex between norepinephrine and native Fe-Tf was occurring from which iron was lost only under denaturing separation conditions (such as urea gel electrophoresis) (Freestone et al. 2000). Addition of ^{55}Fe-labelled Tf to serum-containing culture media in the presence of norepinephrine, and monitoring bacterial uptake of ^{55}Fe showed that catecholamine-mediated growth induction in serum was directly associated with uptake of the catecholamine-released Tf iron (Freestone et al. 2000, 2002). Later work by Freestone et al. (2003), used a dialysis membrane partition methodology to show that norepinephrine could directly shuttle iron from iron replete Tf across a protein-barrier membrane to iron-depleted apo-Tf. This indicates a direct norepinephrine-Fe complex must have occurred, an idea consistent with the chemical studies of Gerard et al. (1999), who showed that mixtures of catecholamine-inorganic iron salts resulted in Fe(III)-catecholamine complex formation.

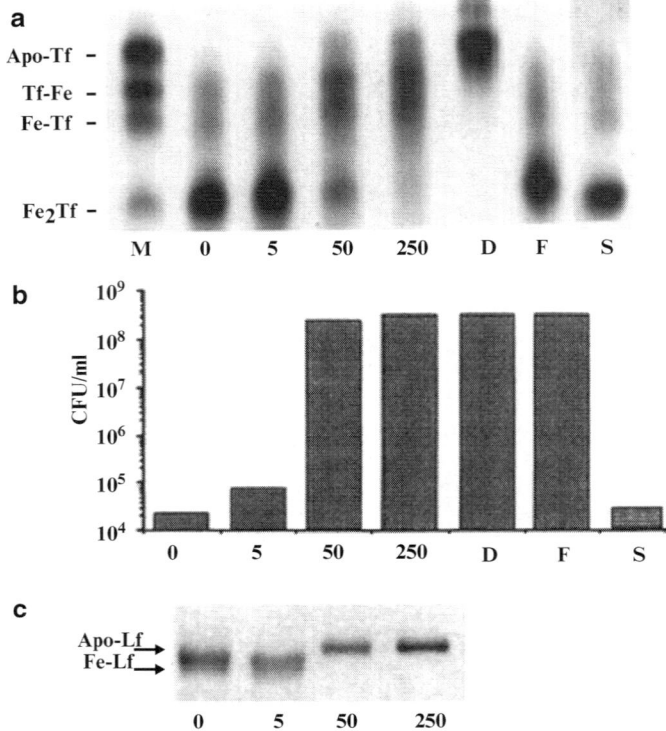

Fig. 3.3 Catecholamines induce growth through provision of Tf and Lf complexed iron. (**a**) Urea-PAGE of iron-saturated Tf after incubation for 18 h at 37°C in the absence of norepinephrine (lane 0) or in the presence of 5, 50 or 250 μM norepinephrine or the iron chelator dihydroxybenzoic acid (DHBA, 10 mM) (tracks 5, 50, 250, and D, respectively). Tracks D, F and S show the effect of incubation with DHBA, or 250 μM norepinephrine plus excess iron (1 mM ferric nitrate) 250 μM NE-S (sulfated norepinephrine), respectively. Lane M contains marker isoforms of Tf, fully iron saturated (Fe_2Tf), monoferric Tf with iron occupying the N-terminal or C-terminal domain (Fe-Tf and Tf-Fe, respectively), plus iron-free Tf. (**b**) Viable counts of *E. coli* strain E2348/69 after 18 h of incubation from an inoculum of 10^2 CFU/ml in serum-SAPI medium supplemented with norepinephrine or NE-S at the concentrations indicated for panel **a** above. (**c**) SDS-PAGE of iron-saturated Lf after incubation for 18 h at 37°C in the absence of norepinephrine (lane 0) or in the presence of 5, 50 or 250 μM NE (lanes 5, 50 and 250, respectively). Fe-Lf and apo-Lf indicate the mobility of iron-bound and iron-free isoforms of Lf, respectively. (This figure was adapted with permission from Freestone et al. (2000))

Although Tf is not a major component of gut and respiratory mucosal secretions, such tissue fluids are rich in a structurally related iron-binding protein, lactoferrin (Lf). We were therefore curious as to whether catecholamines had similar iron removal effects on Lf. Because of the physical characteristics of Lf, it is not possible to analyse iron removal using urea gels. However, mobility changes in SDS-polyacrylamide gels were observed that indicated that the catecholamines also caused loss of iron from Lf (Freestone et al. 2000) (Fig. 3.3c). Use of [55]Fe-Lf showed that the iron released from this protein could be also assimilated by bacteria (Freestone et al. 2000). For both Tf and Lf, catecholamine-mediated iron removal and provision to bacteria could be prevented by the addition of excess iron to the incubation mixtures (Freestone et al. 2000), providing confirmation of the role of iron in the catecholamine growth induction process.

The precise chemical nature of the catecholamine-Tf or Lf-complex and how its formation results in iron release is the subject of a current study, but it is clear from Freestone et al. (2003) and chemical analyses (Gerard et al. 1999) that catecholamines can directly bind ferric iron, and that binding of the Fe(III) in Tf and Lf results in a significant reduction of the iron binding affinity of these host iron binding proteins, such that they then can become bacterial nutrient sources (Freestone et al. 2000, 2003).

3.4 Bacterial Elements Involved in Catecholamine-Mediated Growth Induction

The provision of normally sequestered host Tf or Lf-iron by catecholamines is of considerable significance as it enables pathogenic bacterial species that lack specific binding proteins or uptake systems for Tf or Lf to grow in normally bacteriostatic iron-restricted environments, such as blood or serum (Freestone et al. 2008a, b). As a consequence, a considerable amount of data has been elucidated about the bacterial molecular machinery involved in growth responsiveness to the catecholamines, and to bacterial-Tf- and Lf-catecholamine interactions. For Gramnegative species such as *E. coli* (Freestone et al. 2000; Burton et al. 2002; Freestone et al. 2003), *Salmonella enterica* (Williams et al. 2006) and *Bordetella* spp. (Anderson and Armstrong 2008) siderophore synthesis and uptake systems have both been shown to be the key elements in the catecholamine-mediated growth induction process. This section will therefore examine the genes involved in catecholamine responsiveness for the two most investigated enteric pathogens, *E.coli* O157:H7 and *Salmonella enterica* Sv Typhimurium.

For *E. coli*, the enterobactin synthesis gene *entA* (Freestone et al. 2000, 2003; Burton et al. 2002) was found to be essential for catecholamine growth induction. Later, work from our laboratory demonstrated that the product of the *entF* gene was also required, as were also proteins involved in synthesising amino acid precursors for enterobactin, such as AroK and AroD (Freestone, unpublished data). The presence of a functional uptake system for enterobactin was also found to be necessary, as an *E.coli tonB* mutant failed to grow in the presence of the catecholamine norepinephrine (Freestone et al. 2003) or to dopamine, or a variety of dietary catechol compounds (Freestone et al. 2007c). The response of the *E.coli* siderophore mutants to growth in serum-based medium in the presence of the catecholamine noreipephrine is shown in Fig. 3.4a. In the case of *Salmonella*, initial unpublished work from our laboratory (Fig. 3.4b) that was later extended by Williams et al. (2006) and Bearson et al. (2008) similarly showed that siderophore synthesis and

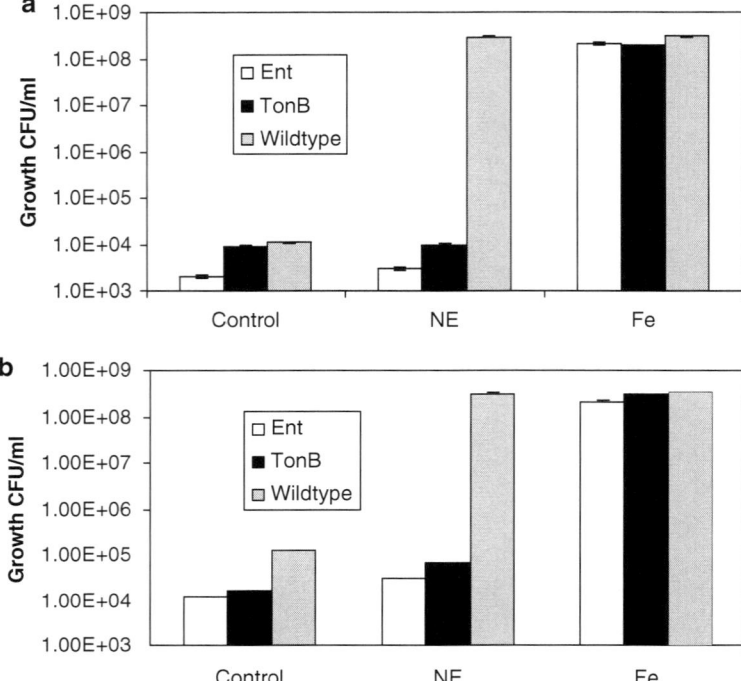

Fig. 3.4 The importance of ferrous iron uptake systems in catecholamine-growth induction. Overnight cultures of *E. coli* O157:H7 (Panel **a**) and *S. enterica* (Panel **b**) were inoculated at approximately 10^2 CFU/ml into triplicate 1 mL aliquots of serum-SAPI medium (Freestone et al. 2000) containing no additions (Control) 100 μM norepinephrine (NE) or 100 μM ferric nitrate (Fe) and incubated statically at 37°C in a 5% CO_2 atmosphere for 18 h. The iron was included to demonstrate that failure to grown in the serum-based medium was due to lack of a ferrous iron uptake system, and not sensitivity to the serum itself. Growth of the cultures was enumerated on luria agar as described in Freestone et al. (2000). The results shown are representative data from at least three separate experiments; all data points showed variation of less than 5%

receptor proteins (*fepA*, *iroN* and *cirA*) were all necessary elements in *Salmonella* catecholamine growth responsiveness. For *Bordetella* species, Anderson and Armstrong (2008) have also demonstrated that the presence of an intact enterobactin production and uptake system were required for growth-related responses to norepinephrine, such as the ability of the bacteria to acquire iron from Tf in the presence of norepinephrine.

Since we had shown that catecholamine growth induction in serum-based medium involved provision of iron from serum-transferrin (Freestone et al. 2000), we also undertook analyses of the importance of ferric uptake (siderophore) systems in the mechanism by which enteric bacteria uptake iron from Tf and other host iron binding proteins; these investigations are illustrated in Figs. 3.5 and 3.6. Figure 3.5 (our unpublished data) shows the ability of norepinephrine to deliver iron

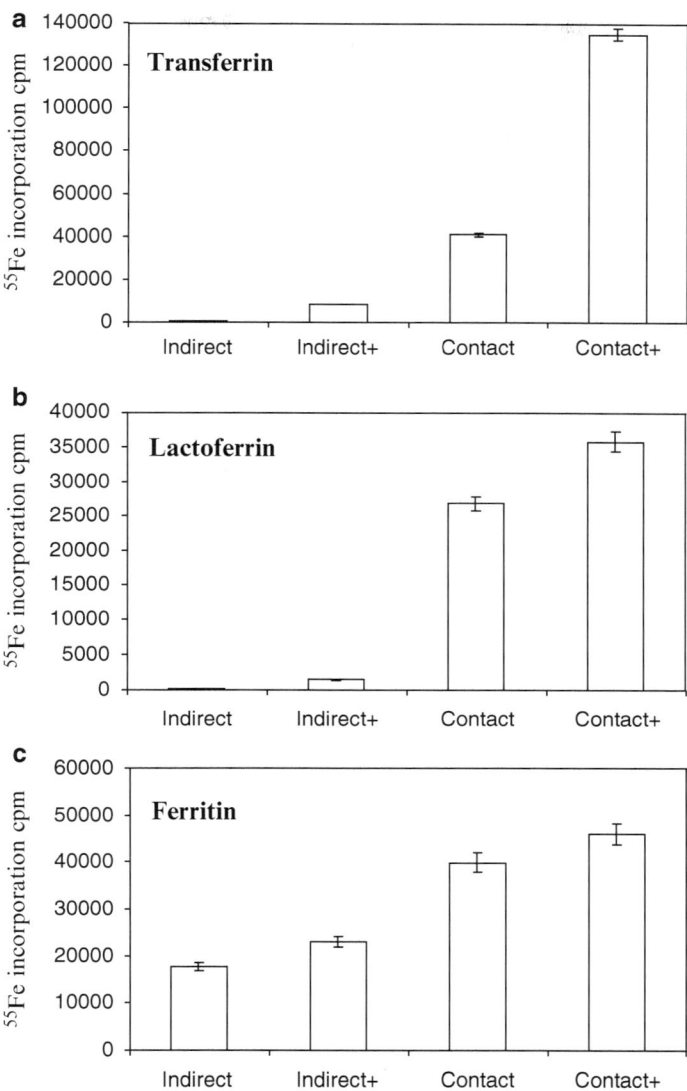

Fig. 3.5 The importance of direct contact in *E. coli* uptake of Tf, Lf and ferritin-complexed iron. Filter-sterilised ^{55}Fe-Tf, ^{55}Fe-Lf and ^{55}Fe-ferritin were prepared as described in Freestone et al. (2000) and added at 2.5×10^5 cpm/ml, either directly into 5 ml of sterile SAPI medium (Freestone et al. 2000) buffered with 50 mM Tris–HCl, pH 7.5 supplemented with 100 µM norepinephrine or an equivalent volume of water. Bacteria were incubated either in direct "contact" with the ^{55}Fe-labelled proteins or enclosed within 1-cm diameter dialysis membrane (4 kDa cut-off) ("indirect" contact). An exponential growing culture of wildtype *E. coli* O157:H7 was added directly to the ^{55}Fe-labelled protein mixtures at 2×10^8 CFU/ml, and incubated at 37°C in a 5% CO_2 atmosphere for 4 h, during which time there was essentially no additional growth. The bacteria were then harvested, washed in PBS and assayed for cell numbers and ^{55}Fe incorporation using scintillation counting as described previously (Freestone et al. 2000). **a–c**: bacterial uptake of ^{55}Fe from ^{55}Fe-Tf (**a**), ^{55}Fe-Lf (**b**) and ^{55}Fe-ferritin (**c**) in the presence and absence of norepinephrine

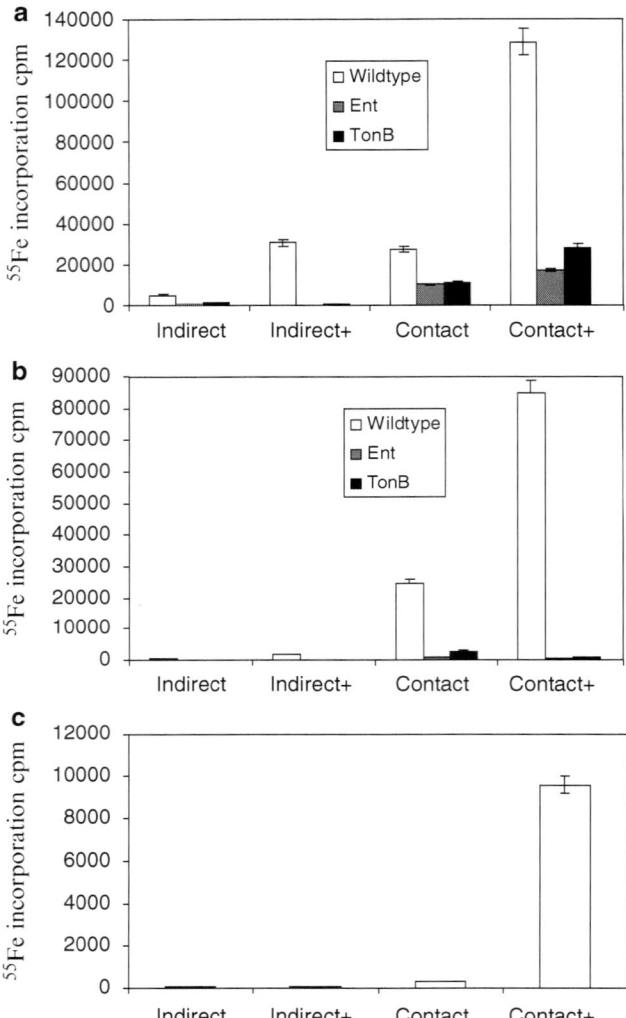

Fig. 3.6 The importance of ferrous iron uptake systems in Tf iron assimilation. Demonstration of the role of ferrous iron uptake systems in the mechanism of Gram-negative uptake of iron from Tf was carried out using enterobactin siderophore synthesis and uptake mutants (*entA* and *tonB*) of *E. coli* O157:H7 (Freestone et al. 2003) and *Salmonella enterica* (current study). To test the ability of the strains to acquire iron from ^{55}Fe-Tf, indirect and direct contact assays were performed as described in the legend to Fig. 3.4. **a** and **b**, uptake of ^{55}Fe- from ^{55}Fe-Tf by *E. coli* O157:H7 (**a**) and *S. enterica* (**b**) wildtype, *entA* and *tonB* mutants in the presence and absence of 100 μM norepinephrine or an equivalent volume of water. (**c**) uptake of ^{55}Fe- from ^{55}Fe-Tf by non-siderophore producing *S. epidermidis* in the presence and absence of 100 μM norepinephrine or an equivalent volume of water

(in the form of ^{55}Fe) from ^{55}Fe-labelled Tf, Lf and ferritin to an enterobactin-producing *E. coli* O157:H7. In the experiment shown, the bacteria were either partitioned into dialysis tubing ("Indirect" contact), and were therefore only able to acquire ^{55}Fe via secreted iron binding molecules or were in direct contact with the ^{55}Fe-protein ("Contact"). It is clear for Tf, Lf and ferritin that bacteria that had direct contact with the iron binding proteins were able to extract the greatest levels of iron (in the form of ^{55}Fe), particularly if the catecholamine was present. However, even when the iron binding protein was spatially distant from the bacteria, the catecholamine still enabled considerable uptake of the host-protein-complexed ^{55}Fe. In the absence of the stress hormone, the secreted factors released by the bacteria were relatively poor at removing ^{55}Fe from the extracellular Tf and Lf, though effective at scavenging ^{55}Fe from the intracellular ferritin, although in the latter case the presence of norepinephrine still enabled greater *E. coli* iron assimilation. Figure 3.5 also shows that bacterial uptake of iron from Lf was less than that of Tf. Wally et al. (2006) showed that differences in Tf and Lf iron binding affinity that were due to the variation in the structure of the iron binding domain's inter-lobe linker within the two proteins existed; this region, which is helical in Lf, is unstructured in Tf, making the removal of iron more difficult from Lf, which in contrast to Tf can retain its iron at acidic pH values (Lambert et al. 2005). This could explain the comparative differences in the potency of norepinephrine effects on Tf- and Lf-iron removal observed.

The ability of the catecholamines to complex and directly remove Tf and Lf iron means that bacteria do not have to closely associate with Tf or Lf in order to acquire this iron. Figure 3.5a and b show the importance of the presence of the ferric iron (siderophore) acquisition system in *E. coli* and *Salmonella enterica* uptake iron from Tf. Siderophore synthesis and uptake mutants (*entA* and *tonB*) along with their wildtype parental strains were incubated in direct contact with ^{55}Fe-Tf, or spatially separated from it within dialysis tubing, and analysed for iron uptake from the proteins (^{55}Fe incorporation) as described in Fig. 3.6. For both *E. coli* and *S. enterica*, whether in direct or indirect contact with the ^{55}Fe-Tf, the *entA* and *tonB* mutants both showed a significant reduction in the uptake of ^{55}Fe when compared with wildtype, indicating the importance of enterobactin synthesis and uptake in the mechanism of catecholamine-mediated host iron acquisition. A similar finding for the role of enterobactin in assimilating catecholamine-mediated Tf iron release by the respiratory pathogen *Bordetella bronchiseptica* has also been made (Anderson and Armstrong 2008).

An example for the ability of Gram-positive bacteria to access Tf iron in the presence of a catecholamine (norepinephrine) is shown in Fig. 3.5c. The microbe shown is *Staphylococcus epidermidis*, a bacterium that does not produce siderophores. The impact of the lack of a siderophore can be clearly seen, as even when in contact with the ^{55}Fe-Tf, the bacteria were able to acquire only very low levels of ^{55}Fe, which is in marked contrast to the Gram-negative bacterial species already considered. However, in the presence of the catecholamine, the *S. epidermidis* culture was able to assimilate nearly 30 times more ^{55}Fe. Significantly, we have shown that this iron provision from Tf can lead to massive growth enhancement

(Freestone et al. 1999; Neal et al. 2001; Freestone et al. 2008b) as well as the enhancement of biofilm formation in intravenous lines (Lyte et al. 2003) (see also Chap. 8).

3.5 Catecholamines Induce Bacterial Growth via Production of Non-Homoserine Lactone Autoinducers

Iron delivery from Tf and Lf is not the only mechanism by which catecholamines can enhance bacterial growth, at least for Gram-negative bacteria. Catecholamine-induced growth of enteric bacteria in a serum-based medium leads to the production of a non-LuxS dependent autoinducer (NE-AI) of growth (Lyte et al. 1996; Freestone et al. 1999). This novel AI is heat stable, very highly cross-species acting, and induces increases in the growth of a magnitude similar to that achievable with the catecholamines (Freestone et al. 1999). Interestingly, this activity works independently of Tf or Lf (Freestone et al. 2003). The NE-AI is also able to rapidly stimulate the recovery to active growth of viable but non-culturable *E. coli* O157:H7 or *Salmonella* (Reissbrodt et al. 2002) as well as increasing the rate of germination of *Bacillus anthrax* spores Reissbrodt et al. (2004). Comparison of NE-AI activity production by various bacteria occupying the human body suggests that it is synthesised in greatest amounts by enteric species (Freestone et al. 1999 and unpublished data), though the NE-AI is also recognised by respiratory pathogens (Freestone et al. 1999) and periodontal bacteria (Roberts et al. 2005). In terms of its synthesis, induction of production of the *E. coli* NE-AI requires only a transient 4–6 h exposure to catecholamines (Lyte et al. 1996; Freestone et al. 1999), after which the NE-AI then induces its own synthesis. This retention of memory of the catecholamine by bacteria indicates that the effects of catecholamine release during acute stress could have lasting and wide acting effects on the gut and other tissue microflora long after catecholamine levels in the tissues of the host animal have returned to normal.

The chemical structure of the NE-AI has yet to be assigned. None the less, Burton et al. (2002) have claimed that the NE-AI is enterobactin, based on the ability of an extract made by growing *E. coli* in dipyridyl-M9 minimal to support growth in serum-based culture medium. The extract made in the iron-limited M9 was of course enterobactin (Burton et al. 2002), whose role in the mechanism of catecholamine-growth induction has already been considered. The claim that the NE-AI was enterobactin is in retrospect surprising, as both Lyte et al. (1996) and Freestone et al. (1999) had previously shown that the NE-AI was made only in serum-based medium. Later work by Freestone et al. (2003) unequivocally showed that the NE-AI could not substitute for enterobactin in ^{55}Fe-Tf iron uptake assays. This indicates the NE-AI does not act in a siderophore-like manner and that it is very unlikely to be enterobactin or any of its breakdown products. Additional work from our laboratory has also shown that the mechanism of growth induction of the NE-AI does not require Tf or Lf (that is, the NE-AI does not remove iron from

^{55}Fe-Tf or Lf) and that it is able to induce bacterial growth in serum or blood based media to a level similar to that attainable by the catecholamines.

3.6   Iron Delivery May Not Be the Whole Story of Catecholamine-Mediated Growth Induction

Norepinephrine- and dopamine-containing sympathetic nerve terminals are distributed throughout the body, including the intestinal tract, where they make up part of the enteric nervous system (ENS). Indeed, around half of all the norepinephrine present within the body is synthesised and utilised within the ENS. In contrast, epinephrine is principally produced by the adrenal glands, and is not made within the ENS since enteric neurons do not synthesise phenyl ethanolamine N-methyltransferase (the enzyme required for epinephrine synthesis). An enteric pathogen is therefore unlikely to be exposed to epinephrine and might not be expected to develop specific sensor systems for it. This makes the suggestion that epinephrine is the catecholamine stress hormone cross-communicating with *E. coli* via the recently recognised AI-3 quorum sensing system (Sperandio et al. 2003; Clarke et al. 2006) of interest. Dopamine, unlike epinephrine is abundant in the mammalian GI tract, as also is norepinephrine, and we have recently shown that at low population densities, reflecting the bacterial numbers likely to be present early in an infection, there exists in Gram-negative enteric bacteria an order of stress hormone preference, with growth responses to the gut catecholamines norepinephrine and dopamine being 10–100 times greater than those to epinephrine (Freestone et al. 2007a). Indeed, in the case of *Y. enterocolitica* (a pathogen which very infrequently invades extraintestinally), none of 17 *Y. enterocolitica* strains tested showed any significant growth response to epinephrine, though all isolates responded to norepinephrine and dopamine with up to 4 log increases in growth. Intriguingly, epinephrine even antagonised *Y. enterocolitica* responses to norepinephrine and dopamine by up to 3 log-orders (Freestone et al. 2007a).

Recently, we have also shown that adrenergic and dopaminergic antagonists can block responses of enteric and staphylococcal bacterial pathogens to catecholamine stress hormones and inotropes (Freestone et al. 2007b, 2008b). Alpha adrenergic antagonist drugs, such as phentolamine, prazosin and phenoxybenzamine, were able to inhibit growth responses of several bacterial species to the adrenergic catecholamines norepinephrine and epinephrine, while beta-receptor specific antagonists such as propranolol or labetalol showed little negative effect. None of the α or β-adrenergic antagonists tested showed any significant inhibition of bacterial responses to dopamine, even though previous reports have shown that this catecholamine is able to induce growth in a similar manner to that shown by norepinephrine and epinephrine through the provision of iron from host iron binding proteins. This is perhaps not surprising given that dopamine does not operate through either α or β-adrenergic receptors, but instead by interaction with specific dopamine receptors.

We found that dopaminergic antagonists such as the anti-psychotic chlorpromazine could inhibit responses to dopamine by over 3 log orders, without substantially affecting either norepinephrine or epinephrine growth induction.

Further investigations from our laboratory also showed that the α- adrenergic and dopaminergic antagonists did not exert their effects through either inhibition of catecholamine-transferrin complex formation and subsequent iron mobilisation, AI synthesis, nor via the siderophore-dependent processes by which Gram-negative and positive bacteria assimilate the Tf-iron release induced by the catecholamine (Freestone et al. 2007a,b, 2008b). Considered collectively, our antagonist findings combined with our demonstration of bacterial catecholamine-specificity point towards the evolution of distinct and specific bacterial catecholamine signalling pathways, although there is as yet no genomic evidence for either an adrenergic or dopaminergic receptor within in any bacterial species.

We also have also recently found that *E. coli* elements proposed to be involved in catecholamine responsiveness (the *Qse* and *luxS* gene products) (Sperandio et al. 2003; Clarke et al. 2006) are not required for catecholamine growth induction (our unpublished data). This suggests that for *E. coli* in the growth context at least, intra-kingdom signalling (quorum sensing) and inter-kingdom signalling (microbial endocrinology) do not overlap.

3.7   Conclusion

Catecholamine-mediated access to the Tf and Lf iron has been shown to significantly enhance the growth of a wide range of commensal and pathogenic bacteria including *E. coli* and the coagulase-negative staphylococci, *S. enterica, Y. enterocolitica, S. aureus* (Freestone et al. 2000, 2002, 2007a, b, 2008a, b; Neal et al. 2001; Lyte et al. 2003, Williams et al. (2006), periodontal pathogens, (Roberts et al. 2002, 2005), and respiratory pathogens (Anderson and Armstrong 2006, 2008). In terms of the relevance of catecholamine-Tf/Lf interactions to the infectious disease process, the gastrointestinal tract provides the most persuasive connection, as bacteria, catecholamines and Lf are all co-localised in the gut (Markel et al. 2007). This could provide an insight into the mechanism of why elevations in catecholamine levels during acute stress can cause the overgrowth and translocation of the gut microflora (Lyte and Bailey 1997; Bailey et al. 2006). This potentially dangerous scenario of stress hormones enabling bacteria to strip iron for growth from a protein that normally by its nature restricts growth may explain why mammals have evolved mechanisms to tightly regulate levels of gut catecholamines. It is not widely appreciated that catecholamine-degrading enzymes are present throughout the entire length of the gastrointestinal tract, with expression levels apparently correlating with bacterial levels: lowest in the stomach and greatest in the large intestine and colon (Harris et al. 2000). Another important clinical consideration, developed further in Chap. 8, is that stress hormone interactions with Tf can enable staphylococci to grow in bacteriostatic tissues such as blood, turning normally harmless skin commensals into highly problematic opportunistic pathogens (Neal et al. 2001; Lyte et al. 2003; Freestone et al. 2008b).

References

Anderson, M. T., and Armstrong, S. K. 2006. The *Bordetella* Bfe system: Growth and transcriptional response to siderophores, catechols, and neuroendocrine catecholamines. J. Bact.188:5731–5740

Anderson, M. T., and Armstrong, S. K. 2008. Norepinephrine mediates acquisition of transferrin-iron on *Bordetella brontiseptica*. J. Bacteriol. 190:3940–3947

Bailey, M., Engler, H., and Sheridan, J. 2006. Stress induces the translocation of cutaneous and gastrointestinal microflora to secondary lymphoid organs of C57BL/6 mice. J. Neuroimmunol. 171:29–37

Bearson, B. L., Bearson, S. M., Uthe, J. J., Dowd, S. E., Houghton, J. O., Lee, I., Toscano, M. J., and Lay Jr., D. C. 2008. Iron regulated genes of *Salmonella enterica* serovar Typhimurium in response to norepinephrine and the requirement of *fepDGC* for norepinephrine-enhanced growth. Microbes Infect. 10:807–816.

Burton, C. L., Chhabra, S. R., Swift, S., Baldwin, T. J., Withers, H., Hill, S. J., and Williams, P. 2002. The growth response of *Escherichia coli* to neurotransmitters and related catecholamine drugs requires a functional enterobactin biosynthesis and uptake system. Infect. Immun. 70:5913–5923

Clarke, M. B., Hughes, D. T., Zhu, C., Boedeker, E. C., and Sperandio, V. 2006. The QseC sensor kinase: a bacterial adrenergic receptor. Proc. Natl. Acad. Sci. U. S. A. 103:10420–10425

Coulanges, V., Andre, P., and Vidon, D. J.-M. 1998. Effect of siderophores, catecholamines, and catechol compounds on *Listeria* spp. growth in iron-complexed medium. Biochem. Biophys. Res. Comm. 249:526–530

Freestone, P. P., Haigh, R. D., Williams, P. H., and Lyte, M. 1999. Stimulation of bacterial growth by heat-stable, norepinephrine-induced autoinducers. FEMS Microbiol. Lett. 172:53–60

Freestone, P. P., Lyte, M., Neal, C. P., Maggs, A. F., Haigh, R. D., and Williams, P. H. 2000. The mammalian neuroendocrine hormone norepinephrine supplies iron for bacterial growth in the presence of transferrin or lactoferrin. J. Bacteriol. 182:6091–6098

Freestone, P. P., Williams, P. H., Haigh, R. D., Maggs, A. F., Neal, C. P., and Lyte, M. 2002. Growth stimulation of intestinal commensal *Escherichia coli* by catecholamines: a possible contributory factor in trauma-induced sepsis. Shock 18:465–470

Freestone, P. P., Haigh, R. D., Williams, P. H., and Lyte, M. 2003. Involvement of enterobactin in norepinephrine-mediated iron supply from transferrin to enterohaemorrhagic *Escherichia coli*. FEMS Microbiol. Lett. 222:39–43

Freestone, P. P., Haigh, R. D., and Lyte, M. 2007a. Blockade of catecholamine-induced growth by adrenergic and dopaminergic receptor antagonists in *Escherichia coli* O157:H7, *Salmonella enterica* and *Yersinia enterocolitica*. BMC Microbiol. 7:8

Freestone, P. P., Haigh, R. D., and Lyte, M. 2007b. Specificity of catecholamine-induced growth in *Escherichia coli* O157:H7, *Salmonella enterica* and *Yersinia enterocolitica*. FEMS Microbiol. Lett. 269:221–228

Freestone, P. P. E., Walton, N., Haigh R. H., and Lyte, M. 2007c. Influence of dietary catechols on the growth of enteropathogenic bacteria. Int. J. Food Micro. 119: 159–169

Freestone, P. P., Sandrini, S. M., Haigh, R. D., and Lyte, M. 2008a. Microbial endocrinology: how stress influences susceptibility to infection. Trends Microbiol. 16:55–64

Freestone, P. P. E., Haigh, R. D., and Lyte, M. 2008b. Catecholamine inotrope resuscitation of antibiotic-damaged staphylococci and its blockade by specific receptor antagonists. J. Infect. Dis. 197:2044–1052

Gerard, C., Chehhal, H., and Aplincourt, M., 1999. Stability of metal complexes with a ligand of biological interest: Noradrenaline. J. Chem. Res. S: 90–91

Harris, R., Picton R., Singh, S., and Waring R. 2000. Activity of phenolsulfotransferases in the human gastrointestinal tract. Life Sci. 67: 2051–2057

Kinney, K. S., Austin, C. E., Morton, D. S., and Sonnenfeld, G. 1999. Catecholamine enhancement of *Aeromonas hydrophila* growth. Microb. Pathog. 26: 85–91

Lacoste, A., Jalabert, F., Malham, S. K., Cueff, A., and Poulet, S. A. 2001. Stress and stress-induced neuroendocrine changes increase the susceptibility of juvenile oysters (*Crassostrea gigas*) to *Vibrio splendidus*. Appl. Environ. Microbiol. 67:2304–2309

Lambert, L., Perri, H., Halbrooks, P.J., and Mason, A.B. 2005. Evolution of the transferrin family: conservation of residues associated with iron and anion binding. Comp. Biochem. Phys. 142: 129–141

Lyte, M., Frank, C. D., and Green, B. T. 1996. Production of an autoinducer of growth by norepinephrine cultured *Escherichia coli* O157:H7. FEMS Microbiol. Lett. 139:155–161

Lyte, M., and Bailey M, T. 1997 Neuroendocrine-bacterial interactions in a neurotoxin-induced model of trauma. J. Surg. Res. 70:195–201.

Lyte, M., Freestone, P. P., Neal, C. P., Olson, B. A., Haigh, R. D., Bayston, R., and Williams, P. H., 2003. Stimulation of *Staphylococcus epidermidis* growth and biofilm formation by catecholamine inotropes. Lancet 361:130–135

Lyte, M. 2004. Microbial endocrinology and infectious disease in the 21st century. Trends Microbiol. 12:14–20.

Markel, T. A., Crisostomo, P. R., Wang, M., Herring, C. M., Meldrum, K. K., Lillemoe, K. D., and Meldrum, D. R., 2007. The struggle for iron: gastrointestinal microbes modulate the host immune response during infection. J. Leukoc. Biol. 81:393–400

Neal, C. P., Freestone, P. P. E., Maggs, A. F., Haigh, R. D., Williams, P. H., and Lyte, M. 2001. Catecholamine inotropes as growth factors for *Staphylococcus epidermidis* and other coagulase-negative staphylococci. FEMS Microbiol. Lett. 194:163–169.

Ratledge, C., and Dover, L.G. 2000. Iron metabolism in pathogenic bacteria. Ann. Rev. Microbiol. 54:881–941

Reissbrodt, R., Rienaecker, I., Romanova, J. M., Freestone, P. P. E., Haigh, R. D., Lyte, M., Tschäpe, H., and Williams, P. H. 2002. Resuscitation of *Salmonella enterica* serovar Typhimurium and Enterohemorrhagic *Escherichia coli* from the viable but nonculturable state by heat-stable enterobacterial autoinducer. Appl. Environ. Microbiol. 68:4788–4794.

Reissbrodt, R., Raßbach, A., Burghardt, B, Rienäcker, I., Mietke, H., Schleif, J., Tschäpe H., Lyte, M., and Williams, P. H. 2004. Assessment of a new selective chromogenic *Bacillus cereus* group plating medium, and use of enterobacterial autoinducer of growth for cultural identification of *Bacillus* species. J. Clin. Microbiol. 42:3795–3798.

Roberts, A., Matthews, J. B., Socransky, S. S., Freestone, P. P., Williams, P. H., and Chapple, I. L. 2002. Stress and the periodontal diseases: effects of catecholamines on the growth of periodontal bacteria in vitro. Oral Microbiol. Immunol. 17:296–303

Roberts, A., Matthews, J., Socransky, S., Freestone, P., Williams, P., and Chapple, I. 2005. Stress and the periodontal diseases: growth responses of periodontal bacteria to *Escherichia coli* stress-associated autoinducer and exogenous Fe. Oral Microbiol. Immunol. 20:147–153.

Sperandio, V., Torres, A. G., Jarvis, B., Nataro, J. P., and Kaper, J. B. 2003. Bacteria–host communication: the language of hormones. Proc. Natl. Acad. Sci. U. S. A. 100:8951–8956.

Wally, J., Halbrooks, P., Vonrhein, C., Rould, M., Everse, S., Mason, A., and Buchanan, S.K. 2006. The crystal structure of iron-free human serum transferrin provides insight into inter-lobe communication and receptor binding. J. Biol. Chem. 281:24934–24944

Williams, P. H., Rabsch, W., Methner, U., Voigt, W., Tschäpe, H., and Reissbrodt, R. (2006) Catecholate receptor proteins in *Salmonella enterica*: role in virulence and implications for vaccine development. Vaccine 24:3840–3844

Chapter 4
Dietary Catechols and their Relationship to Microbial Endocrinology

Neil Shearer and Nicholas J. Walton

4.1 Introduction

The effects of catecholamines on bacterial growth and behaviour are well documented (see other chapters in this volume) and appear to occur principally through facilitation of iron acquisition from the host glycoproteins, transferrin and lactoferrin. Nevertheless, it is clear that mechanisms other than simple iron acquisition are responsible for some of the effects of catecholamines. The question that naturally arises, however, is how far these phenomena are more generally characteristic of compounds with a catecholic structure, rather than being specific to the neuroendocrine catecholamines, norepinephrine (noradrenaline), epinephrine (adrenaline) and dopamine, or their analogues. This is important, since catechols are common; in particular, they are ingested routinely as components of plant-based diets and often in substantial amounts. A number of subsidiary questions then arise: what are these other catecholic compounds and what are the prevalent dietary sources; what levels of these compounds or of their relevant metabolites might exist in plasma, organs or tissues; and are there any specific effects of catecholamines upon bacteria that are not observed with other catechols, and vice versa? This account attempts to provide some insights.

4.2 Dietary Sources and Distribution of Catechols

Catechols bind iron, and this property is exploited in catecholic siderophores (Crosa and Walsh 2002) and, by contrast, in strategies to prevent bacterial growth by restricting the supply of iron (Scalbert 1991; Mila et al. 1996). Further, since free Fe^{2+} ions participate in the Fenton reaction with H_2O_2, which produces highly

N. Shearer (✉) and N.J. Walton
Institute of Food Research, Norwich Research Park, Colney, Norwich NR4 7UA, UK
e-mail: Neil.Shearer@bbsrc.ac.uk

M. Lyte and P.P.E. Freestone (eds.), *Microbial Endocrinology*,
Interkingdom Signaling in Infectious Disease and Health,
DOI 10.1007/978-1-4419-5576-0_4, © Springer Science+Business Media, LLC 2010

reactive hydroxyl radicals, the sequestration of free iron by catechols and other iron-chelating compounds can restrict the potential for free-radical generation and protect biological molecules from oxidative damage (Rice-Evans et al. 1995; Khokhar and Apenten 2003). Catechols also possess an autoxidation activity that oxidises Fe^{2+} to Fe^{3+}, although conversely, reduction of Fe^{3+} to Fe^{2+} can also occur, particularly at low pH (Moran et al. 1997; Chvátalová et al. 2008). Catechols with an additional neighbouring hydroxyl group, such as gallic acid (3,4,5-trihydroxy-benzoic acid), appear to bind iron and also to oxidise Fe^{2+} to Fe^{3+} somewhat less effectively than simple catechols such as protocatechuic acid (3,4-dihydroxybenzoic acid) (Khokhar and Apenten 2003; Andjelković et al. 2006; Chvátalová et al. 2008). As discussed later, a study of the effects of tea catechins revealed that only compounds possessing a 3,4-dihydroxyphenyl group in the B-ring stimulated bacterial growth under iron-restrictive conditions; compounds with an additional neighbouring hydroxyl group in the 5-position failed to stimulate growth (Freestone et al. 2007c). In this chapter, the term catechol is used to denote compounds with the *o*-dihydroxyphenyl, as distinct from a trihydroxyphenyl, grouping.

Catechols in the diet are most often plant-derived. The large majority are eventual products of the central phenylpropanoid pathway from phenylalanine via cinnamic acid. In plants, phenylpropanoid-pathway derivatives fulfil diverse functions in defence, signalling, protection against ultraviolet light and insect attraction (Parr and Bolwell 2000; Boudet 2007). Those present in the largest amounts are likely to serve a broad-spectrum defence function as anti-feedants or as antimicrobial agents. Catechols are rarely a specific focus of attention, being more often considered nonspecifically within other secondary-product classes, especially the hydroxycinnamic acids and the various sub-classes of the flavonoids. Many plant polyphenolic substances exhibit a broad-spectrum antibacterial activity and catechols are not necessarily among the most potent (Taguri et al. 2006); however, studies of siderophore mutants of *Erwinia chrysanthemi* demonstrate that plant polyphenols containing catechol groups can act to prevent bacterial growth by sequestering Fe^{3+}, although sequestration of other metal ions, notably Cu^{2+} and Zn^{2+}, might also occur (Mila et al. 1996). As discussed further, plant catechols include, in particular, catecholic representatives of the benzoic and cinnamic acids and their derivatives, notably protocatechuic acid (3,4-dihydroxybenzoic acid), caffeic acid (3,4-dihydroxy-*trans*-cinnamic acid) and the chlorogenic acids (principally 5-*O*-caffeoylquinic acid); oleuropein, a hydroxytyrosol ester found in olives; catecholic flavonols, notably quercetin; catecholic flavanols, for example (−)-epicatechin and (−)-epicatechin gallate; many anthocyanins, which are glycosides of anthocyanidins such as the catechol, cyanidin, and which are widespread as blue, purple and red plant pigments; and finally, many proanthocyanidins or condensed tannins, formally polymerised flavanols. Flavonols, flavones and anthocyanins, though not flavanols, are all generally not found free in plants but are typically found as *O*-glycosides. Figure 4.1 shows the structures of the principal flavonoid sub-classes.

Many of these compounds attract considerable interest on account of their reported effects in relation to a range of cancers, inflammatory conditions and cardiovascular diseases, which to varying extents (and not in every case) may be a

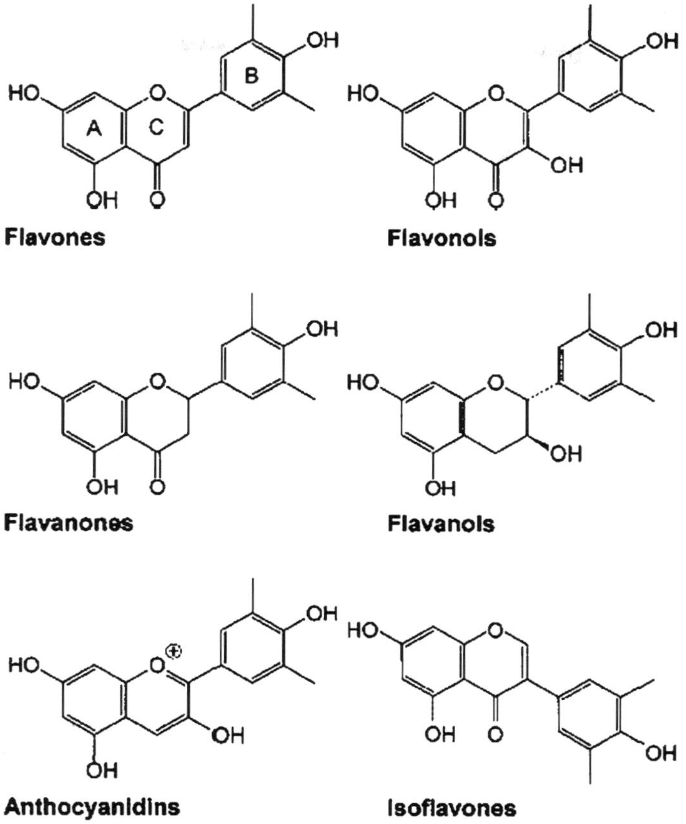

Fig. 4.1 Subclasses of the flavonoids. Classification is based on variations in the heterocyclic C-ring. Reproduced from Hollman and Arts (2000), © Society of Chemical Industry, with permission

result of their radical-scavenging and antioxidant properties (Ross and Kassum 2002; Cooper et al. 2005; Evans et al. 2006; Hodgson and Croft 2006; Prior et al. 2006; Rahman 2006; Schroeter et al. 2006; Espín et al. 2007; Khan et al. 2008; Loke et al. 2008). One consequence of this is a concern to acquire reliable data on dietary intakes. Extensive information on the contents of phenylpropanoid metabolites in fruits, vegetables and beverages has been compiled (Hollman and Arts 2000; Tomás-Barberán and Clifford 2000; Clifford 2000a, b; Santos-Buelga and Scalbert 2000; Manach et al. 2004). Levels can vary widely on account of varietal characteristics, cultural conditions, developmental stage, position on the plant and storage – and very often within an individual harvested fruit or vegetable (Manach et al. 2004).

Catecholic benzoic and cinnamic acids are amongst the simplest catecholic compounds found in plant foods. They and their derivatives are widespread and may sometimes be present in appreciable amounts. Thus, blackberry fruits may contain ca. 0.07–0.2 g of protocatechuic acid kg⁻¹ fresh weight (Tomás-Barberán and Clifford 2000) and potato tubers may contain around 1.2 g kg⁻¹ fresh weight of chlorogenic acid (principally 5-O-caffeoylquinic acid), though much is likely to be lost during cooking (Clifford 2000a). Coffee also contains large amounts of chlorogenic acid and Clifford (2000a) has estimated that 200 ml of instant coffee brew (2% w/v) may provide 50–150 mg of the compound (equivalent to ca. 25–75 mg of caffeic acid).

The occurrence of oleuropein, a secoiridoid glucoside ester of the catechol, hydroxytyrosol, is restricted to olives, where in young fruits it may account for 14% of dry matter, although in processed fruit and in olive oil, levels are lower than this, partly on account of hydrolysis, including hydrolysis to hydroxytyrosol. Olives also contain verbascoside, a compound containing two catecholic residues, hydroxytyrosol and caffeic acid (Soler-Rivas et al. 2000).

In most fruits, vegetables and beverages the levels of flavonols, flavones and flavanols are below about 0.015 g kg⁻¹ fresh weight, although there are conspicuous exceptions (Hollman and Arts 2000). Thus, the catecholic flavonol, quercetin, occurs in the form of glycosides in onions at 0.35 g kg⁻¹ fresh weight and in kale at around 0.11 g kg⁻¹ fresh weight (and in each case possibly considerably more, depending upon variety and cultural conditions); the flavone, luteolin, may reach 0.2 g kg⁻¹ fresh weight in celery leaves; levels of the catecholic flavanols, (−)-epicatechin and (−)-epicatechin gallate, can reach ca. 20–150 µg ml⁻¹ in brewed tea; and levels of (+)-catechin and (−)-epicatechin can reach 100–200 µg ml⁻¹ in some red wines. Cocoa and chocolate are also rich sources of flavanols (Schroeter et al. 2006). Levels of anthocyanins are greatest in those fruits and vegetables that are highly pigmented; for example, blueberries contain ca. 0.8–4.2 g of anthocyanins kg⁻¹ fresh weight and aubergines may contain 7.5 g kg⁻¹ fresh weight (Clifford 2000b). Not surprisingly, red wines, and port wine in particular, contain appreciable levels of anthocyanins, within the range 140–1,100 µg ml⁻¹ (Clifford 2000b). Proanthocyanidins or condensed tannins are abundant in many common or staple foods or beverages (Santos-Buelga and Scalbert 2000), notably black tea, in which two groups of these compounds, the theaflavins and thearubigins, are derived from the flavanols of green tea during processing. Levels of proanthocyanidins of ca. 3–10 g kg⁻¹ dry weight in lentils, of up to 7.4 g kg⁻¹ dry weight in faba beans and of up to 39 g kg⁻¹ dry weight in sorghum, have been determined. High concentrations of proanthocyanidins are often present in red wine and apple juices, and particularly in cider, where levels are reported to range between 2,300 and 3,700 µg ml⁻¹ (Santos-Buelga and Scalbert 2000).

In addition to the catechols discussed above, catecholamines also occur in many plants, and there is evidence for their involvement in defence against pathogens, in responses to plant growth substances and in carbohydrate metabolism, but details of the mechanisms involved still remain uncertain (Kulma and Szopa 2007). Where determined, levels of catecholamines have been found to be low (below 1 mg kg⁻¹ fresh weight), except for bananas, plantains and avocados. Thus, levels in excess of 40 mg kg⁻¹ of dopamine were found in the fruit pulp of red banana and yellow banana, whilst the peel of Cavendish banana contained 100 mg kg^{-1} (Kulma and Szopa 2007). These values are similar to those of, for example, protocatechuic acid in blackberry fruits (Tomás-Barberán and Clifford 2000). Fruit pulp of Fuerte avocado contained smaller amounts of catecholamines: 4 mg kg^{-1} of dopamine and <3.5 mg kg^{-1} of norepinephrine. The relatively high content of dopamine in banana pulp is of particular interest in view of the ability of banana pulp to promote the growth of Gram-negative bacteria in iron-restricted medium, as reported by Lyte (1997).

Dietary catechols also arise from non-plant sources. In particular, tyramine arises in cheeses and other fermented foods as a result of the bacterial decarboxylation of tyrosine (Santos 1996). Concentrations exceeding 0.1 g kg^{-1} can be present in matured cheeses (Komprda et al. 2008). Micromolar concentrations of tyramine have been found to increase the adherence of *Escherichia coli* O157:H7 to murine intestinal mucosa (Lyte 2004b).

4.3 Dietary Intakes of Catechols

Manach et al. (2004) reviewed and summarised data from several countries, including the United States, Denmark, the Netherlands and Spain and concluded that the total intake of polyphenols probably reached 1,000 mg d^{-1} in individuals who ate several portions of fruit and vegetables each day. The consumption of flavonols probably accounted for 20–50 mg d^{-1}, only a proportion of which would be the glycosides of quercetin and other catechols. Levels of anthocyanins consumed were estimated to be broadly similar (though somewhat higher in Finland where appreciable amounts of berries are commonly eaten). However, depending upon individual dietary habits, the intakes of both total and specific polyphenols are likely to be highly variable. In some individuals, chlorogenic acid may predominate as a result of coffee consumption and could amount to more than 200 mg d^{-1} (Clifford 2000a), whereas drinkers of tea may well ingest comparable amounts of catecholic flavanols and proanthocyanidins (Hollman and Arts 2000; Santos-Buelga and Scalbert 2000; Cooper et al. 2005). Taking all principal sources into account, daily intakes of catechols therefore seem likely to exceed 500 mg d^{-1} in some individuals.

4.4 Absorption and Availability of Catechols

Although substantial amounts of catechols may be consumed in the diet, many intervening factors and processes will determine whether appreciable concentrations are to be found in the plasma and in tissues. These include the structure of the particular compound, notably the extent and type of glycosylation or esterification;

the extent and mechanisms of absorption; the degree of conjugation or derivatisation during or following absorption; potential food-matrix effects; the extent of any metabolism by the gastrointestinal flora, dependent in part upon the degree to which the compound remains unabsorbed (or is possibly re-secreted) in the small intestine; and finally, absorption and metabolism by individual tissues and organs (Manach et al. 2004, 2005; Prior et al. 2006).

Structural effects on absorption are readily apparent, for example, in the case of quercetin glycosides. The 4'-glucoside of quercetin is absorbed more rapidly than the 3β-rutinoside (rutin); maximal absorption of the former occurs about 30–40 min post-ingestion in humans, whereas maximal absorption of the latter requires 6–9 h. Thus, quercetin is found to be absorbed more rapidly from onions, which contain the flavonol predominantly in the form of glucosides, than from apples, which contain it in the form of both glucosides and other glycosides (Hollman et al. 1997). Enzyme systems of the small intestine that can perform the uptake and hydrolysis of quercetin glucosides have been identified and comprise the sodium-dependent glucose transporter, SGLT1, followed by a cytosolic β-glucosidase, or alternatively lactase phorizin hydrolase, a glucosidase of the brush border membrane (Day et al. 2003; Németh et al. 2003; Sesink et al. 2003); glycosides other than glucosides may be absorbed relatively poorly in the small intestine and absorption may be appreciably dependent upon bacterial hydrolysis in the colon. In contrast to flavonols, flavanols are not glycosylated and therefore require no hydrolysis step (Manach et al. 2004). Simple hydroxy-benzoic or hydroxycinnamic acids, such as caffeic acid, also appear to be readily absorbed (Clifford 2000a; Cremin et al. 2001), although chlorogenic acid (5-O-caffeoylquinic acid) is not, and hydrolysis by a number of colonic bacteria (strains of E. coli, Bifidobacterium lactis and Lactobacillus gasseri) and by human faecal microbiota has been demonstrated (Couteau et al. 2001; Gonthier et al. 2006). Anthocyanins differ from flavonols in that the native glycosides can apparently be absorbed and appear in the plasma, although bioavailability appears comparatively low (Wu et al. 2002; Manach et al. 2005). The major metabolite of cyanidin glucosides in humans is protocatechuic acid (Vitaglione et al. 2007). Proanthocyanidins, particularly those of higher molecular weight, are poorly absorbed, and large proportions are likely to reach the colon (Manach et al. 2004, 2005).

Following absorption, most catechols and polyphenols become derivatised by glucuronidation, sulphation or methylation, or by a combination of these, which are reactions that are undergone by many xenobiotics (Wu et al. 2002; Manach et al. 2004, 2005; Prior et al. 2006). The UDP-glucuronosyltransferase, sulpho-transferase and catechol-O-methyltransferase (COMT) activities that are responsible are active in intestinal enterocytes and in the liver; COMT is also active in kidney and has a wide tissue distribution. Prior et al. (2006) have summarised the principal patterns of conjugation that are observed for flavonoids. Thus, fla-vonoids with a catecholic B-ring, such as quercetin and cyanidin, are derivatised mainly to 3'-O-methyl derivatives, with smaller amounts of 4'-O-methyl deriva-tives. The catecholic B-ring also promotes glucuronidation. For quercetin, the principal compounds detected in human plasma after the ingestion of quercetin glucosides contained in onion were found to be 3'-O-methylquercetin-3-O-glucuronide and quercetin-3'-O-sulphate (Day et al. 2001). Extensive methylation and glucuronidation of the 3'- and 4'-positions also occur in the case of catechin and epicatechin (Natsume et al. 2003).

The bioavailability of polyphenols, including their pharmacokinetics in plasma has been comprehensively reviewed (Manach et al. 2005). For example, meals containing the equivalent of 80–100 mg of quercetin gave rise to plasma concentrations of quercetin conjugates of around 0.3–0.75 μM (Hollman et al. 1997; Manach et al. 1998; Graefe et al. 2001); consumption of 80 g of chocolate (containing a total of 5.3 mg of procyanidins g^{-1}) could produce a plasma epicatechin concentration of as much as 0.36 μM (Rein et al. 2000; Wang et al. 2000); and a glass of red wine containing 35 mg of catechin could produce a plasma catechin concentration of 0.09 μM (Donovan et al. 1999). As reviewed by Manach et al. (2004, 2005), half-lives in plasma have been found to be variable, from as little as ca. 2 h for anthocyanins and flavanols to 11–28 h for quercetin derivatives. Therefore, unless dietary catechols are ingested at frequent intervals, their plasma levels, or those of their derivatives, may decline rapidly, and only in the case of compounds such as quercetin glycosides can successive intakes lead to some degree of accumulation of plasma derivatives (Manach et al. 2005). Elimination by both biliary and urinary routes has been demonstrated, with the biliary route likely to be predominant for larger and multiple-conjugated metabolites (Manach et al. 2004).

Besides appearing in plasma, labelled dietary phenolic compounds can give rise to derivatives in a wide range of tissues and organs, especially the digestive organs and the liver. From a number of studies in different laboratories, levels in rats and mice varied between 30 ng and 3,000 ng (as aglycone) g^{-1} of tissue, depending upon the compound and the dose administered (Manach et al. 2004). It is likely that accumulation in any given organ is non-homogeneous, although it is not clear to what extent organ- or tissue-specific uptake mechanisms may exist.

4.5 Bacterial Growth May be Stimulated Experimentally by a Range of Catechols

It is clear that catechols ingested in the diet, or their derivatives, may be found in plasma and tissues at submicromolar and micromolar levels. These are comparable with, or higher than, normal plasma concentrations of norepinephrine and epinephrine, which are in the nanomolar range (Benedict and Grahame-Smith 1978); and, against this background, we shall shortly examine the evidence from two studies (Coulanges et al. 1998; Freestone et al. 2007c) that have shown that dietary catechols can stimulate bacterial growth in a manner similar to that observed with the neuroendocrine catecholamines, norepinephrine and epinephrine.

However, it is important to keep in mind that a substantial proportion, possibly approaching 50% in some cases (see Prior et al. 2006), of dietary catechol derivatives in the plasma and tissues may be conjugates, in which one of the catecholic hydroxyl groups is methylated, glucuronidated or sulphated. Furthermore, these compounds may not be free in solution. Quercetin incubated with human plasma becomes almost entirely bound to plasma proteins, chiefly albumin (Boulton et al. 1998) and other flavonoids behave similarly, although sulphation and glycosidation may substantially reduce the binding affinity (Dufour and Dangles 2005). The potential effect of protein binding is important because the stimulation of bacterial growth in iron-restricted medium that occurs in response to norepinephrine appears to occur concomitantly with uptake of the catecholamine into the bacterial cell (Freestone et al. 2000). However, the role of norepinephrine uptake in relation to both iron uptake and growth stimulation still remains unclear (Freestone et al. 2003, 2007b). It is not yet known whether the iron uptake and growth stimulation that occur in response to dietary catechols, and which are discussed below, are associated with the bacterial uptake of these compounds. In any event, it remains to be established how far the reported effects of catechols on bacterial growth and behaviour are affected by protein-binding of the catechol. However, the iron-restricted, serum-SAPI medium employed by Freestone et al. (2000, 2007c) contains, by definition, the proteins present in adult bovine serum, suggesting that the binding of catechols to proteins is not an issue, at least in this experimental set-up.

In the first of the two studies examining the effects of catechols on bacterial growth, Coulanges et al. (1998) examined the growth-promoting effects of a range of catechols upon a number of *Listeria* species, which (as far as is known) are unable to biosynthesise siderophores. Their ability to overcome growth inhibition induced by the iron-chelator, tropolone, was measured in disk diffusion assays. A number of compounds possessing a catechol grouping were effective in relieving growth inhibition. These included dopamine, epinephrine and norepinephrine and DL-DOPA, the siderophores pyoverdine and rhodotorulic acid, and plant-derived catechols including caffeic acid, esculetin, quercetin and rutin. Salicylic acid (*o*-hydroxybenzoic acid) was ineffective, as was dihydroxybenzoic acid (the isomer was not specified). This study therefore demonstrated that the relief of tropolone-induced growth inhibition in *Listeria monocytogenes* that had previously been observed to occur with catecholamines (Coulanges et al. 1997) was not restricted to these compounds but could occur also with a range of non-amine catechols.

In the second study, and following work summarised by Freestone et al. (2002) and Lyte (2004a), Freestone et al. (2007c) examined the growth of *E. coli* O157:H7 and *Salmonella enterica* SV Enteriditis in response to a number of catechols commonly consumed in the diet, and to fruit and vegetable extracts known to contain catechols. In the case of the individual catechols, they determined growth responses both in iron-restricted medium and in iron-replete medium and followed up these experiments with measurements of the uptake of [55]Fe from [55]Fe-labelled transferrin and lactoferrin. Marked differences in growth were observed, depending upon whether a rich medium (Luria broth) or an iron-restricted medium (serum-SAPI) was employed (Figs. 4.2 and 4.3). In the rich medium, none of the compounds tested – catechin, caffeic acid, chlorogenic (5-O-caffeoylquinic) acid and tannic acid – had any significant effect upon growth, with the exception of tannic acid, which became progressively more inhibitory to growth with increasing concentration (up to 200 μg ml^{-1}). In contrast, in the iron-restricted medium (and hence broadly consistent with the findings in *Listeria* of Coulanges et al. 1998), all four catechol compounds tested promoted growth, by a factor of around 4 log-orders, with saturation occurring at 50–100 μM (50 μg ml^{-1} for tannic acid). Similarly, all the fruit and vegetable extracts tested (apple, carrot, grape, pear, plum, orange and strawberry), and infusions of tea and coffee, promoted growth in the iron-restricted medium. Study of the effects of tea flavanols (see Fig. 4.4) revealed that the ability to promote growth in the iron-restricted medium was restricted to those compounds possessing a 3,4-dihydroxyphenyl (i.e. catecholic) B-ring ((+)–catechin, (–)-catechin gallate and (–)-epicatechin gallate); trihydroxy-compounds with an additional hydroxyl group in the 5-position of the B-ring ((–)-epigallocatechin, (–)-epigallocatechin gallate and (–)-gallocatechin gallate) failed to promote growth.

The work of Freestone et al. (2007c) revealed important mechanistic similarities between the responses to dietary catechols and the responses to neuroendocrine catecholamines. All of the catechols and plant extracts that promoted growth in the iron-restricted medium were also able to stimulate the uptake of ^{55}Fe from ^{55}Fe-transferrin or ^{55}Fe-lactoferrin in both organisms studied (Fig. 4.5). Their activity in this respect was therefore in general comparable with that of the neuroendocrine catecholamines, typified by norepinephrine. Further similarity with the behaviour of the catecholamines was shown by the absence of any growth promotion by dietary catechols, or by fruit or vegetable extracts, in *E. coli* siderophore biosynthesis (*entA*) and transport (*tonB*) mutants. Mutations in these genes were previously shown to prevent norepinephrine-stimulated growth of *E. coli* (Burton et al. 2002; Freestone et al. 2003).

The evidence from these studies by Freestone et al. (2007c) and Coulanges et al. (1998) therefore show conclusively that the behaviour of the catecholamines in promoting growth in iron-restricted medium is not unique but instead is common to a diverse range of compounds all possessing the catechol structure. Thus, any consideration of the role of catecholamines in the promotion of bacterial growth (and potentially of virulence) needs to be broadened to take account of the occurrence and effects of dietary catechols, but subject to the caveats outlined at the beginning of this section.

4.6 Are any of the Effects of Catechols on Bacteria Catecholamine-Specific?

Although a wide range of catechols are able to stimulate the growth of certain bacteria in minimal, iron-complexed media, there is at least one report of heightened specificity. The growth of *Yersinia enterocolitica* is stimulated by norepinephrine

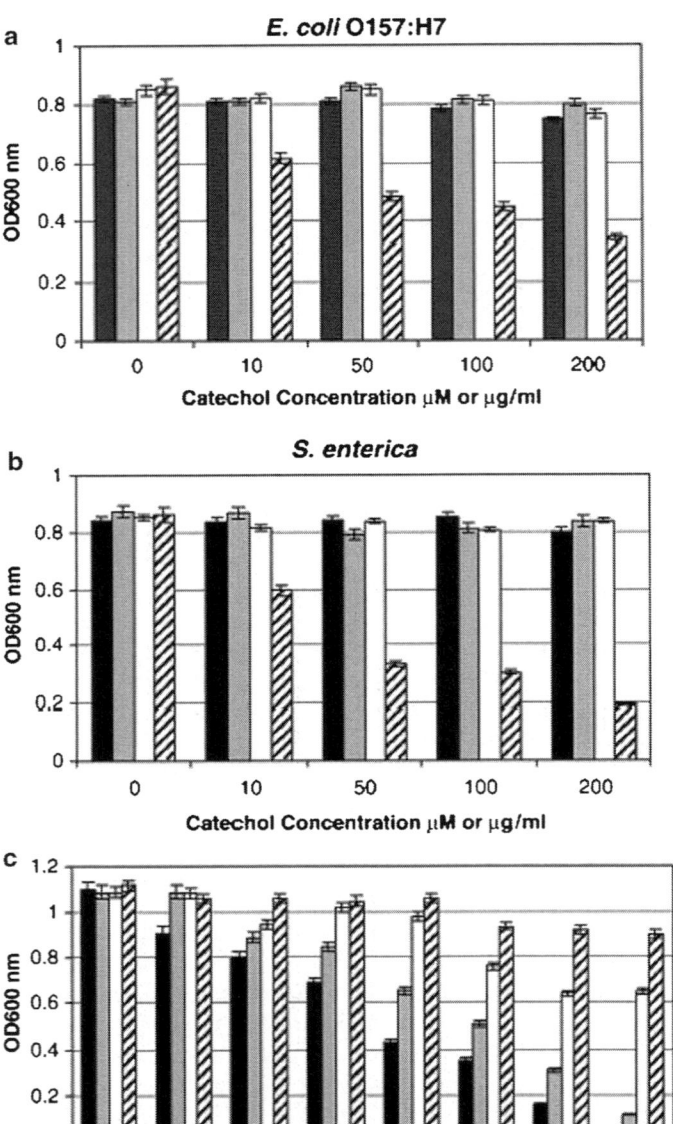

Fig. 4.2 Growth modulation by dietary catechols in laboratory culture media. *E. coli* O157:H7 (**a**) and *S. enterica* (**b**) were inoculated at approximately 10^2 CFU ml^{-1} into duplicate 1 ml volumes of Luria broth containing the concentrations of the catechols shown, incubated for 10 h and enumerated for growth by measurement of the absorbance of the cultures at 600 nm. Catechin, caffeic and chlorogenic acids are measured in μM units, while tannic acid is measured in units of μg ml^{-1}. The results shown are representative data from two separate experiments; data points showed

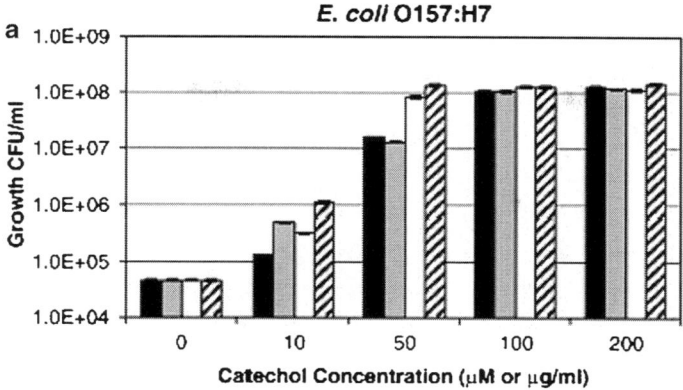

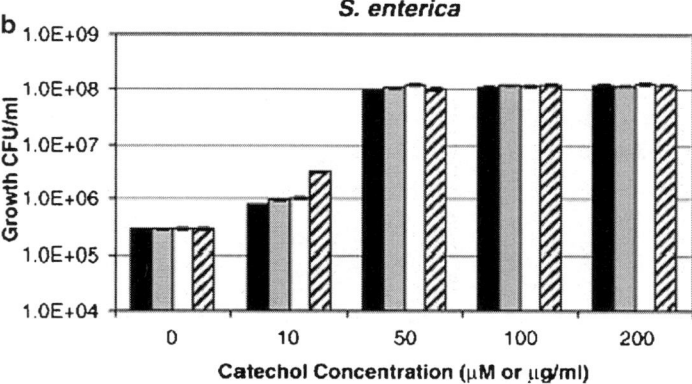

Fig. 4.3 Growth modulation by dietary catechols in serum-based media. *E. coli* O157:H7 and *S. enterica* were inoculated at approximately 10^2 CFU ml^{-1} into duplicate 1 ml aliquots of serum-SAPI containing the concentrations of the catechols shown, incubated for 18 h, and enumerated for growth (CFU ml^{-1}). Catechin, caffeic and chlorogenic acids are measured in μM units, while tannic acid is measured in units of μg ml^{-1}. The results shown are representative data from two separate experiments; data points showed variation of less than 5%. Catechin (*black bar*); Caffeic acid (*grey bar*); Chlorogenic acid (*white bar*); Tannic acid (*diagonal hatch*). Reproduced from Freestone et al. (2007c), with permission

Fig. 4.2 (continued)   variation of less than 5%. Catechin (*black bar*); Caffeic acid (*grey bar*); Chlorogenic acid (*white bar*); Tannic acid (*diagonal hatch*). (**c**) examines the mechanistic basis of the growth inhibition by tannic acid of *E. coli* O157:H7. A similar methodology to that used in (**a**) and (**b**) was employed, except that the culture media were supplemented with either no additions (*black bar*), 50 mM Tris–HCl, pH 7.5 (*grey bar*), 100 μM ferric nitrate (*white bar*) and 50 mM Tris–HCl, pH 7.5 and 100 μM ferric nitrate (*diagonal hatch*). Cultures were corrected for absorbance due to media components. The results shown are representative data from two separate experiments; data points showed variation of less than 5%. Reproduced from Freestone et al. (2007c), with permission

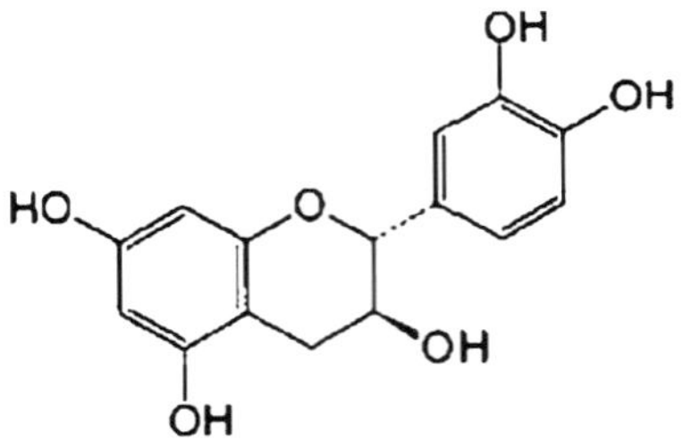

Fig. 4.4 Structures of flavanols: (+)-catechin (I), (−)-epicatechin (II, R =H), (−)-epigallocatechin (II, R =OH), (−)-epicatechin-3-gallate (III, R = H) and (−)-epigallocatechin-3-gallate (III R = OH). Reproduced from Hollman and Arts (2000), © Society of Chemical Industry, with permission

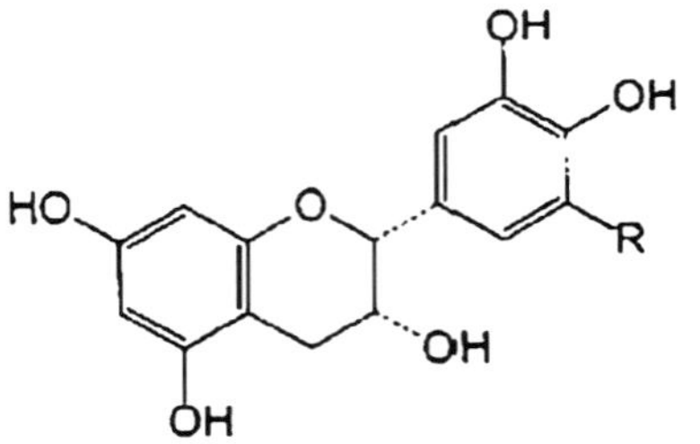

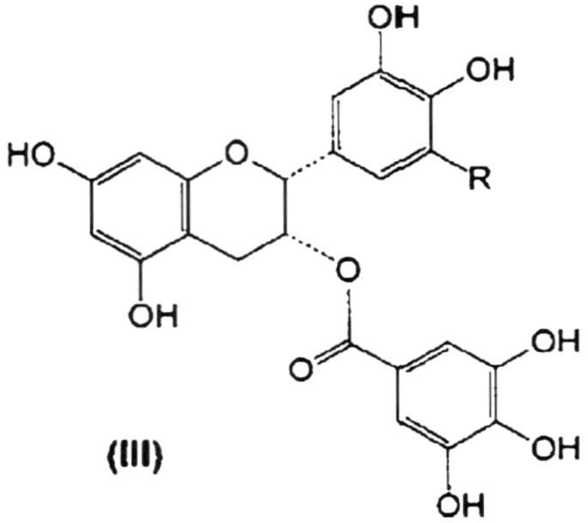

and dopamine but not by epinephrine (Freestone et al. 2007a). It has been postulated that this specificity reflects the fact that *Y. enterocolitica* infection is principally limited to the gut where epinephrine-containing neurones are not found. This study also showed that epinephrine was a less potent growth inducer for *E. coli* 0157:H7, with 50 μM epinephrine being required to elicit the stimulation of growth produced by 20 μM norepinephrine or dopamine. Thus, although a general growth response to catecholic substances appears to be common, more specificity may exist than is presently apparent.

In addition to the growth stimulation by neuroendocrine catecholamines, there are also several reports of bacteria utilising these catecholamines as signalling molecules in the activation of expression of colonisation and/or virulence factors. The question therefore arises as to whether these effects are specific to catecholamines or whether similar responses might also be elicited by dietary catechols. Expression of the outer-membrane enterobactin transporter BfeA of *Bordetella bronchiseptica* is activated in response to its substrate by the AraC family transcriptional regulator, BfeR. Anderson and Armstrong (2006) demonstrated that BfeA expression could

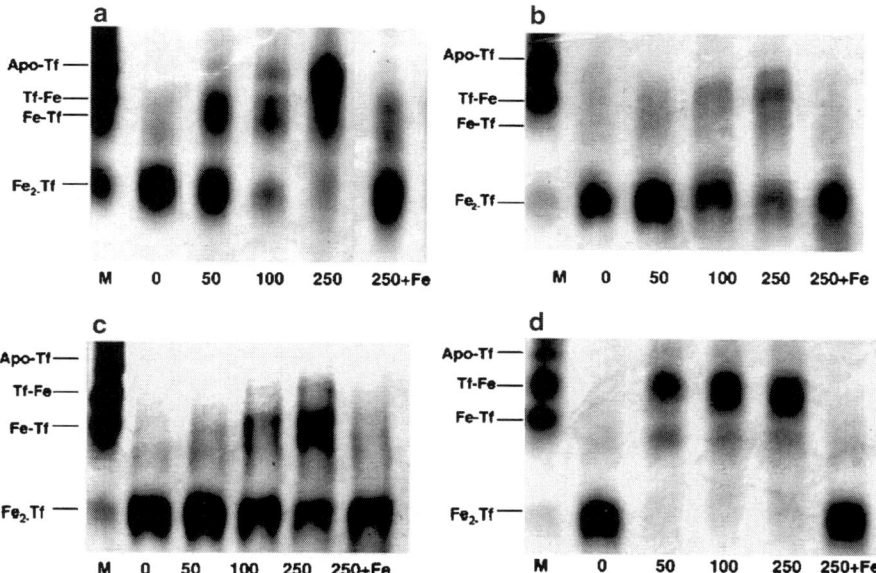

Fig. 4.5 Dietary catechols are able to remove Tf-complexed iron. Urea gels illustrate the effect on the iron-binding status of Tf incubated for 18 h at 37°C in the presence of increasing concentrations of the catechol compounds shown. (**a**) Caffeic acid (lower legend μM); (**b**) catechin (lower legend μM); (**c**) chlorogenic acid (lower legend μM); (**d**) tannic acid (lower legend μg ml^{-1}). *Lane M* contains iron-free (Apo-Tf), monoferric with iron in the N-terminal or C-terminal domains (Fe-Tf and Tf-Fe, respectively), and saturated (Fe$_2$-Tf) isoforms as markers. +Fe (1 mM ferric nitrate). Reproduced from Freestone et al. (2007c), with permission

also be activated by norepinephrine as well as by several other catecholamines and by the non-amine catechols, pyrocatechol and 3,4-dihydroxymandelic acid. It remains to be established whether catechols derived from the diet might also activate BfeA expression.

Although the activation of *B. bronchiseptica* BfeA does not appear to be catecholamine-specific, there is some evidence of catecholamine specificity in *Borrelia burgdorferi*, the causative agent of Lyme disease, in which expression of outer-surface protein A (OspA) is induced by epinephrine and norepinephrine (Scheckelhoff et al. 2007). This induction is blocked by a competitive inhibitor of human β-adrenergic receptors, suggesting that the mechanism is catecholamine-specific. However, since no non-amine catechols were tested, there remains in principle the possibility that other catechols could be bound by, and activate, the as-yet-uncharacterised adrenergic receptor. Importantly, no effect on *B. burgdorferi* growth was seen with either epinephrine or norepinephrine, suggesting that this signalling pathway is independent of iron acquisition.

Additional evidence for the presence of bacterial adrenergic receptors comes from work with *E. coli* 0157:H7. In this bacterium, two response regulators, QseA and QseB, activate virulence gene expression in response to epinephrine, norepinephrine and the bacterial autoinducer AI-3. Binding of epinephrine directly activates QseC, the cognate histidine kinase for QseB, and this activation is blocked by the α-adrenergic antagonist, phentolamine, but not by the β-adrenergic antagonist, propanolol (Clarke et al. 2006). However, both α- and β-adrenergic receptor antagonists block the activation of LEE (Locus of Enterocyte Effacement) gene expression by QseA (Sperandio et al. 2003). This difference in antagonist specificity implies that *E. coli* 0157:H7 possesses an additional, as yet undiscovered, adrenergic-receptor, which is responsible for activating QseA.

Clarke et al. (2006) established that the periplasmic signal-sensing domain of QseC is strongly conserved across a range of bacteria, including *Shigella* sp., *Salmonella* sp., *Erwinia carotovora*, *Haemophilus influenzae*, *Pasteurella multocida*, *Actinobacillus pleuropneumoniae* and *Psychrobacter* sp., and that it shares no primary sequence homology with classical G-protein-coupled adrenergic receptors. It will now be of particular interest to establish the structural features and binding properties of this domain and to discover the extent to which it might also bind, if at all, other catecholic molecules of biological interest. However, the observation that phentolamine antagonizes the effects of norepinephrine upon QseC (Clarke et al. 2006), yet does not block the growth stimulation of *E. coli* 0157-H7 that is seen in response to several dietary catechols when they are provided under iron-restricted conditions (Freestone, unpublished), suggests that growth stimulation by catechols may be mechanistically separate from effects that are mediated via QseC.

The evidence discussed in the last two chapters suggests that catechols in general can stimulate growth of bacteria through their ability to increase the bioavailability of iron. However, studies with inhibitors of adrenergic receptors indicate that there are additional bacterial signalling pathways that are likely to be specific for neuroendocrine catecholamines. There is also evidence that these catecholamines may stimulate growth in a manner independent of iron release from transferrin. It was recently shown in *E. coli* 0157:H7 that adrenergic and dopaminergic receptor antagonists that block growth stimulation by norepinephrine, epinephrine and dopamine do not block [55]Fe uptake from [55]Fe transferrin (Freestone et al. 2007b). The presence of the α-adrenergic receptor antagonists did, however, block norepinephrine uptake. However, these effects may still be linked to iron provision since the addition of ferric nitrate overcame the antagonist blockade of growth induction.

4.7   Concluding Remarks

It is apparent that catechols derived from the diet can be present in plasma and in a range of tissues and organs at concentrations that are comparable with those of the neuroendocrine catecholamines and that they both have the potential to promote bacterial growth through the relief of iron restriction. It seems much less probable, on the other hand, that catechols other than catecholamines can mediate responses that require bacterial adrenergic receptors, although definitive experiments have yet to be carried out. However, diet-derived catechols and related compounds may indirectly affect plasma levels of neuroendocrine catecholamines, through the inhibition of catechol-O-methyltransferase activity, as demonstrated particularly for (−)-epigallocatechin-3-O-gallate (Shixian et al. 2006; Zhu et al. 2008).

Whilst the emphasis in the foregoing discussion has been upon the possible effects on bacteria of dietary catechols in the plasma, catechols and other polyphenols will also influence the behaviour and composition of the colonic microflora. It has been argued that (in plants) a major function of tannins may be to prevent or limit the growth of pathogens and degradative microorganisms, especially (though not exclusively) through the complexation of iron (Scalbert 1991; Mila et al. 1996). It is also well established that a tannin-rich diet has the potential to restrict iron availability in humans (Santos-Buelga and Scalbert 2000; Mennen et al. 2005). It is not certain how far tannins and other catechols present in the gut lumen inhibit bacterial growth in vivo (Chung et al. 1998), or whether in contrast they might, in particular circumstances, *promote* bacterial growth, as shown to occur for *E. coli* O157:H7 and *S. enterica* SV Enteriditis in the serum-SAPI model (Freestone et al. 2007c) and for commensal *E. coli* in an in vivo rat model (Samanta 2004). One factor likely to be important is the low oxygen tension of the gut lumen, which may favour the operation of bacterial iron-uptake mechanisms for Fe^{2+} (Andrews et al. 2003; Naikare et al. 2006). A second factor to be taken into account is the availability and effect of transferrin and lactoferrin, especially given that Freestone et al. (2007c) have demonstrated that dietary catechols can release iron from these glycoproteins and promote its uptake by *E. coli* O157:H7 and *S. enterica* SV Enteriditis. However (in the absence of catechols), both apo-lactoferrin and lactoferrin partially saturated with iron have been found to inhibit the growth of *E. coli* O157:H7 (Griffiths et al. 2003), presumably by chelating any free iron in the media. The growth of the probiotic organism, *Bifidobacterium infantis*, on the other hand, was unaffected. Many lactic-acid bacteria, including *Bifidobacterium* and *Lactobacillus* species, which are widely regarded as beneficial members of the colonic microflora, are notable for their almost complete lack of dependence upon iron (Pandey et al. 1994; Bruyneel et al. 1989). Therefore, in the gut-lumen environment, these organisms might benefit from factors such as tannins that could inhibit the growth of their iron-requiring competitors.

Acknowledgements N.S. is supported by BBSRC responsive mode grant No. BB/D013135/1 awarded under the BBSRC's GPS scheme and co-funded by the FSA.

References

Anderson, M. T., and Armstrong, S. K. 2006. The *Bordetella* Bfe system: growth and transcriptional response to siderophores, catechols, and neuroendocrine catecholamines. J. Bacteriol. 188:5731–5740.

Andjelković, M., Van Camp, J., De Meulenaer, B., Depaemelaere, G., Socaciu, C., Verloo, M., and Verhe, R. 2006. Iron-chelation properties of phenolic acids bearing catechol and galloyl groups. Food Chem. 98:23–31.

Andrews, S. C., Robinson, A. K., and Rodriguez-Quinones, F. 2003. Bacterial iron homeostasis. FEMS Microbiol. Rev. 27:215–237.

Benedict, C. R., and Grahame-Smith, D. G. 1978. Plasma noradrenaline and adrenaline concentrations and dopamine-β-hydroxylase activity in patients with shock due to septicaemia, trauma and haemorrhage. Q. J. Med.1–20.

Boudet, A.-M. 2007. Evolution and current status of research in phenolic compounds. Phytochemistry 68:2722–2735.

Boulton, D. W., Walle, U. K., and Walle, T. 1998. Extensive binding of the bioflavonoid quercetin to human plasma proteins. J. Pharm. Pharmacol. 50:243–249.

Bruyneel, B., Vande Woestyne, M., and Verstraete, W. 1989. Lactic-acid bacteria – microorganisms able to grow in the absence of available iron and copper. Biotechnol. Lett. 11:401–406.

Burton, C. L., Chhabra, S. R., Swift, S., Baldwin, T. J., Withers, H., Hill, S. J., and Williams, P. 2002. The growth response of *Escherichia coli* to neurotransmitters and related catecholamine drugs requires a functional enterobactin biosynthesis and uptake system. Infect. Immun. 70:5913–5923.

Chvátalová, K., Slaninová, I., Březinová, L., and Slanina, J. 2008. Influence of dietary phenolic acids on redox status of iron: ferrous iron autoxidation and ferric iron reduction. Food Chem. 106:650–660.

Chung, K.-T., Lu, Z., and Chou, M. W. 1998. Mechanism of inhibition of tannic acid and related compounds on the growth of intestinal bacteria. Food Chem. Toxicol. 36:1053–1060.

Clarke, M. B., Hughes, D. T., Zhu, C., Boedeker, E. C., and Sperandio, V. 2006. The QseC sensor kinase: a bacterial adrenergic receptor. Proc. Natl. Acad. Sci. USA 103:10420–10425.

Clifford, M. N. 2000a. Chlorogenic acids and other cinnamates – nature, occurrence, dietary burden, absorption and metabolism. J. Sci. Food. Agric. 80:1033–1043.

Clifford, M. N. 2000b. Anthocyanins – nature, occurrence and dietary burden. J. Sci. Food. Agric. 80:1063–1072.

Cooper, R., Morré, D. J., and Morré, D. M. 2005. Medicinal benefits of green tea: part I. Review of noncancer health benefits. J. Alternat. Complement. Med. 11:521–528.

Coulanges, V., Andre, P., and Vidon, D. J.-M. 1998. Effect of siderophores, catecholamines, and catechol compounds on *Listeria* spp. growth in iron-complexed medium. Biochem. Biophys. Res. Commun. 249:526–530.

Coulanges, V., Andre, P., Ziegler, O., Buchheit, L., and Vidon, D. J.-M. 1997. Utilization of iron-catecholamine complexes involving ferric reductase activity in *Listeria monocytogenes*. Infect. Immun. 65:2778–2785.

Couteau, D., McCartney, A. L., Gibson, G. R., Williamson, G., and Faulds, C. B. 2001. Isolation and characterization of human colonic bacteria able to hydrolyse chlorogenic acid. J. Appl. Microbiol. 90:873–881.

Cremin, P., Kasim-Karakas, S., and Waterhouse, A. L. 2001. LC/ES-MS detection of hydroxycinnamates in human plasma and urine. J. Agric. Food Chem. 49:1747–1750.

Crosa, J. H., and Walsh, C. T. 2002. Genetics and assembly line enzymology of siderophore biosynthesis in bacteria. Microbiol. Mol. Biol. Rev. 66:223–249.

Day, A. J., Mellon, F., Barron, D., Sarrazin, G., Morgan, M. R., and Williamson, G. 2001. Human metabolism of dietary flavonoids: identification of plasma metabolites of quercitin. Free Radic. Res. 35:941–952.

Day, A. J., Gee, J. M., DuPont, M. S., Johnson, I. T., and Williamson, G. 2003. Absorption of quercetin-3-glucoside and quercetin-4'-glucoside in the rat small intestine: the role of lactase phlorizin hydrolase and the sodium-dependent glucoside transporter. Biochem. Pharmacol. 65:1199–1206.

Donovan, J. L., Bell, J. R., Kasim-Karakas, S., German, J. B., Walzem, R. L., Hansen, R. J., and Waterhouse, A. L. 1999. Catechin is present as metabolites in human plasma after consumption of red wine. J. Nutr. 129:1662–1668.

Dufour, C., and Dangles, O. 2005. Flavonoid-serum albumin complexation: determination of binding constants and binding sites by fluorescence spectroscopy. Biochim. Biophys. Acta 1721:164–173.

Espín, J. C., García-Conesa, M. T., and Tomás-Barberán, F. A. 2007. Nutraceuticals: facts and fiction. Phytochemistry 68:2986–3008.

Evans, D. A., Hirsch, J. B., and Dushenkov, S. 2006. Phenolics, inflammation and nutrigenomics. J. Sci. Food Agric. 86:2503–2509.

Freestone, P. P. E., Lyte, M., Neal, C. P., Maggs, A. F., Haigh, R. D., and Williams, P. H. 2000. The mammalian neuroendocrine hormone norepinephrine supplies iron for bacterial growth in the presence of transferrin or lactoferrin. J. Bacteriol. 182:6091–6098.

Freestone, P. P. E., Williams, P. H., Haigh, R. D., Maggs, A. F., Neal, C. P., and Lyte, M. 2002. Growth stimulation of intestinal commensal *Escherichia coli* by catecholamines: a possible contributory factor in trauma-induced sepsis. Shock 18:465–470.

Freestone, P. P. E., Haigh, R. D., Williams, P. H., and Lyte, M. 2003. Involvement of enterobactin in norepinephrine-mediated iron supply from transferrin to enterohaemorrhagic *Escherichia coli*. FEMS Microbiol. Lett. 222:39–43.

Freestone, P. P. E., Haigh, R. D., and Lyte, M. 2007a. Specificity of catecholamine-induced growth in *Escherichia coli* O157:H7, *Salmonella enterica* and *Yersinia enterocolitica*. FEMS Microbiol. Lett. 269:221–228.

Freestone, P. P. E., Haigh, R. D., and Lyte, M. 2007b. Blockade of catecholamine-induced growth by adrenergic and dopaminergic receptor antagonists in *Escherichia coli* O157:H7, *Salmonella enterica* and *Yersinia enterocolitica*. BMC Microbiol. 7:8.

Freestone, P. P. E., Walton, N. J., Haigh, R. D., and Lyte, M. 2007c. Influence of dietary catechols on the growth of enteropathogenic bacteria. Int. J. Food Microbiol. 119:159–169.

Gonthier, M.-P., Remesy, C., Scalbert, A., Cheynier, V., Souquet, J.-M., Poutanen, K., and Aura, A.-M. 2006. Microbial metabolism of caffeic acid and its esters chlorogenic and caftaric acids by human faecal microbiota in vitro. Biomed. Pharmacother. 60:536–540.

Graefe, E. U., Wittig, J., Mueller, S., Riethling, A. K., Uehleke, B., Drewelow, B., Pforte, H., Jacobasch, G., Derendorf, H., and Veit, M. 2001. Pharmacokinetics and bioavailability of quercetin glycosides in humans. J. Clin. Pharmacol. 41:492–499.

Griffiths, E. A., Duffy, L. C., Schanbacher, F. L., Dryja, D., Leavens, A., Neiswander, R. L., Qiao, H. P., DiRienzo, D., and Ogra, P. 2003. In vitro growth responses of bifidobacteria and enteropathogens to bovine and human lactoferrin. Dig. Dis. Sci. 48:1324–1332.

Hodgson, J. M., and Croft, K. D. 2006. Dietary flavonoids: effects on endothelial function and blood pressure. J. Sci. Food Agric. 86:2492–2498.

Hollman, P. C. H., and Arts, I. C. W. 2000. Flavonols, flavones and flavanols – nature, occurrence and dietary burden. J. Sci. Food Agric. 80:1081–1093.

Hollman, P. C. H., van Trijp, J. M. P., Buysman, M. N. C. P., Gaag, M. S. v. d., Mengelers, M. J. B., de Vries, J. H. M., and Katan, M. B. 1997. Relative bioavailability of the antioxidant flavonoid quercetin from various foods in man. FEBS Lett. 418:152–156.

Khan, N., Afaq, F., and Mukhtar, H. 2008. Cancer chemoprevention through dietary antioxidants: progress and promise. Antioxid. Redox Signal. 10:475–510.

Khokhar, S., and Apenten, R. K. O. 2003. Iron binding characteristics of phenolic compounds: some tentative structure–activity relations. Food Chem. 81:133–140.

Komprda, T., Burdychová, R., Dohnal, V., Cwiková, O., Sládková, P., and Dvořáčková, H. 2008. Tyramine production in Dutch-type semi-hard cheese from two different producers. Food Microbiol. 25:219–227.

Kulma, A., and Szopa, J. 2007. Catecholamines are active compounds in plants. Plant Sci. 172:433–440.

Loke, W. M., Proudfoot, J. M., Stewart, S., McKinley, A. J., Needs, P. W., Kroon, P. A., Hodgson, J. M., and Croft, K. D. 2008. Metabolic transformation has a profound effect on anti-inflammatory activity of flavonoids such as quercetin: lack of association between antioxidant and lipoxygenase inhibitory activity. Biochem. Pharmacol. 75:1045–1053.

Lyte, M. 1997. Induction of Gram-negative bacterial growth by neurochemical containing banana (*Musa × paradisiacal*) extracts. FEMS Microbiol. Lett. 154:245–250.

Lyte, M. 2004a. Microbial endocrinology and infectious disease in the 21st century. Trends Microbiol. 12:14–20.

Lyte, M. 2004b. The biogenic amine tyramine modulates the adherence of *Escherichia coli* O157:H7 to intestinal mucosa. J. Food Protect. 67:878–883.

Manach, C., Morand, C., Crespy, V., Demigné, C., Texier, O., Régérat, F., and Rémésy, C. 1998. Quercetin is recovered in human plasma as conjugated derivatives which retain antioxidant properties. FEBS Lett. 426:331–336.

Manach, C., Scalbert, A., Morand, C., Rémésy, C., and Jiménez, L. 2004. Polyphenols: food sources and bioavailability. Am. J. Clin. Nutr. 79:727–747.

Manach, C., Williamson, G., Morand, C., Scalbert, A., and Rémésy, C. 2005. Bioavailability and bioefficacy of polyphenols in humans. I. Review of 97 bioavailability studies. Am. J. Clin. Nutr. 81(suppl.):230S–242S

Mennen, L. I., Walker, R., Bennetau-Pelissero, C., and Scalbert, A. 2005. Risks and safety of polyphenol consumption. Amer. J. Clin. Nutr. 81(suppl.):326S–329S.

Mila, I., Scalbert, A., and Expert, D. 1996. Iron withholding by plant polyphenols and resistance to pathogens and rots. Phytochemistry 42:1551–1555.

Moran, J. F., Klucas, R. V., Grayer, R. J., Abian, J., and Becana, M. 1997. Complexes of iron with phenolic compounds from soybean nodules and other legume tissues: prooxidant and antioxidant properties. Free Radic. Biol. Med. 22:861–870.

Naikare, H., Palyada, K., Panciera, R., Marlow, D., and Stintzi, A. 2006. Major role for FeoB in *Campylobacter jejuni* ferrous iron acquisition, gut colonization, and intracellular survival. Infect. Immun. 74:5433–5444.

Natsume, M., Osakabe, N., Oyama, M., Sasaki, M., Baba, S., Nakamura, Y., Osawa, T., and Terao, J. 2003. Structures of (−)-epicatechin glucuronide identified from plasma and urine after oral ingestion of (−)-epicatechin: differences between human and rat. Free Radic. Biol. Med. 34:840–849.

Németh, K., Plumb, G. W., Berrin, J. G., Juge, N., Jacob, R., Naim, H. Y., Williamson, G., Swallow, D. M. and Kroon, P. A. 2003. Deglycosylation by small intestinal epithelial cell beta-glucosidases is a critical step in the absorption and metabolism of dietary flavonoid glycosides in humans. Eur. J. Nutr. 42:29–42.

Pandey, A., Bringel, F., and Meyer, J.-M. 1994. Iron requirement and search for siderophores in lactic-acid bacteria. Appl. Microbiol. Biotechnol. 40:735–739.

Parr, A. J., and Bolwell, G. P. 2000. Phenols in the plant and in man. The potential for possible nutritional enhancement of the diet by modifying the phenols content or profile. J. Sci. Food Agric. 80:985–1012.

Prior, R. L., Wu, X., and Gu, L. 2006. Flavonoid metabolism and challenges to understanding health effects. J. Sci. Food Agric. 86:2487–2491.

Rahman, I. 2006. Beneficial effects of dietary polyphenols using lung inflammation as a model. J. Sci. Food Agric. 86:2499–2502.

Rein, D., Lotito, S., Holt, R. R., Keen, C. L., Schmitz, H. H., and Fraga, C. G. 2000. Epicatechin in human plasma: in vivo determination and effect of chocolate consumption on plasma oxidation status. J. Nutr. 130:2109S–2114S.

Rice-Evans, C. A., Miller, N. J., Bolwell, P. G., Bramley, P. M., and Pridham, J. B. 1995. The relative antioxidant activities of plant-derived polyphenolic flavonoids. Free Radic. Res. 22:375–379.

Ross, J. A., and Kassum, C. M. 2002. Dietary flavonoids: bioavailability, metabolic effects, and safety. Annu. Rev. Nutr. 22:19–34.

Samanta, S. 2004. Impact of tannic acid on the gastrointestinal microflora. Microb. Ecol. Health Dis. 16:32–34.

Santos, M. H. S. 1996. Biogenic amines: their importance in foods. Int. J. Food Microbiol. 29:213–231.

Santos-Buelga, C., and Scalbert, A. 2000. Proanthocyanidins and tannin-like compounds – nature, occurrence, dietary intake and effects on nutrition and health. J. Sci. Food Agric. 80:1094–1117.

Scalbert, A. 1991. Antimicrobial properties of tannins. Phytochemistry 30:3875–3883.

Scheckelhoff, M. R., Telford, S. R., Wesley, M., and Hu, L. T. 2007. *Borrelia burgdorferi* intercepts host hormonal signals to regulate expression of outer surface protein A. Proc. Natl. Acad. Sci. USA 104:7247–7252.

Schroeter, H., Heiss, C., Balzer, J., Kleinbongard, P., Keen, C. L., Hollenberg, N. K, Sies, H., Uribe-Kwik, C., Schmitz, H. H., and Kelm, M. 2006. (–)-Epicatechin mediates beneficial effects of flavanol-rich cocoa on vascular function in humans. Proc. Natl. Acad. Sci. USA 103:1024–1029.

Sesink, A. L. A., Arts, I. C. W., Faassen-Peters, M., and Hollman, P. C. H. 2003. Intestinal uptake of quercetin-3-glucoside in rats involves hydrolysis by lactase phlorizin hydrolase. J. Nutr. 133:773–776.

Shixian, Q., VanCrey, B., Shi, J., Kakuda, Y., and Jiang, Y. 2006. Green tea extract thermogenesis-induced weight loss by epigallocatechin gallate inhibition of catechol-O-methyltransferase. J. Med. Food 9: 451–458.

Soler-Rivas, C., Espín, J. C., and Wichers, H. J. 2000. Oleuropein and related compounds. J. Sci. Food. Agric. 80:1013–1023.

Sperandio. V., Torres, A. G., Jarvis, B., Nataro, J. P., and Kaper, J. B. 2003. Bacterial-host communication: The language of hormones. Proc. Natl. Acad. Sci. USA 100:9851–8956.

Taguri, T., Tanaka, T., and Kouno, I. 2006. Antibacterial spectrum of plant polyphenols and extracts depending upon hydroxyphenyl structure. Biol. Pharm. Bull. 29:2226–2235.

Tomás-Barberán, F. A., and Clifford, M. N. 2000. Dietary hydroxybenzoic acid derivatives – nature, occurrence and dietary burden. J. Sci. Food. Agric. 80:1024–1032.

Vitaglione, P., Donnarumma, G., Napolitano, A., Galvano, F., Gallo, A., Scalfi, L., and Fogliano, V. 2007. Protocatechuic acid is the major human metabolite of cyanidin-glucosides. J. Nutr. 137:2043–2048.

Wang, J. F., Schramm, D. D., Holt, R. R., Ensunsa, J. L., Fraga, C. G., Schmitz, H. H., and Keen, C. L. 2000. A dose-response effect from chocolate consumption on plasma epicatechin and oxidative damage. J. Nutr. 130:2115S–2119S.

Wu, X., Cao, G., and Prior, R. L. 2002. Absorption and metabolism of anthocyanins in elderly women after consumption of elderberry or blueberry. J. Nutr. 132:1865–1871.

Zhu, B. T., Shim, J.-Y., Nagai, M., and Bai, H.-W. 2008. Molecular modelling study of the mechanism of high-potency inhibition of human catechol-O-methyltransferase by (–)-epigallocate-chin-3-O-gallate. Xenobiotica 38: 130–146.

Chapter 5
Interactions Between Bacteria and the Gut Mucosa: Do Enteric Neurotransmitters Acting on the Mucosal Epithelium Influence Intestinal Colonization or Infection?

Benedict T. Green and David R. Brown

5.1 Introduction

The mechanisms governing the ability of bacteria to adhere to and colonize human and animal hosts in health and disease are still incompletely understood. Throughout the extensive mucosal surfaces of the body that are in contact with the external environment, epithelial cells represent the first point of cellular contact between bacteria and the host. In the intestinal tract, the colonization of the mucosal epithelium by bacteria has become increasingly recognized as an important determinant in the maintenance and protection of health (Ley et al. 2006). Prokaryotic factors, such as flagellin or intimin, play important roles in epithelial adherence or invasion by commensal or pathogenic bacteria; physicochemical factors, such as ambient temperature and pH, contribute to bacterial colonization as well. The roles served by other, host-related factors in microbe–host interactions, such as host regulatory molecules, have recently been discovered. In this chapter, we discuss the nature of intercellular communication that occurs among four key cells, i.e., intestinal epithelial cells (IECs), enteroendocrine cells, neurons, and gut immunocytes, which participate in modulating interactions of bacteria with the intestinal mucosa. We pay special consideration to the emerging role of host-derived biogenic amines in this process. One class of biogenic amines, the catecholamines, epinephrine, norepinephrine (NE) and dopamine (DA) have been extensively studied over the past two decades for their direct effects on the growth and virulence properties of enteric bacteria. This rapidly increasing body of information is discussed elsewhere in this book (cf. Chaps. 3, 6, 9, and 12).

B.T. Green
Research Pharmacologist, Agricultural Research Service, United States Department of Agriculture, 1150 E, 1400 N, Logan, UT 84341, USA
e-mail: ben.green@ars.usda.gov

D.R. Brown (✉)
Department of Veterinary and Biomedical Sciences, University of Minnesota, St. Paul, MN 55108, USA
e-mail: brown013@umn.edu

M. Lyte and P.P.E. Freestone (eds.), *Microbial Endocrinology*, Interkingdom Signaling in Infectious Disease and Health, DOI 10.1007/978-1-4419-5576-0_5, © Springer Science+Business Media, LLC 2010

5.2 Gut Bacteria and Conversations Among Cells of the Intestinal Mucosa–Submucosa

A growing body of data indicates that there is extensive intercellular crosstalk among host cells in the intestinal mucosa. Intestinal epithelial cells and enteroendocrine cells, which are situated to first encounter ingested bacteria, chemically communicate with immunocytes and nerves that are located below the mucosal surface. The products secreted from these diverse classes of mucosal and submucosal cells, i.e., cytokines, bioactive peptides, and biogenic amines, affect mucosal defense functions, initiate and regulate inflammatory responses, and alter the outcome of microbe–host associations in health and disease.

5.2.1 Intestinal Epithelial Cells

Intestinal epithelial cells function as accessory immune cells in humans and other mammalian species. They express several types of pathogen-recognition receptors, including Toll-like receptors (TLRs). These receptors are coupled through intracellular signaling pathways and trigger proinflammatory and host defense responses in IECs. TLRs recognize a variety of pathogen-associated molecular patterns, such as lipopolysaccharide (LPS) and flagellin, which respectively stimulate TLR types 4 and 5. The protective and defensive responses triggered by TLRs and other pathogen-recognition receptors are highly modulated in IECs, probably to permit the coexistence of commensal microflora with the intestinal mucosa. For example, alkaline phosphatase expressed by IECs catalyzes the degradation of LPS, which would otherwise exist at high levels in the intestinal lumen and persistently activate pro-inflammatory signaling cascades in IECs (Bates et al. 2007). Proinflammatory substances, such as interleukin 1-beta or tumor necrosis factor-alpha, act in turn to decrease alkaline phosphatase gene expression in IECs (Malo et al. 2006). In addition, IEC expression of some TLRs appears to be down-regulated, a phenomenon which may serve to limit inflammatory reactions to commensal bacteria contacting these cells (Shibolet and Podolsky 2007). Commensal flora plays an important role in the growth and differentiation of IECs (Hooper 2004).

5.2.2 Enteroendocrine Cells

Mucosal enteroendocrine cells constitute the largest mass of endocrine cells in the body. They are distributed diffusely along the length of the intestine where they are poised to sample luminal contents and come into contact with mucosa-associated bacteria. Compared to animals with normal gut microbiota, germ-free animals manifest differences in the numbers and hormonal contents of enteroendocrine cells

(Uribe et al. 1994; Sharma and Schumacher 1996). Recent evidence indicates that some enteroendocrine cells express functional taste and olfactory receptors (Braun et al. 2007; Sternini et al. 2008) as well as TLRs (Bogunovic et al. 2007; Palazzo et al. 2007). The secreted products of enteroendocrine cells, including serotonin and a diverse array of gut peptide hormones, act directly and through neurons and most likely gut leukocytes to control aspects of epithelial secretion, growth, and defense.

5.2.3 Enteric Neurons

Serotonin and gut peptide hormones released from enteroendocrine cells interact with nerves and leukocytes in the intestinal submucosa as well. There are in excess of 100 million intrinsic neurons surrounded by supporting glial cells that reside within the intestinal wall (Fig. 5.1). In addition, neurons lying outside the intestine send projections to target cells in the intestinal wall and mucosa. Neurons projecting to the mucosa participate in the monitoring of intestinal contents and regulation of mucosal defense function; their activity is modulated by mediators derived from IECs, enteroendocrine cells, and immunocytes (Levite and Chowers 2001; Downing and Miyan 2000; Lundgren 2004). Like these other cell types, enteric neurons appear to express TLRs capable of detecting bacteria-associated molecular patterns (Barajon et al. 2009; Rumio et al. 2006; Arciszewski et al. 2005). There is also evidence that commensal flora is necessary for proper enteric nervous system (ENS) structure and function (Dupont et al. 1965).

5.2.4 Diffuse and Organized Gut-Associated Lymphoid Tissue

The gut-associated lymphoid tissue (GALT) represents the largest component of the common mucosal immune system and functions to control intestinal infections. It consists of a diffuse lymphoid compartment containing large populations of lymphocytes and antigen-presenting cells (e.g., macrophages, dendritic cells) in the intestinal lamina propria and Peyer's patches, which are organized lymphoid follicles covered by a single layer of specialized epithelial cells (i.e., M cells). Studies of germ-free and gnotobiotic rodents have shown that microflora play a role in the development of the GALT (Bauer et al. 2006). As the inductive site for mucosal immunity, Peyer's patches play a critical role in sampling of luminal contents and initiating adaptive immune responses towards potentially harmful microorganisms and antigenic materials (Mowat 2003). This includes the generation of immunoglobulin A (IgA)-producing lymphoblasts. These cells mature in the system circulation and then traffic as plasma cells to mucosal effector sites in the gut lamina propria. Neurons and nerve fibers exist in close proximity to lamina propria leukocytes, including mast cells (Wood 2007) and lymphocytes (Downing and Miyan 2000).

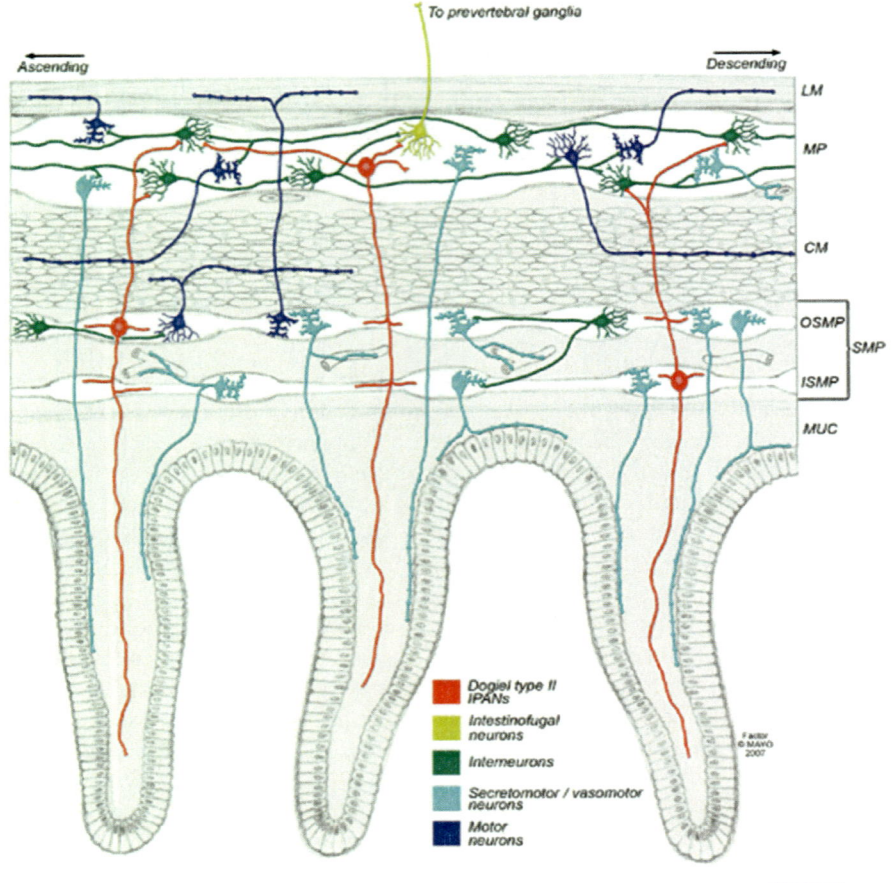

Fig. 5.1 Functional classes of enteric neurons and their major projections to the mucosa, submucosal neurons, and smooth muscle coats in the intestinal wall of a large mammal, such as a pig. Note that the mucosa receives both afferent and efferent innervation. *CM* circular muscle; *IPAN* intrinsic primary afferent neuron; *ISMP* and *OSMP* inner and outer submucosal plexuses; *LM* longitudinal muscle; *MP* myenteric plexus; *MUC* mucosa (from Linden and Farrugia 2008)

Peyer's patches are highly innervated (Defaweux et al. 2005; Vulchanova et al. 2007; Chiocchetti et al. 2008). Antigen-specific secretory IgA synthesized in these plasma cells is the major immunoglobulin secreted onto mucosal surfaces and plays an important role in mucosal protection; furthermore, constitutively produced secretory IgA is thought to regulate the gut microfloral population (Suzuki et al. 2007; Macpherson and Slack 2007). In addition to their important immunological role, Peyer's patches are exploited as portals of entry into the body for several species of enteropathogenic bacteria (Clark and Jepson 2003).

5.3 The Enteric Nervous System, Mucosally Directed Nerves, and Gut Bacteria

5.3.1 Organization of the Enteric Nervous System

The mammalian ENS originates mainly from the vagal neural crest during gestation and constitutes two or more distinct ganglionated plexuses (Burns and Thapar 2006). These include the submucosal (Meissner) and myenteric (Auerbach) plexuses, which are organized into an interconnected neural network of reflex arcs containing intrinsic primary afferent neurons, interneurons, and excitatory or inhibitory, motor and secretomotor neurons, each with distinct plurichemical coding (Furness 2006). Thus, the intestine is the only visceral organ capable of executing complex preprogrammed behaviors that can occur independently of the central nervous system.

The myenteric plexus, which is located between the longitudinal and circular smooth muscle layers, coordinates intestinal propulsion and segmentation. It is well known that neurally mediated disruptions in myoelectrical activity and mechanical functions of the intestine can alter the enteric content of microorganisms (Scott and Cahall 1982).

Neurons in the submucosal plexus(es) regulate active ion transport and paracellular permeability of the intestinal epithelium, as well as relay sensory information from the mucosa to the myenteric plexus and central nervous system (Furness 2006). Species differences have been observed in the structure and chemical coding of submucosal neurons. Rodents possess a single submucosal ganglionated plexus whereas larger mammals such as cattle and pigs have both inner (Meissner's) and outer submucosal (Schabadasch) plexuses (Fig. 5.1). Humans possess a third submucosal plexus, which appears to be similar in neurochemical coding to the outer submucosal plexus (Timmermans et al. 2001).

5.3.2 The ENS and Gut Bacteria

Submucosal and myenteric nerves play an important defensive role in the initial stages of bacterial infection, by coordinating the intestinal secretory and propulsive functions necessary to dilute and purge enteropathogens (Spiller 2002). Ingested pathogenic bacteria can alter enteric neural activity and plasticity, either through (1) direct interactions with nerve cell bodies and fibers, or (2) induction of inflammatory responses in neurons and neighboring cells, such as enteric glia, or (3) the release of neuroactive exotoxins (Lundgren 2002). Enteric neuropeptides mediate or mitigate these bacterial effects on the ENS. Some gut neuropeptides, including SP, neuropeptide Y, and neurotensin, even possess inherent antimicrobial activity (Brogden et al. 2005).

LPS at high concentrations (100 ng/ml) produces death in cultured myenteric neurons, an effect that can be prevented by the enteric neurotransmitter vasoactive intestinal peptide (VIP) (Arciszewski et al. 2005, 2008). *Bacteroides fragilis* infections alter the relative proportions of substance P- and somatostatin-expressing colonic neurons and fibers (Gonkowski et al. 2003). Intestinal infection with the causative agent in porcine proliferative enteropathy, *Lawsonia intercellularis*, is associated with an increase in the number of submucosal neurons immunoreactive for the gut peptides VIP, galanin, somatostatin, and calcitonin gene-related peptide (Pidsudko et al. 2008). The probiotic bacterium Lactobacillus reuteri appears to modulate enteric neurotransmission linked to colonic motility by affecting the gating properties of neuronal cation channels (Wang et al. 2010). Several disease-causing bacteria, including enteropathogenic *Escherichia coli*, *Salmonella Typhimurium*, and *Shigella dysenteriae*, induce the expression of receptors for the enteric neuropeptide galanin on colonic epithelial cells; this effect is not produced by normal colonic microflora. Galanin released by colonic submucosal nerves acts upon these epithelial cell receptors to stimulate active transepithelial anion secretion, which in turn contributes to secretory diarrhea (Hecht et al. 1999; Matkowskyj et al. 2000). The gut neuropeptide substance P (SP), which activates intestinal defenses against bacteria, has been shown to reduce host susceptibility to enteric *Salmonella* infections (Pascual 2004). This phenomenon appears to be due to the involvement of this neuropeptide and its cognate receptors in Salmonella-induced gut inflammation (Walters et al. 2005).

Toxins from *Clostridium difficile*, *Shigella* spp., *Campylobacter* spp., and other enteropathogens can evoke the release of inflammatory mediators, which can acutely alter neuronal activity and chronically produce structural and chemical changes in the ENS (Vasina et al. 2006; Lomax et al. 2006). The copious diarrhea associated with *Vibrio cholerae* infections has been attributed in part to cholera toxin-evoked increases in the activity of enteric neurons expressing VIP, a potent secretogogue (Mourad and Nassar 2000). *Clostridium difficile* toxin B appears to activate VIPergic submucosal neurons via an interleukin 1-dependent mechanism (Neunlist et al. 2003). Fluid secretion in diarrhea associated with *Salmonella enterica* and enterotoxigenic *E. coli* infections may involve enteric neural circuits as well (Lundgren 2002).

5.3.3   Catecholamines in the ENS

Both DA and NE are synthesized in and released from enteric nerves innervating the intestinal mucosa (Wu and Gaginella 1981; Llewellyn-Smith et al. 1981, 1984; Eisenhofer et al. 1997; Vieira-Coelho and Soares-da-Silva 1993; Wang et al. 1997; Li et al. 2004; Lomax et al. 2010). Enteric dopaminergic neurons appear to reside within the intestinal wall, but the noradrenergic innervation of the intestine originates in neurons lying outside the gut wall are located in prevertebral ganglia (Anlauf et al. 2003; Li et al. 2004; Furness 2006). These "extrinsic" sympathetic nerve fibers may co-contain peptide neurotransmitters such as neuropeptide Y or somatostatin

(Timmermans et al. 1997; Straub et al. 2006, 2008). Butyrate, which is produced by colonic bacteria, has been found to transcriptionally activate the gene encoding tyrosine hydroxylase, the rate-limiting enzyme in catecholamine synthesis (Patel et al. 2005). Epinephrine is not synthesized in enteric nerves because phenyletha-nolamine *N*-methyltransferase, the enzyme catalyzing epinephrine synthesis from norepinephrine, does not appear to be expressed in the digestive tract (Black et al. 1981; Bäck et al. 1995; Kennedy and Ziegler 2000; Costa et al. 2000). Interestingly, the growth of enteropathogens such as enterohemorrhagic *Escherichia coli* O157:H7 (EHEC) and *Salmonella enterica* is preferentially enhanced by catecholamines that are normally present in the GI tract (NE and DA) in comparison to epinephrine, and growth of the more exclusive enteric pathogen *Yersinia enterocolitica* is stimulated by NE and DA but not epinephrine (Freestone et al. 2007a). The direct action of NE on bacterial growth is unlikely to be mediated by conventional adrenergic receptors, which recognize both NE and epinephrine (cf. Sect. 5.3.4 below).

In addition to neuronal cells, there is accumulating evidence that immune cells are capable of synthesizing, releasing, and degrading catecholamines. As there is an abundance of immunocytes within the GALT, these cells may represent an alternate, non-neuronal source of NE and DA in the intestinal wall (Flierl et al. 2008).

5.3.4 Catecholamine Receptor Pharmacology

Norepinephrine and DA activate their cognate G protein-coupled receptors expressed on closely apposed neurons, IECs, and other target cells, which influence overall mucosal function and alter intestinal susceptibility to infection. Receptors for NE and DA have been defined over approximately four decades through the development and use of highly selective receptor agonists and antagonists in functional pharmacological investigations and, more recently, through molecular cloning and structure-function studies in isolated cells and transgenic animals.

The receptor concept was based at the turn of the last century on the powerful and physiologically relevant approach of defining binding interactions of endogenous substances or their synthetic homologs with specific receptors that are linked to a biological response. Biochemical analyses of selective ligand-binding site interactions that were developed some 60 years later provide valuable information on the affinities (K_d, dissociation constant) and competitive interactions of ligands at specific binding sites as well as the relative density (B_{max}) of the binding sites. Because they do not measure the biological activity of the ligands examined, however, they do not truly define a "receptor," which is an entity that is functionally coupled to intracellular signal transduction pathways and mediates a biological function. Binding studies, if carefully executed, provide information on a specific binding site for an endogenous or synthetic ligand, which can serve as supporting evidence for the presence of the receptor in a biological system. Through GTPγ^{35}S binding assays (Harrison and Traynor 2003), it is now possible to assess the effectiveness of ligands to activate G proteins coupled to a particular receptor and this approach affords a better approximation of drug activity (e.g., it is possible to distinguish an agonist from an antagonist). There is generally a good concordance in the results of studies determining the affinities and competitive interactions among ligands for adrenergic receptors (ARs) and dopaminergic receptors (DRs) by functional means with affinities and interactions of the same ligands determined biochemically at specific binding sites.

Presently, nine ARs are classified into alpha- and beta-AR types; there are two subtypes of alpha-ARs (alpha$_1$ and alpha$_2$) having three isoforms each, and three subtypes of beta-ARs. Compared to epinephrine, NE has relatively higher binding affinity for alpha-ARs and beta$_1$- and beta$_3$-ARs, but lower affinity for beta$_2$-ARs. Two main DR types exist through gene duplication events in the vertebrate lineage (D$_1$R and D$_2$R); from these, five receptor subtypes have been defined. Our understanding of the nature of these receptors and their relationships to G proteins and downstream intracellular signaling cascades is evolving (Strange 2008). For example, there is accumulating evidence that these receptors may form homo- or heterodimeric complexes on cell membranes that possess a pharmacological profile that differs from that of the monomeric receptor(s) or may be involved in the process of receptor down-regulation. With respect to ARs, this dimerization phenomenon appears to occur between different ARs as well between ARs and other classes of G protein-coupled receptors, including chemokine receptors (Milligan et al. 2005; Hague et al. 2006); dopamine receptors are known to form heterodimeric complexes with adenosine receptors (Fuxe et al. 2005).

5.3.5   Catecholamine Receptors and Mucosal Function

Enteric dopaminergic and alpha$_1$-, alpha$_2$-, or beta-adrenergic binding sites have been detected on submucosal nerves or IECs in some species, notably the guinea pig and rat (Chang et al. 1983; Cotterell et al. 1984; Senard et al. 1990; Valet et al. 1993; Vieira-Coelho and Soares-da-Silva 2001; Baglole et al. 2005). Catecholamine receptors are probably expressed by immunocytes and enteroendocrine cells as well. NE alters immunocyte function (Elenkov et al. 2000; Meredith et al. 2005; Kin and Sanders 2006) and modulates serotonin release from enterochromaffin cells in the intestinal mucosa (Pettersson 1979; Simon and Ternaux 1990; Schäfermeyer et al. 2004). In summary, enteric catecholamines, NE in particular, can potentially modulate crosstalk between several different types of cells in the intestinal mucosa and submucosa.

Norepinephrine alters active, transepithelial ion transport in the intestinal mucosa through interactions with functionally defined alpha-ARs and to a lesser extent the beta-ARs. Depending upon the animal species and intestinal segment examined, this action is mediated indirectly through enteric nerves or by direct effects on IECs (Brown and O'Grady, 1997; Horger et al. 1998). In addition, NE modulates epithelial growth and turnover (Tutton and Helme 1974; Tutton and

Barkla 1977; Olsen et al. 1985), paracellular permeability (Lange and Delbro 1995), and the vectorial secretion of secretory IgA towards the gut lumen (Schmidt et al. 1999, 2007). Dopamine affects active ion transport through direct and indirect actions on enteric adrenergic and dopaminergic receptors (Donowitz et al. 1982, 1983; Vieira-Coelho and Soares-da-Silva 1998; Al-Jahmany et al. 2004).

5.3.6 Enteric Nerves, Catecholamines, and IEC:Bacteria Interactions

Both NE and DA have been shown to alter the mucosal attachment or invasiveness of bacterial pathogens such as EHEC or serovars of *Salmonella enterica* not always through direct contact with these bacteria, but rather by acting on cells of the intestinal mucosa (Table 5.1). The actions of these catecholamines on bacteria–mucosa interactions have been examined in mucosal explants mounted in Ussing chambers (Brown and O'Grady 2008). This apparatus has been used for decades in studies of transepithelial ion transport, and more recently in investigations of bacteria–host interactions (Ding et al. 2001; Crane et al. 2006). This system extends the viability of mucosal explants under quasi-physiological conditions, allows for continuous, tangential flow of bacteria across a fixed mucosal surface area, and permits the selective contact of drugs and bacteria with the luminal or contraluminal surfaces of intestinal tissues (Fig. 5.2).

Table 5.1 Functional evidence for mucosal *alpha*-adrenergic receptors influencing EHEC adherence to explants of porcine cecal and colonic mucosae

Pharmacological characteristic	Supporting evidence	References
Selective agonism	At equimolar concentrations, UK14,304 (alpha$_2$-adrenoceptor agonist), but neither phenylephrine (alpha$_1$) nor isoproterenol (beta) increase EHEC adherence to cecal mucosa	Chen et al. (2006)
Selective antagonism	At equimolar concentrations in both cecum or colon and phentolamine (alpha-adrenoceptor antagonist), but not propranolol (beta), inhibits NE action. Furthermore, yohimbine (alpha$_2$), but not prazosin (alpha$_1$), inhibits NE action	Green et al. (2004) and Chen et al. (2006)
Laterality of drug action	NE at low (μM) concentrations effective only when added to the contraluminal, but not luminal bathing medium, consistent with submucosal localization of adrenergic receptors	Green et al. (2004)

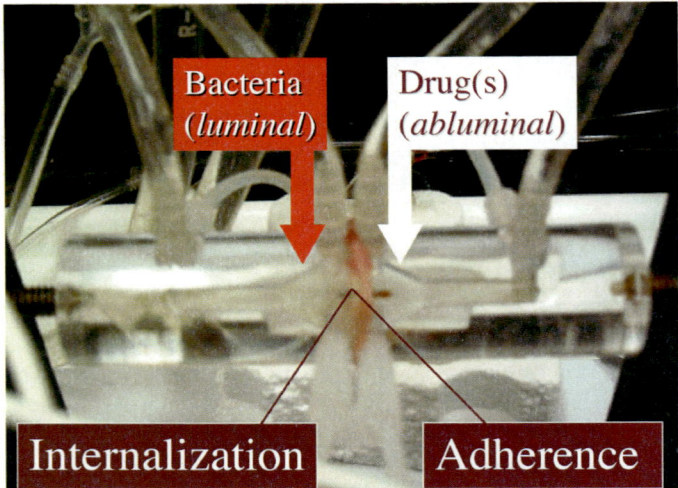

Fig. 5.2 Photograph of an Ussing chamber containing a porcine intestinal mucosa explant. *Arrows* indicate the locations of bacterial inoculations and drug additions as performed in many experiments such as those summarized in Table 5.1

5.3.6.1   Role of the ENS and Catecholamines in Bacterial Internalization into the Mucosa of the Small Intestine

The attachment and invasion of enteropathogenic bacteria to the intestinal mucosa appears to be modulated by the ENS. Inhibition of neural conduction by the serosal side addition of the neuronal sodium channel blocker saxitoxin increases internalization of luminally inoculated *Salmonella enterica* serovar Choleraesuis and EHEC by >6-fold in Peyer's patch explants from the porcine jejunum. Internalization of a rodent commensal *E. coli* strain is unaffected by the toxin (Green et al. 2003) and that of *S. enterica* serovar Typhimurium is decreased by threefold (Brown and Price 2008). Serosal application of the neurotoxin or the local anesthetic lidocaine decreased *S. Typhimurium* internalization by three- to fourfold in explants of non-Peyer's patch absorptive mucosa from porcine jejunum. In contrast, electrical stimulation of enteric nerves in this preparation increased *S. Typhimurium* internalization by 2.5-fold and this effect was inhibited by saxitoxin or lidocaine (Schreiber et al. 2007). Although these neurally mediated effects on *Salmonella* internalization may appear to be small, it should be emphasized that they were measured over a surface area of 2 cm^2 in isolated tissues over a relatively short (90 min) time period. If extrapolated to the large surface area encompassed by the small intestine or even a segment thereof, these changes in Salmonella uptake are likely to be biomedically significant.

The cellular mechanisms underlying these neurally mediated effects on Salmonella internalization in the porcine small intestinal mucosa are undefined at present.

There is evidence that they differ for different serovars Choleraesuis and Typhimurium of *S. enterica* as saxitoxin increases internalization of the former and decreases that of the latter. Moreover, Salmonella internalization is inhibited by the actin polymerization inhibitor cytochalasin D in the nonfollicular absorptive mucosa and monolayers of the porcine enterocyte cell line, IPEC J2, a result that is in agreement with other studies of actin-dependent Salmonella invasion in epithelial preparations (Schreiber et al. 2007; Brown and Price 2007). In contrast, cytochalasin D has no effect on the uptake of *S. enterica* serovars Cholerasuis and Typhimurium into jejunal Peyer's patch mucosa explants (Green and Brown 2006; Brown and Price 2008).

Norepinephrine and DA have also been implicated in *Salmonella* internalization, especially in porcine jejunal Peyer's patches, where there is strong immunohistochemical evidence for catecholaminergic innervation (Kulkarni-Narla et al. 1999) At these inductive sites for mucosal immunity, nerve fibers immunoreactive for the catecholamine synthetic enzymes tyrosine hydroxylase and dopamine beta-hydroxylase can be seen terminating beneath epithelial cells. Enteric nerves near Peyer's patch follicles express immunoreactivities for the type 2 vesicular monoamine transporter, which transports catecholamines into synaptic vesicles, and the norepinephrine transporter NET, a target of cocaine action (Kulkarni-Narla et al. 1999; Green et al. 2003).

The serosal application of NE at a bath concentration of 10 μM produced a six- to ninefold increase in luminal *S. enterica* serovar Choleraesuis and EHEC internalization in porcine Peyer's patch explants. This effect was not mimicked by luminally-applied NE, but was inhibited in tissues pretreated with the alpha-AR antagonist phentolamine. These results indicate that this NE action is mediated by alpha-ARs which are likely localized to the basolateral aspect of Peyer's patch epithelial cells (Green et al. 2003). In a study of *S. enterica* serovar Typhimurium internalization, the serosal administration of DA or the sympathomimetic drugs cocaine and methamphetamine decreased Salmonella recovery from Peyer's patch explants (Brown and Price 2008). It is not known if these effects of NE and DA extend to species other than swine. Although the underlying cellular mechanisms for them must be investigated further, it is tempting to hypothesize that catecholamines may regulate the sampling function of Peyer's patches to control the entry or immune processing of pathogenic microbes at these intestinal sites.

5.3.6.2   Catecholamines and EHEC Adherence to the Mucosa of the Large Intestine

When added to the medium bathing the contraluminal surface of cecal explants from mice, NE and DA increase the number of EHEC adhering to the mucosal surface. They do so at 50% effective concentrations (EC_{50}) of 3.8 and 4.2 μM, respectively. The concentrations of NE applied to the basolateral aspect of the intestinal epithelium that are sufficient to promote EHEC adherence are somewhat lower than those necessary to promote epithelial EHEC adherence when incubated directly with the bacterium (Vlisidou et al. 2004; Bansal et al. 2007). The adherence-enhancing actions of NE and DA on the epithelium are inhibited respectively by AR and DR receptor antagonists, a result indicating that they are mediated by specific catecholamine receptors (Chen et al. 2003). This appears to differ from the mechanism by which these catecholamines produce their direct effects on bacterial function (Freestone et al. 2007b).

This phenomenon extends to species other than the mouse. Indeed, in mucosal explants of porcine cecum and colon, NE increases mucosal adherence of EHEC through interactions with $alpha_2$-ARs that were characterized by conventional receptor criteria (Table 5.1). Increases in active anion secretion across the porcine colonic mucosa are in comparison mediated by $alpha_1$-ARs (Brown and O'Grady 1997). Therefore, it appears that the actions of NE on ion transport and EHEC adherence are not linked through a common cellular mechanism. $Alpha_2$-ARs are negatively coupled to cyclic AMP production and a concomitant decrease in intracellular protein kinase A activity. In support of this receptor–effector association, the adherence-promoting action of NE in the porcine colonic mucosa is inhibited by the protein kinase A activator Sp-cAMPS and mimicked by the protein kinase A inhibitor Rp-cAMPs (Green et al. 2004). The effects of NE in the mouse and pig cecal mucosae are relatively rapid (≤ 90 min), and experiments with EHEC eae and EspA deletion mutants strongly suggest that NE and other sympathomimetic drugs enhance early, nonintimate bacterial adherence (Chen et al. 2003, 2006). As with their effects on *Salmonella* internalization, although the effects of NE and other sympathomimetic drugs on EHEC adherence may appear small (<1.0 log unit increase in the number of adherent bacteria in mucosal explants with an exposed surface area of 1 or 2 cm^2), they may assume considerable medical importance when extrapolated over the extensive surface area of the cecal or colonic mucosa (Snipes 1997).

The mechanisms underlying this unique catecholamine action remain to be further defined through the identification of epithelial surface factors that mediate bacterial adherence and the receptor–effector pathways that are linked to their rapid expression. Beta1-integrins are IEC surface receptors implicated in aspects of EHEC adherence (Sinclair et al. 2006). By blocking epithelial beta1-integrins, heparin has been shown to inhibit EHEC adherence to human colonic epithelial cells (Gu et al. 2008). Norepinephrine is known to enhance interactions between blood cells and the vascular endothelium by stimulating the rapid expression of beta1-integrins (Levite et al. 2001; Butta et al. 2004; Delahunty et al. 2006), and it is tempting to speculate that it may similarly do so in promoting IEC interactions with luminal bacteria. In addition to dissecting the cellular and molecular mechanisms underlying this phenomenon, studies of catecholamine action on bacterial adherence in vitro should be extended to investigations of the role of endogenous and exogenous DA and NE in isolated intestinal loops and intact animal models which encompass larger surface areas and have greater translational relevance.

Norepinephrine may play a physiological role in promoting bacterial colonization of the large intestine, perhaps as an element in host defense. This hypothesis is based in part on a finding that NE increases cecal adherence of a non-O157 strain of *E. coli*, which was isolated from the porcine colonic mucosa (Chen et al. 2006). One interpretation of this result is that the action of NE is not limited to a particular bacterial strain or species. Presumptive NE nerve fibers immunoreactive for dopamine *beta*-hydroxylase are present throughout the submucosa and appear to terminate near the basal membranes of crypt and surface epithelial cells of the porcine distal colon and cecum (Green et al. 2004; Chen et al. 2006). Drugs capable of inhibiting the degradation (such as the monoamine oxidase inhibitor, pargyline) and neural reuptake (desipramine, cocaine) of NE at neuroepithelial junctions mimic the EHEC adherence-promoting action of NE, and their effects are inhibited by phentolamine (Green et al. 2004; Chen et al. 2006). In the porcine colonic mucosa, dopamine beta-hydroxylase-immunoreactive nerves terminate near IgA-positive B lymphocytes and neighboring IECs immunoreactive for the polymeric immunoglobulin receptor (Schmidt et al. 2007). Norepinephrine stimulates the vectorial secretion of secretory IgA in porcine colonic mucosa explants. This effect has been attributed to an alpha-AR-mediated increase in the luminally directed transport of secretory factor, a component of the polymeric Ig receptor (Schmidt et al. 2007). As noted above, constitutively produced secretory IgA is hypothesized to modulate colonization of the intestinal mucosa by commensal bacteria (Suzuki et al. 2007; Macpherson and Slack 2007).

5.4   Other Biogenic Amines and Gut Bacteria

Histamine is another biogenic amine transmitter that has been implicated in intestinal host–pathogen interactions. Its synthesis from the amino acid histidine is catalyzed by histidine decarboxylase. It has long been known that the expression of the inducible isoform of this enzyme is increased by LPS (Oh et al. 1988). Indeed, ingestion of LPS increases the histamine content of the intestinal tract (Aschenbach et al. 2003). Histamine generated by histidine decarboxylase induced by *E. coli* acts via H_1- and H_2-histamine receptors to reduce the clearance of *E. coli* from the peritoneal cavity in a murine experimental peritonitis model (Hori et al. 2002). Selective agonists at each receptor mimicked the effects of *E. coli* and exogenous histamine; on the other hand, the selective H_1- and H_2-histamine receptor antagonists pyrilamine and cimetidine, respectively, accelerated the peritoneal clearance of this microorganism. *Yersinia enterocolitica* infection of murine Peyer's patches is similarly associated with the induction of histidine decarboxylase and consequent generation of histamine. For example, cimetidine, but not pyrilamine, decreased the survival of *Yersinia*-infected mice, an effect that may be related to a suppression of H_2-histamine receptor-induced innate immune responses to this pathogenic bacterium (Handley et al. 2006).

The indoleamine serotonin (5-hydroxytryptamine) is released from enteric neurons and mucosal enterochromaffin cells. Although the gut contains most of the body's serotonin content, little is known of the potential effects of this biogenic amine on microbial interactions with the intestinal mucosa. Infections produced by *Citrobacter rodentium*, a Gram-negative bacterial pathogen that adheres to the colonic epithelium of mice, are associated with an increase in the stimulated release of serotonin from the colonic mucosa, but reduced numbers of enterochromaffin cells exhibiting immunoreactivity for serotonin and its uptake transporter (O'Hara et al. 2006). Enteropathogenic E. coli also appears to inhibit the serotonin transporter (Esmalli et al. 2009). These changes in mucosal serotonin signaling appear to be a consequence of mucosal immune responses to this pathogen. Indeed, there is recent evidence that enterochromaffin cell number and serotonin content may be influenced by the types of cytokines released in the course of enteric infections (Wang et al. 2007; Motomura et al. 2008).

5.5 Conclusions

The potential involvement of biogenic amines in intestinal interactions with bacteria and their connection to organismal biology is of emerging interest. The biological significance of the effects of the catecholamines, histamine, and serotonin on mucosal interactions with bacteria remains a mystery. Despite the roles of the ENS and the diverse array of neurotransmitters it contains in responding to bacterial infection, their involvement in host–pathogen interactions at the cellular level has yet to be firmly established.

In addition to their direct effects on bacterial growth and virulence, catecholamines released from enteric nerves appear to selectively act upon host receptors to alter mucosal internalization or adherence of intestinal bacteria. Their actions are exerted at mucosal inductive and effector sites in the intestinal tract and underscore the complex interactions of these neurotransmitter substances with IECs, enteroendocrine cells, neurons, and immunocytes in the intestinal mucosa and submucosa. Indeed, NE and possibly DA may serve roles in the regulation of bacterial sampling at immune recognition and processing sites and in the establishment or maintenance of mucosa-associated microfloral populations. The role of neuropeptides including neuropeptide Y and somatostatin or other substances (e.g., adenosine 5′-triphosphate) coreleased with NE from enteric sympathetic nerve terminals remains unknown, but they may serve to modulate NE actions. Neuropeptide Y and ATP, for example, have been shown to modulate the vasomotor effects of NE on blood vessels receiving adrenergic innervation (Huidobro-Toro and Donoso 2004).

With particular reference to NE, it is tempting to speculate that stress-induced sympathetic neural outflow may influence mucosal interactions with intestinal bacteria. This notion is supported by the finding that another stress-evoked neuroactive substance, adrenocorticotrophic hormone, acts like NE to enhance the colonic adherence of EHEC (Schreiber and Brown 2005). Sympathetic nerves are also implicated in intestinal inflammatory states, and the relationships of noradrenergic activity with intestinal inflammation, colonization of mucosa-associated commensal bacteria, and predisposition to enteric infections offers a fruitful area for future investigations (Irving and Gibson 2008; Lawley et al. 2008; Straub et al. 2008).

References

Al-Jahmany, A.A., Schultheiss, G., and Diener, M. 2004. Effects of dopamine on ion transport across the rat distal colon. Pflugers Arch. 448:605–612.

Anlauf, M., Schäfer, M.K., Eiden, L., and Weihe, E. 2003. Chemical coding of the human gastrointestinal nervous system: cholinergic, VIPergic, and catecholaminergic phenotypes. J. Comp. Neurol. 459:90–111.

Arciszewski, M., Pierzynowski, S., and Ekblad, E. 2005. Lipopolysaccharide induces cell death in cultured porcine myenteric neurons. Dig. Dis. Sci. 50:1661–1668.

Arciszewski, M.B., Sand, E., Ekblad, E. 2008. Vasoactive intestinal peptide rescues cultured rat myenteric neurons from lipopolysaccharide induced cell death. Regul Pept. 146:218–223.

Aschenbach, J.R., Seidler, T., Ahrens, F., Schrödl, W., Buchholz, I., Garz, B., Krüger, M., and Gäbel, G. 2003. Luminal salmonella endotoxin affects epithelial and mast cell function in the proximal colon of pigs. Scand. J. Gastroenterol. 38:719–726.

Bäck, N., Ahonen, M., Soinila, S., Kivilaakso, E., and Kiviluoto, T. 1995. Catecholamine-synthesizing enzymes in the rat stomach. Histochem. Cell Biol. 104:63–67.

Baglole, C.J., Davison, J.S., and Meddings, J.B. 2005. Epithelial distribution of neural receptors in the guinea pig small intestine. Can. J. Physiol. Pharmacol. 83:389-395.

Bansal, T., Englert, D., Lee, J., Hegde, M., Wood, T.K., and Jayaraman, A. 2007. Differential effects of epinephrine, norepinephrine, and indole on *Escherichia coli* O157:H7 chemotaxis, colonization, and gene expression. Infect. Immun. 75:4597–4607.

Barajon, I., Serrao, G., Arnaboldi, F., Opizzi, E., Ripamonti, G., Balsari, A., and Rumio, C. 2009. Toll-like receptors 3, 4, and 7 are expressed in the enteric nervous system and dorsal root ganglia. J. Histochem. Cytochem. 57:1013-1023.

Bates, J.M., Akerlund, J., Mittge, E., and Guillemin, K. 2007. Intestinal alkaline phosphatase detoxifies lipopolysaccharide and prevents inflammation in zebrafish in response to the gut microbiota. Cell Host Microbe 2:371–382.

Bauer, E., Williams, B.A., Smidt, H., Verstegen, M.W., and Mosenthin, R. 2006. Influence of the gastrointestinal microbiota on development of the immune system in young animals. Curr. Issues Intest. Microbiol. 7:35–51.

Black, I.B., Bohn, M.C., Jonakait, G.M., and Kessler, J.A. 1981. Transmitter phenotypic expression in the embryo. Ciba Found. Symp. 83:177–193.

Bogunovic, M., Davé, S.H., Tilstra, J.S., Chang, D.T., Harpaz, N., Xiong, H., Mayer, L.F., and Plevy, S.E. 2007. Enteroendocrine cells express functional Toll-like receptors. Am. J. Physiol. Gastrointest. Liver Physiol. 292:G1770–G1783.

Braun, T., Voland, P., Kunz, L., Prinz, C., and Gratzl, M. 2007. Enterochromaffin cells of the human gut: sensors for spices and odorants. Gastroenterology 132:1890–1901.

Brogden, K.A., Guthmiller, J.M., Salzet, M., and Zasloff, M. 2005. The nervous system and innate immunity: the neuropeptide connection. Nat. Immunol. 6:558–564.

Brown, D.R., and O'Grady, S.M. 1997. Regulation of ion transport in the porcine intestinal tract by enteric neurotransmitters and hormones. Comp. Biochem. Physiol. 118A:309–317.

Brown, D.R., and O'Grady, S.M. 2008. The Ussing chamber and measurement of drug actions on mucosal ion transport. In: *Current Protocols in Pharmacology*, vol 41. Wiley, NY, pp. 7.12.1–7.12.17.

Brown, D.R., and Price, L.D. 2007. Characterization of *Salmonella enterica* serovar Typhimurium DT104 invasion in an epithelial cell line (IPEC J2) from porcine small intestine. Vet. Microbiol. 120:328–333.

Brown, D.R., and Price, L.D. 2008. Catecholamines and sympathomimetic drugs decrease early *Salmonella* Typhimurium uptake into porcine Peyer's patches. FEMS Immunol. Med. Microbiol 52:29–35.

Burns, A.J., and Thapar, N. 2006. Advances in ontogeny of the enteric nervous system. Neurogastroenterol. Motil. 18:876–887.

Butta, N., Larrucea, S., Gonzalez-Manchon, C., Alonso, S., and Parrilla, R. 2004. *alpha*-Adrenergic-mediated activation of human reconstituted fibrinogen receptor (integrin alphaIIbbeta3) in Chinese hamster ovary cells. Thromb. Haemost. 92:1368–1376.

Chang, E.B., Field, M., and Miller, R.J. 1983. Enterocyte alpha 2-adrenergic receptors: yohimbine and p-aminoclonidine binding relative to ion transport. Am. J. Physiol. 244:G76-G82.

Chen, C., Brown, D.R., Xie, Y., Green, B.T., and Lyte, M. 2003. Catecholamines modulate *Escherichia coli* O157:H7 adherence to murine cecal mucosa. Shock 20:183–188.

Chen, C., Lyte, M., Stevens, M.P., Vulchanova, L., and Brown, D.R. 2006. Mucosally-directed adrenergic nerves and sympathomimetic drugs enhance non-intimate adherence of *Escherichia coli* O157:H7 to porcine cecum and colon. Eur. J. Pharmacol. 539:116–124.

Chiocchetti, R., Mazzuoli, G., Albanese, V., Mazzoni, M., Clavenzani, P., Lalatta-Costerbosa, G., Lucchi, M.L., Di Guardo, G., Marruchella, G., and Furness, J.B. 2008. Anatomical evidence for ileal Peyer's patches innervation by enteric nervous system: a potential route for prion neuroinvasion? Cell Tissue Res. 332:185–194.

Clark, M.A., and Jepson, M.A. 2003. Intestinal M cells and their role in bacterial infection. Int. J. Med. Microbiol. 293:17–39.

Costa, M., Brookes, S.J., and Hennig, G.W. 2000. Anatomy and physiology of the enteric nervous system. Gut 47:15–19.

Cotterell, D.J., Munday, K.A., and Poat, J.A. The binding of [3H]prazosin and [3H]clonidine to rat jejunal epithelial cell membranes. Biochem Pharmacol. 33:751-756.

Crane, J.K., Choudhari, S.S., Naeher, T.M., and Duffey, M.E. 2006. Mutual enhancement of virulence by enterotoxigenic and enteropathogenic *Escherichia coli*. Infect. Immun. 74:1505–1515.

Defaweux, V., Dorban, G., Demonceau, C., Piret, J., Jolois, O., Thellin, O., Thielen, C., Heinen, E., and Antoine, N. 2005. Interfaces between dendritic cells, other immune cells, and nerve fibres in mouse Peyer's patches: potential sites for neuroinvasion in prion diseases. Microsc. Res. Tech. 66:1–9.

Delahunty, M., Zennadi, R., and Telen, M.J. 2006. LW protein: a promiscuous integrin receptor activated by adrenergic signaling. Transfus. Clin. Biol. 13:44–49.

Ding, J., Magnotti, L.J, Huang, Q., Xu, D.Z., Condon, M.R., and Deitch, E.A. 2001. Hypoxia combined with *Escherichia coli* produces irreversible gut mucosal injury characterized by increased intestinal cytokine production and DNA degradation. Shock 16:189–195.

Donowitz, M., Cusolito, S., Battisti, L., Fogel, R., and Sharp, G.W. 1982. Dopamine stimulation of active Na and Cl absorption in rabbit ileum: interaction with alpha 2-adrenergic and specific dopamine receptors. J. Clin. Invest. 69:1008–1016.

Donowitz, M., Elta, G., Battisti, L., Fogel, R., and Label-Schwartz, E. 1983. Effect of dopamine and bromocriptine on rat ileal and colonic transport. Stimulation of absorption and reversal of cholera toxin-induced secretion. Gastroenterology 84:516–523.

Downing, J.E., and Miyan, J.A. 2000. Neural immunoregulation: emerging roles for nerves in immune homeostasis and disease. Immunol. Today 21:281–289.

Dupont, J.R., Jervis, H.R., and Sprinz, H. 1965. Auerbach's plexus of the rat cecum in relation to the germfree state. J. Comp. Neurol. 125:11–18.

Eisenhofer, G., Aneman, A., Friberg, P., Hooper, D., Fåndriks, L., Lonroth, H., Hunyady, B., and Mezey, E. 1997. Substantial production of dopamine in the human gastrointestinal tract. J. Clin. Endocrinol. Metab. 82:3864–3871.

Elenkov, I.J., Wilder, R.L., Chrousos, G.P., and Vizi, E.S. 2000. The sympathetic nerve – an integrative interface between two supersystems: the brain and the immune system. Pharmacol. Rev. 52:595–638.

Esmaili, A., Nazir, S.F., Borthakur, A., Yu, D., Turner, J.R., Saksena, S., Singla, A., Hecht, G.A., Alrefai, W.A., and Gill, R.K. 2009. Enteropathogenic *Escherichia coli* infection inhibits intestinal serotonin transporter function and expression. Gastroenterology 137:2074-2083.

Flierl, M.A., Rittirsch, D., Huber-Lang, M., Sarma, J.V., and Ward, P.A. 2008. Catecholamines – Crafty weapons in the inflammatory arsenal of immune/inflammatory cells or opening Pandora's box? Mol. Med. 14:195–204.

Freestone, P.P., Haigh, R.D., and Lyte, M. 2007a. Specificity of catecholamine-induced growth in *Escherichia coli* O157:H7, *Salmonella enterica* and *Yersinia enterocolitica*. FEMS Microbiol. Lett. 269:221–228.

Freestone, P.P., Haigh, R.D., and Lyte, M. 2007b. Blockade of catecholamine-induced growth by adrenergic and dopaminergic receptor antagonists in *Escherichia coli* O157:H7, *Salmonella enterica* and *Yersinia enterocolitica*. BMC Microbiol. 7:8.

Furness, J.B. 2006. *The Enteric Nervous System*, Blackwell, Malden, MA.

Fuxe, K., Ferré, S., Canals, M., Torvinen, M., Terasmaa, A., Marcellino, D., Goldberg, S.R., Staines, W., Jacobsen, K.X., Lluis, C., Woods, A.S., Agnati, L.F., and Franco, R. 2005. Adenosine A2A and dopamine D2 heteromeric receptor complexes and their function. J. Mol. Neurosci. 26:209–220.

Gonkowski, S., Kaminska, B., Bossowska, A., Korzon, M., Landowski, P., and Majewski, M. 2003. The influence of experimental *Bacteroides fragilis* infection on substance P and somatostatin-immunoreactive neural elements in the porcine ascending colon – a preliminary report. Folia Morphol. (Warsz) 62:455–457.

Green, B.T., and Brown, D.R. 2006. Differential effects of clathrin and actin inhibitors on internalization of *Escherichia coli* and *Salmonella choleraesuis* in porcine jejunal Peyer's patches. Vet. Microbiol. 113:117–122.

Green, B.T., Lyte, M., Kulkarni-Narla, A., and Brown, D.R. 2003. Neuromodulation of enteropathogen internalization in Peyer's patches from porcine jejunum. J. Neuroimmunol. 141:74–82.

Green, B.T., Lyte, M., Chen, C., Xie, Y., Casey, M.A., Kulkarni-Narla, A., Vulchanova, L., and Brown, D.R. 2004. Adrenergic modulation of *Escherichia coli* O157:H7 adherence to the colonic mucosa. Am. J. Physiol. Gastrointest. Liver Physiol. 287:G1238–G1246.

Gu, L., Wang, H., Guo, Y-L., and Zen, K. 2008. Heparin blocks the adhesion of E. *coli* O157:H7 to human colonic epithelial cells. Biochem. Biophys. Res. Comm. 369:1061–1064.

Hague, C., Lee, S.E., Chen, Z., Prinster, S.C., Hall, R.A., and Minneman, K.P. 2006. Heterodimers of alpha1B- and alpha1D-adrenergic receptors form a single functional entity. Mol. Pharmacol. 69:45–55.

Handley, S.A., Dube, P.H., and Miller, V.L. 2006. Histamine signaling through the H_2 receptor in the Peyer's patch is important for controlling *Yersinia enterocolitica* infection. Proc. Natl. Acad. Sci. USA 103:9268–9273.

Harrison, C., and Traynor, J.R. 2003. The [^{35}S]GTP*gamma*S binding assay: approaches and applications in pharmacology. Life Sci. 74:489–508.

Hecht, G., Marrero, J.A., Danilkovich, A., Matkowskyj, K.A., Savkovic, S.D., Koutsouris, A., and Benya, R.V. 1999. Pathogenic *Escherichia coli* increase Cl⁻ secretion from intestinal epithelia by upregulating galanin-1 receptor expression. J. Clin. Invest. 104:253–262.

Hooper, L.V. 2004. Bacterial contributions to mammalian gut development. Trends Microbiol. 12:129–134.

Horger, S., Schultheiss, G., and Diener, M. 1998. Segment-specific effects of epinephrine on ion transport in the colon of the rat. Am. J. Physiol. 275:G1367–G1376.

Hori, Y., Nihei, Y., Kurokawa, Y., Kuramasu, A., Makabe-Kobayashi, Y., Terui, T., Doi, H., Satomi, S., Sakurai, E., Nagy, A., Watanabe, T., and Ohtsu, H. 2002. Accelerated clearance of *Escherichia coli* in experimental peritonitis of histamine-deficient mice. J. Immunol. 169:1978–1983.

Huidobro-Toro, J.P., and Donoso, M.V. 2004. Sympathetic co-transmission: the coordinated action of ATP and noradrenaline and their modulation by neuropeptide Y in human vascular neuroeffector junctions. Eur. J. Pharmacol. 500:27–35.

Irving, P.M., and Gibson, P.R. 2008. Infections and IBD. Nat. Clin. Pract. Gastroenterol. Hepatol. 5:18–27.

Kennedy, B., and Ziegler, M.G. 2000. Ontogeny of epinephrine metabolic pathways in the rat: role of glucocorticoids. Int. J. Dev. Neurosci. 18:53–59.

Kin, N.W., and Sanders, V.M. 2006. It takes nerve to tell T and B cells what to do. J. Leukoc. Biol. 79:1093–1104.

Kulkarni-Narla, A., Beitz, A.J., and Brown, D.R. 1999. Catecholaminergic, cholinergic and peptidergic innervation of gut-associated lymphoid tissue in porcine jejunum and ileum. Cell Tissue Res. 298:275–286.

Lange, S., and Delbro, D.S. 1995. Adrenoceptor-mediated modulation of Evans blue dye permeation of rat small intestine. Dig. Dis. Sci. 40:2623–2629.

Lawley, T.D., Bouley, D.M., Hoy, Y.E., Gerke, C., Relman, D.A., and Monack, D.M. 2008. Host transmission of *Salmonella enterica* serovar Typhimurium is controlled by virulence factors and indigenous intestinal microbiota. Infect. Immun. 76:403–416.

Levite, M., and Chowers, Y. 2001. Nerve-driven immunity: neuropeptides regulate cytokine secretion of T cells and intestinal epithelial cells in a direct, powerful and contextual manner. Ann. Oncol. 12 Suppl 2:S19–S25.

Levite, M., Chowers, Y., Ganor, Y., Besser, M., Hershkovits, R., and Cahalon L. 2001. Dopamine interacts directly with its D3 and D2 receptors on normal human T cells, and activates *beta*1 integrin function. Eur. J. Immunol. 31:3504–3512.

Ley, R.E., Peterson, D.A., and Gordon, J.I. 2006. Ecological and evolutionary forces shaping microbial diversity in the human intestine. Cell 124:837–848.

Li, Z.S., Pham, T.D., Tamir, H., Chen, J.J., and Gershon, M.D. 2004. Enteric dopaminergic neurons: definition, developmental lineage, and effects of extrinsic denervation. J. Neurosci. 24:1330–1339.

Linden, D.R., and Farrugia, G. 2008. Autonomic control of gastrointestinal function. In: *Clinical Autonomic Disorders*, 3rd ed. Low, P.A., and Benarroch, E.E. (eds.), Lippincott Williams and Wilkins, Baltimore, pp. 88–105.

Llewellyn-Smith, I.J., Wilson, A.J., Furness, J.B., Costa, M., and Rush, R.A. 1981. Ultrastructural identification of noradrenergic axons and their distribution within the enteric plexuses of the guinea-pig small intestine. J. Neurocytol. 10:331–352.

Llewellyn-Smith, I.J., Furness, J.B., O'Brien, P.E., and Costa, M. 1984. Noradrenergic nerves in human small intestine. Distribution and ultrastructure. Gastroenterology 87:513–529.

Lomax, A.E., Linden, D.R., Mawe, G.M., and Sharkey, K.A. 2006. Effects of gastrointestinal inflammation on enteroendocrine cells and enteric neural reflex circuits. Auton. Neurosci. 126–127:250–257.

Lomax, A.E., Sharkey, K.A., and Furness, J.B. 2010. The participation of the sympathetic innervation of the gastrointestinal tract in disease states. Neurogastroenterol. Motil. 22:7-18.

Lundgren, O. 2002. Enteric nerves and diarrhoea. Pharmacol. Toxicol. 90:109–120.

Lundgren, O. 2004. Interface between the intestinal environment and the nervous system. Gut 53 Suppl 2:ii16–ii18.

Macpherson, A.J., and Slack, E. 2007. The functional interactions of commensal bacteria with intestinal secretory IgA. Curr. Opin. Gastroenterol. 23:673–678.

Malo, M.S., Biswas, S., Abedrapo, M.A., Yeh, L., Chen, A., and Hodin, R.A. 2006. The proinflammatory cytokines, IL-1*beta* and TNF-*alpha*, inhibit intestinal alkaline phosphatase gene expression. DNA Cell Biol. 25:684–695.

Matkowskyj, K.A., Danilkovich, A., Marrero, J., Savkovic, S.D., Hecht, G., and Benya, R.V. 2000. Galanin-1 receptor up-regulation mediates the excess colonic fluid production caused by infection with enteric pathogens. Nat. Med. 6:1048–1051.

Meredith, E.J., Chamba, A., Holder, M.J., Barnes, N.M., and Gordon, J. 2005. Close encounters of the monoamine kind: immune cells betray their nervous disposition. Immunology 115:289–295.

Milligan, G., Wilson, S., and López-Gimenez, J.F. 2005. The specificity and molecular basis of $alpha_1$-adrenoceptor and CXCR chemokine receptor dimerization. J. Mol. Neurosci. 26:161–168.

Motomura, Y., Ghia, J.E., Wang, H., Akiho, H., El-Sharkawy, R.T., Collins, M., Wan, Y., McLaughlin, J.T., and Khan, W.I. 2008. Enterochromaffin cell and 5-hydroxytryptamine responses to the same infectious agent differ in Th1 and Th2 dominant environments. Gut 57:475–481.

Mourad, F.H., and Nassar, C.F. 2000. Effect of vasoactive intestinal polypeptide (VIP) antagonism on rat jejunal fluid and electrolyte secretion induced by cholera and *Escherichia coli* enterotoxins. Gut 47:382–386.

Mowat, A.M. 2003. Anatomical basis of tolerance and immunity to intestinal antigens. Nat. Rev. Immunol. 3:331–341.

Neunlist ,M., Barouk, J., Michel, K., Just, I., Oreshkova, T., Schemann, M., and Galmiche, J.P. 2003. Toxin B of *Clostridium difficile* activates human VIP submucosal neurons, in part via an IL-1*beta*-dependent pathway. Am. J. Physiol. Gastrointest. Liver Physiol. 285:G1049–G1055.

Oh, C., Suzuki, S., Nakashima, I., Yamashita, K., and Nakano, K. 1988. Histamine synthesis by non-mast cells through mitogen-dependent induction of histidine decarboxylase. Immunology 65:143–148.

O'Hara, J.R., Skinn, A.C., MacNaughton, W.K., Sherman, P.M., and Sharkey, K.A. 2006. Consequences of *Citrobacter rodentium* infection on enteroendocrine cells and the enteric nervous system in the mouse colon. Cell. Microbiol. 8:646–660.

Olsen, P.S., Poulsen, S.S., and Kirkegaard, P. 1985. Adrenergic effects on secretion of epidermal growth factor from Brunner's glands. Gut 26:920–927.

Palazzo, M., Balsari, A., Rossini, A., Selleri, S., Calcaterra, C., Gariboldi, S., Zanobbio, L., Arnaboldi, F., Shirai, Y.F., Serrao, G., and Rumio, C. 2007. Activation of enteroendocrine cells via TLRs induces hormone, chemokine, and defensin secretion. J. Immunol. 178:4296–4303.

Pascual, D.W. 2004. The role of tachykinins on bacterial infections. Front. Biosci. 9:3209–3217.

Patel, P., Nankova, B.B., LaGamma, E.F. 2005. Butyrate, a gut-derived environmental signal, regulates tyrosine hydroxylase gene expression via a novel promoter element. Brain Res. Dev. Brain Res.160:53–62.

Pettersson, G. 1979. The neural control of the serotonin content in mammalian enterochromaffin cells. Acta Physiol. Scand. Suppl. 470:1–30.

Pidsudko, Z., Kaleczyc, J., Wasowicz, K., Sienkiewicz, W., Majewski, M., Zajac, W., and Lakomy, M. 2008. Distribution and chemical coding of intramural neurons in the porcine ileum during proliferative enteropathy. J. Comp. Pathol. 138:23–31.

Rumio, C., Besusso, D., Arnaboldi, F., Palazzo, M., Selleri, S., Gariboldi, S., Akira, S., Uematsu, S., Bignami, P., Ceriani, V., Ménard, S., and Balsari, A. 2006. Activation of smooth muscle and myenteric plexus cells of jejunum via Toll-like receptor 4. J. Cell Physiol. 208:47–54.

Schäfermeyer, A., Gratzl, M., Rad, R., Dossumbekova, A., Sachs, G., and Prinz, C. 2004. Isolation and receptor profiling of ileal enterochromaffin cells. Acta Physiol. Scand. 182:53–62.

Schmidt, P.T., Eriksen, L., Loftager, M., Rasmussen, T.N., and Holst, J.J. 1999. Fast acting nervous regulation of immunoglobulin A secretion from isolated perfused porcine ileum. Gut 45:679–685.

Schmidt, L.D., Xie, Y., Lyte, M., Vulchanova, L., and Brown, D.R. 2007. Autonomic neurotransmitters modulate immunoglobulin A secretion in porcine colonic mucosa. J. Neuroimmunol. 185:20–28.

Schreiber, K.L., and Brown, D.R. 2005. Adrenocorticotrophic hormone modulates *Escherichia coli* O157:H7 adherence to porcine colonic mucosa. Stress 8:185–190.

Schreiber, K.L., Price, L.D., and Brown, D.R. 2007. Evidence for neuromodulation of enteropathogen invasion in the intestinal mucosa. J. Neuroimmune Pharmacol. 2:329–337.

Scott, L.D., and Cahall, D.L. 1982. Influence of the interdigestive myoelectric complex on enteric flora in the rat. Gastroenterology 82:737–745.

Senard, J.M., Langin, D., Estan, L., and Paris, H. 1990. Identification of alpha 2-adrenoceptors and non-adrenergic idazoxan binding sites in rabbit colon epithelial cells. Eur. J. Pharmacol. 191:59-68.

Sharma, R., and Schumacher, U. 1996. The diet and gut microflora influence the distribution of enteroendocrine cells in the rat intestine. Experientia 52:664–670.

Shibolet, O., and Podolsky, D.K. 2007. TLRs in the Gut. IV. Negative regulation of Toll-like receptors and intestinal homeostasis: addition by subtraction. Am. J. Physiol. Gastrointest. Liver Physiol. 292:G1469–G1473.

Simon, C., and Ternaux, J.P.1990. Regulation of serotonin release from enterochromaffin cells of rat cecum mucosa. J. Pharmacol. Exp. Ther. 253:825–832.

Sinclair, J.F., Dean-Nystrom, E.A., and O'Brien, A.D. 2006. The established intimin receptor Tir and the putative eucaryotic intimin receptors nucleolin and beta1 integrin localize at or near the site of enterohemorrhagic *Escherichia coli* O157:H7 adherence to enterocytes in vivo. Infect. Immun. 74:1255–1265.

Snipes, R.L. 1997. Intestinal absorptive surface in mammals of different sizes. Adv. Anat. Embryol. Cell Biol. 138:III–VIII, 1–90.

Spiller, R.C. 2002. Role of nerves in enteric infection. Gut 51:759–762.

Sternini, C., Anselmi, L., and Rozengurt, E. 2008. Enteroendocrine cells: a site of 'taste' in gastrointestinal chemosensing. Curr. Opin. Endocrinol. Diabetes Obes. 15:73–78.

Strange, P.G. 2008. Signaling mechanisms of GPCR ligands. Curr. Opin. Drug Discov. Devel. 11:196–202.

Straub, R.H., Wiest, R., Strauch, U.G., Härle, P., and Schölmerich, J. 2006. The role of the sympathetic nervous system in intestinal inflammation. Gut 55:1640–1649.

Straub, R.H., Grum, F., Strauch, U.G., Capellino, S., Bataille, F., Bleich, A., Falk, W., Schölmerich, J., and Obermeier, F. 2008. Anti-inflammatory role of sympathetic nerves in chronic intestinal inflammation. Gut 57:911–921.

Suzuki, K., Ha, S.A., Tsuji, M., and Fagarasan, S. 2007. Intestinal IgA synthesis: a primitive form of adaptive immunity that regulates microbial communities in the gut. Semin. Immunol. 19:127–135.

Timmermans, J.P., Adriaensen, D., Cornelissen, W., and Scheuermann, D.W. 1997. Structural organization and neuropeptide distribution in the mammalian enteric nervous system, with special attention to those components involved in mucosal reflexes. Comp. Biochem. Physiol. A Physiol. 118:331–340.

Timmermans, J.P., Hens, J., and Adriaensen, D. 2001. Outer submucous plexus: an intrinsic nerve network involved in both secretory and motility processes in the intestine of large mammals and humans. Anat. Rec. 262:71–78.

Tutton, P.J., and Barkla, D.H. 1977. The influence of adrenoceptor activity on cell proliferation in colonic crypt epithelium and in colonic adenocarcinomata. Virchows Arch. B Cell. Pathol. 24:139–146.

Tutton, P.J., and Helme, R.D. 1974. The influence of adrenoreceptor activity on crypt cell proliferation in the rat jejunum. Cell Tissue Kinet. 7:125–136.

Uribe, A., Alam, M., Johansson, O., Midtvedt, T., and Theodorsson, E. 1994. Microflora modulates endocrine cells in the gastrointestinal mucosa of the rat. Gastroenterology 107:1259–1269.

Valet, P., Senard, J.M., Devedjian, J.C., Planat, V., Salomon, R., Voisin, T., Drean, G., Couvineau, A., Daviaud, D., Denis, C., Laburthe, M., and Paris, H. 1993. Characterization and distribution of alpha 2-adrenergic receptors in the human intestinal mucosa. J. Clin. Invest. 91:2049-2057.

Vasina, V., Barbara, G., Talamonti, L., Stanghellini, V., Corinaldesi, R., Tonini, M., De Ponti, F., and De Giorgio, R. 2006. Enteric neuroplasticity evoked by inflammation. Auton. Neurosci. 126–127:264–272.

Vieira-Coelho, M.A., and Soares-da-Silva, P. 1993. Dopamine formation, from its immediate precursor 3,4-dihydroxyphenylalanine, along the rat digestive tract. Fundam. Clin. Pharmacol. 7:235–243.

Vieira-Coelho, M.A., and Soares-da-Silva, P. 1998. Alpha$_2$-adrenoceptors mediate the effect of dopamine on adult rat jejunal electrolyte transport. Eur. J. Pharmacol. 356:59–65.

Vieira-Coelho, M.A., and Soares-da-Silva, P. 2001. Comparative study on sodium transport and Na+,K+-ATPase activity in Caco-2 and rat jejunal epithelial cells: effects of dopamine. Life Sci. 69:1969-1981.

Vlisidou, I., Lyte, M., van Diemen, P.M., Hawes, P., Monaghan, P., Wallis, T.S., and Stevens, M.P. 2004. The neuroendocrine stress hormone norepinephrine augments *Escherichia coli* O157:H7-induced enteritis and adherence in a bovine ligated ileal loop model of infection. Infect. Immun. 72:5446–5451.

Vulchanova, L., Casey, M.A., Crabb, G.W., Kennedy, W.R., and Brown, D.R. 2007. Anatomical evidence for enteric neuroimmune interactions in Peyer's patches. J. Neuroimmunol. 185:64–74.

Walters, N., Trunkle, T., Sura, M., and Pascual, D.W. 2005. Enhanced immunoglobulin A response and protection against *Salmonella enterica* serovar Typhimurium in the absence of the substance P receptor. Infect. Immun. 73:317–324.

Wang, B., Mao, Y.K., Diorio, C., Wang, L., Huizinga, J.D., Bienenstock, J., Kunze, W. 2010. *Lactobacillus reuteri* ingestion and IK(Ca) channel blockade have similar effects on rat colon motility and myenteric neurones. Neurogastroenterol. Motil. 22:98-107.

Wang, Y.F., Mao, Y.K., Xiao, Q., Daniel, E.E., Borkowski, K.R., McDonald, T.J. 1997. The distribution of NPY-containing nerves and the catecholamine contents of canine enteric nerve plexuses. Peptides 18:221–234.

Wang, H., Steeds, J., Motomura, Y., Deng, Y., Verma-Gandhu, M., El-Sharkawy, R.T., McLaughlin, J.T., Grencis, R.K., and Khan, W.I. 2007. CD4+ T cell-mediated immunological control of enterochromaffin cell hyperplasia and 5-hydroxytryptamine production in enteric infection. Gut 56:949–957.

Wood, J.D. 2007. Effects of bacteria on the enteric nervous system: implications for the irritable bowel syndrome. J. Clin. Gastroenterol. 41(5 Suppl 1):S7–S19.

Wu, Z.C., and Gaginella, T.S. 1981. Release of [^{3}H]norepinephrine from nerves in rat colonic mucosa: effects of norepinephrine and prostaglandin E_2. Am. J. Physiol. 241:G416–G421.

Chapter 6
Modulation of the Interaction of Enteric Bacteria with Intestinal Mucosa by Stress-Related Catecholamines

Mark P. Stevens

6.1 Introduction

Stress and susceptibility to microbial infection have long been correlated. A plausible explanation for this link is that persistent activation of adrenal axes under chronic stress or depression leads to the release of soluble mediators that may impair innate and adaptive immunity (reviewed in Nance and Sanders 2007). Indeed, neurotransmitters, neuropeptides and adrenal hormones modulate specific and nonspecific activities of the cell-mediated immune response by binding to cellular receptors and sympathetic nerve fibres, densely innervate lymphoid organs and terminate in the proximity of immune cell populations. In recent years however, it has become clear that many bacteria are able to sense mediators of the host stress response and respond by activating growth and the expression of virulence factors. Particular emphasis has been placed on the ability of Gram-negative enteric pathogens to respond to stress-related catecholamine hormones. This article reviews evidence that stress is correlated to the outcome of enteric bacterial infections and examines the molecular mechanisms that may be at work.

6.1.1 Stress and Enteric Bacterial Infections in Animals

In food-producing animals, stress associated with social interaction, handling and transport has been correlated with increased excretion of pathotypes of *Escherichia coli* and *Salmonella enterica*. An understanding of the dialogue between such pathogens and the stressed host is important, not only in the interests of animal welfare, but because the organisms may transmit through the food chain and environment to humans, where they may cause acute enteritis and severe systemic sequelae.

M.P. Stevens (✉)
Division of Microbiology, Institute for Animal Health, Compton, Berkshire RG20 7NN, UK
e-mail: mark-p.stevens@bbsrc.ac.uk

M. Lyte and P.P.E. Freestone (eds.), *Microbial Endocrinology*,
Interkingdom Signaling in Infectious Disease and Health,
DOI 10.1007/978-1-4419-5576-0_6, © Springer Science+Business Media, LLC 2010

Studies in animals may also inform us of events relevant in the pathogenesis of human bacterial diseases and aid the development of strategies to interfere with host-pathogen signalling.

In young piglets, isolation from the sow, short-lived cold stress or mixing of litters has been reported to increase faecal excretion of enterotoxigenic *E. coli* (ETEC) relative to control piglets (Jones et al. 2001). Measurement of antibody and T-cell proliferation against a model antigen (ovalbumin) administered intramuscularly at the time of oral ETEC challenge revealed no differences in the immune response between control and stressed piglets, indicating that enhanced faecal excretion of ETEC under stress may not involve modulation of host adaptive immunity. Even mild physical handling of pigs, involving a daily weight measurement requiring movement to-and-from scales in a proximal pen, increased the faecal excretion of *E. coli* and total coliforms relative to control pigs (Dowd et al. 2007). Conclusive data that transport or social stress influence the carriage and virulence of other *E. coli* pathotypes is lacking, however, elevated growth of enterohemorrhagic *E. coli* (EHEC) serotype O157:H7 could be detected in semi-permeable chambers implanted in the peritoneal cavity of mice in a social conflict model, relative to control nonstressed mice (Dréau et al. 1999).

In the case of *Salmonella*, transportation has been correlated with reactivation of subacute *S. enterica* serovar Typhimurium infections in pigs (Isaacson et al. 1999), and with increased hide and faecal *Salmonella* contamination in transported beef cattle (Barham et al. 2002). Social stress caused by mixing has also been found to increase faecal excretion and invasion of *S.* Typhimurium in early-weaned pigs (Callaway et al. 2006). Mixed pigs also exhibited elevated cecal coliform counts and increased translocation of *Salmonella* to intestinal lymph nodes (Callaway et al. 2006). Although studies in poultry are lacking, evidence is emerging that stress associated with stocking density, access to feed and water, physical disturbance and depopulation (thinning) may be correlated with the carriage of zoonotic pathogens and disease susceptibility (reviewed in Humphrey 2006).

6.1.2   Response of the Enteric Nervous System to Stress and Implications for Enteric Bacteria

A key response of the enteric nervous system (ENS) to stress is the release of the catecholamine hormones. Catecholamines are derived from tyrosine and possess a benzene ring with adjacent hydroxyl moieties and an opposing amine side chain. Norepinephrine (NE) is released from sympathetic nerve fibres originating in the prevertebral ganglia that innervate the gut mucosa (reviewed in Furness 2006). In contrast, dopamine (DA) is expressed in a subpopulation of nonsympathetic enteric neurons residing in the intestinal wall. It has been estimated that the human ENS may comprise at least 500 million neurons (Furness 2006). Nerves reactive to antibody against tyrosine hydroxylase (which catalyses the rate-limiting step in NE synthesis), alone or in combination with dopamine β-hydroxylase (which mediates the terminal step in NE synthesis), have been found to innervate porcine intestinal mucosa with some terminating close to the lumen (Fig. 6.1; Kulkarni-Narla et al. 1999; Green et al. 2003). Neurons containing phenylethanolamine N-methyltransferase, which is required for the synthesis of epinephrine from NE, are lacking in the intestinal

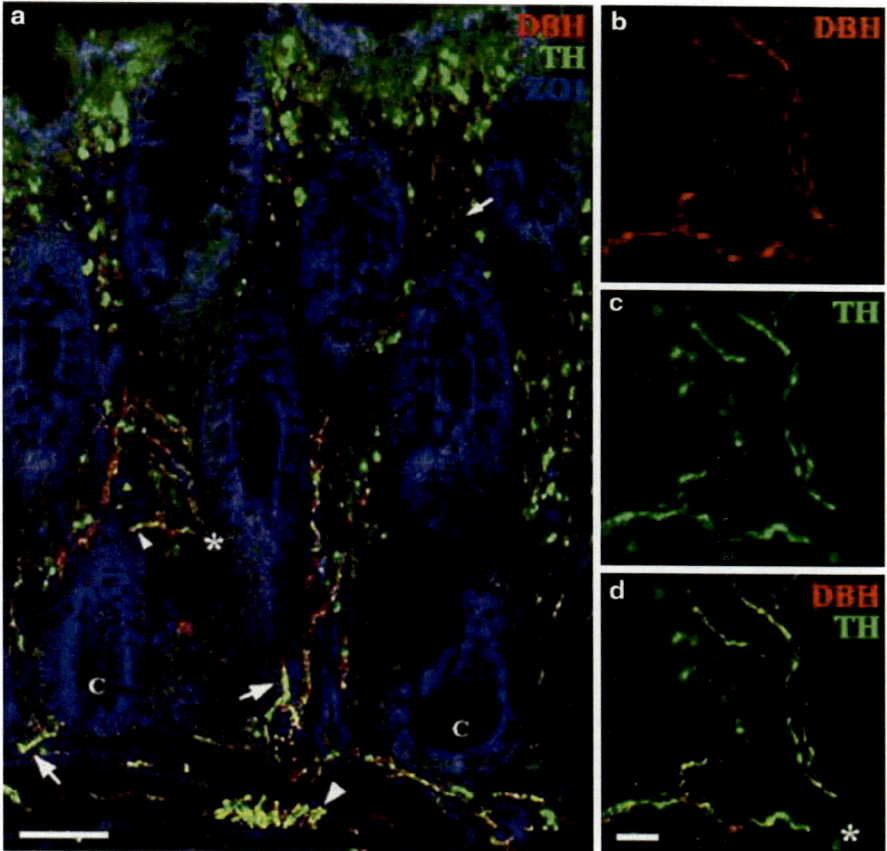

Fig. 6.1 Noradrenergic innervation in an oblique longitudinal section of porcine distal colon. (**a**) Colonic mucosa was triple-labelled for dopamine β-hydroxylase (DBH; *red*), tyrosine hydroxylase (TH; *green*), and the tight-junction protein zonula occludens-1 (ZO1; *blue*) to visualise epithelial cells. Fibres reactive to antibody against DBH and TH were present in the colonic villi and often terminated in close proximity to epithelial cells (*small arrowhead*). Only very fine fibres could be distinguished in the most apical portion of the villi (*small arrow*). Colocalisation of DBH and TH (*yellow*) was evident in submucosal ganglia (*large arrowhead*) and nerve bundles (*large arrow*) near crypts (**c**) at the base of the villi. (**b–d**) Imaging at high magnification demonstrated reactivity to antibodies against DBH (**b**) and TH (**c**) colocalised in varicose nerve fibres (**d**). *Asterisks* in **a** and **d** indicate the region of **a** (scale bar 100 μm) presented in **b–d** (scale bar 10 μm). The image is a projection of six optical sections collected at 1 μm intervals. Reprinted from Green et al. (2004)

mucosa (Costa et al. 2000), and it is therefore considered unlikely that epinephrine would be found in significant quantities in the gut, except perhaps when the mucosal barrier is substantially damaged.

Significant spillover of catecholamines from the systemic circulation into the intestines can occur during episodes of acute stress (Aneman et al. 1996; Eisenhofer et al. 1997) and enhanced release of catecholamines by the ENS under stress has been demonstrated experimentally. Intestinal expression of tyrosine hydroxylase is upregulated in response to surgical perforation of the bowel and gut-derived sepsis in rats (Zhou et al. 2004) and catabolic stress induced by partial hepatectomy in mice has been reported to result in elevated levels of NE in faecal pellets compared to those from mice subject to sham-laparotomy (Alverdy et al. 2000). Evidence that such events may be significant in the progression of enteric disease was provided by the finding that NE and DA are able to induce the growth of commensal *E. coli* in vitro (Freestone et al. 2002). It was subsequently shown that NE stimulates the growth of Gram-negative enteric pathogens of various genera, including *Salmonella* (Freestone et al. 1999, 2007a, b; Williams et al. 2006), *Shigella* (O'Donnell et al. 2006), *Yersinia* (Freestone et al. 1999, 2007a, b; Lyte and Ernst 1992), *Vibrio* (Nakano et al. 2007b), *Campylobacter* (Cogan et al. 2007) as well as pathotypes of *E. coli* such as EHEC and ETEC (Lyte et al. 1997a). Norepinephrine can also resuscitate *E. coli* and *Salmonella* from a viable but nonculturable state (Reissbrodt et al. 2002), indicating that it may aid the outgrowth of physiologically-stressed bacteria that may enter mammalian hosts from the environment.

Systemic translocation of gut-derived bacteria and sepsis are well-known complications of surgery (reviewed in Nieuwenhuijzen and Goris 1999), and it is believed that outgrowth of enteric bacteria in response to the release of stress-related catecholamines may be significant in this process. Increased translocation of gut-derived bacteria to inguinal and mesenteric lymph nodes and the liver has also been observed during social conflict and restraint in C57BL/6 mice (Bailey et al. 2006). In addition, mice subject to partial hepatectomy are more susceptible to gut-derived sepsis caused by the opportunistic pathogen *Pseudomonas aeruginosa* (Laughlin et al. 2000) and exhibit elevated levels of NE in faeces (Alverdy et al. 2000). This was correlated with increased expression of the *Ps. aeruginosa* PA-I lectin in vivo, which is vital for pathogenesis (Alverdy et al. 2000; Laughlin et al. 2000). Expression of PA-I lectin is sensitive to NE in vitro (Alverdy et al. 2000) suggesting a molecular link between NE release under stress and the activity of bacteria in the intestinal lumen. Soluble factors in filtrates of faeces from mice subject to partial hepatectomy are able to activate PA-I lectin expression in vitro (Wu et al. 2003), though the role of NE in such stimulation has yet to be formally proven. The endogenous opioid dynorphin, which is released from the intestinal mucosa in mice subject to ischemia/reperfusion injury, also enhanced *Ps. aeruginosa* virulence in this model and virulence gene expression in vitro and in vivo (Zaborina et al. 2007). A detailed review of the ability of *Ps. aeruginosa* to respond to cues from the host may be found in Chap. 9. The following sections will focus on the impact and mode of action of stress-related catecholamines on the interaction of *E. coli* and *Salmonella* with intestinal mucosa.

6.2 Impact of Catecholamines on the Interaction of *E. coli* with Intestinal Mucosa

As indicated earlier, increased faecal excretion of *E. coli* has been noted in pigs subject to social stress (Jones et al. 2001) or physical handling (Dowd et al. 2007). Evidence that such events may be correlated with the release of stress-related catecholamines is afforded by the observation that the selective neurotoxin 6-hydroxy-dopamine rapidly and dramatically increases the number of *E. coli* in the cecum of mice (Lyte and Bailey 1997). 6-hydroxydopamine destroys noradrenergic nerve terminals without transit across the blood–brain barrier and elicits an immediate systemic release of NE. Within 14 days, during which time noradrenergic nerves regenerate, coliform counts returned to normal (Lyte and Bailey 1997). The effect could be inhibited by prior administration of desipramine hydrochloride, which specifically inhibits catecholamine uptake into noradrenergic nerve terminals, implying that damage to such neurons is required for bacterial outgrowth (Lyte and Bailey 1997). In murine models, stress induced by partial hepatectomy or short-term starvation also caused a profound increase in the number of *E. coli* adhering to the cecal mucosa compared to control mice (Hendrickson et al. 1999).

In addition to evidence that release of endogenous catecholamines may alter the activities of luminal *E. coli*, direct instillation of NE into the gut lumen has been reported to enhance adherence of *E. coli* O157:H7 to the mucosal surface in a bovine ligated ileal loop model (Vlisidou et al. 2004). *E. coli* O157:H7 can rarely be found in association with the mucosa 12 h post-inoculation of bovine ileal loops (Fig. 6.2a), however, if combined with 5 mM NE immediately prior to inoculation,

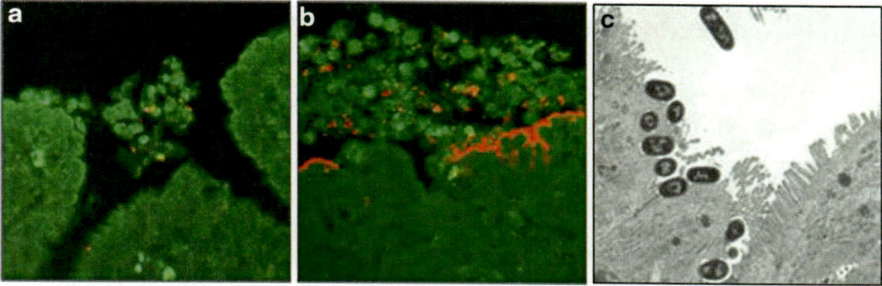

Fig. 6.2 Confocal laser scanning micrographs of bovine mid-ileal mucosa from ligated loops inoculated with *E. coli* O157:H7 strain 85–170 nal[R] in the presence of diluent (**a**), or 5 mM NE (**b**). F-actin was stained with fluorescein isothiocyanate-conjugated phalloidin (*green*), and bacteria detected with rabbit anti-O157 typing serum and anti-rabbit Ig-Alexa[568] (*red*). Dense microcolonies of intimately attached bacteria were seen only in the presence of NE. Magnification ×630. Increased recruitment of neutrophils in the presence of NE was confirmed by analysis of γ-emissions associated with ^{111}In-labelled neutrophils in the same tissues as used for microscopy (Vlisidou et al. 2004). (**c**) Shows a transmission electron micrograph of AE lesions induced by *E. coli* O157:H7 in the presence of 5 mM NE. Scale bar = 5 μm. Reprinted from Vlisidou et al. (2004)

dense microcolonies of adherent bacteria could be observed (Fig. 6.2b). Adherent bacteria were found to have formed "attaching and effacing" (AE) lesions that are known to be vital for bacterial persistence in the bovine intestines (Fig. 6.2c; Dziva et al. 2004; van Diemen et al. 2005). NE also significantly increased fluid accumulation and the recruitment of ^{111}In oxinate-labelled neutrophils in response to *E. coli* O157:H7 compared to control loops filled with bacteria and diluent, albeit only when NE used at 5 mM and not 50 μM (Vlisidou et al. 2004). NE alone did not induce enteritis in this model, indicating that the effect was not the result of outgrowth of resident enteric bacteria or damage to intestinal mucosa.

The relevance of observations in the ligated loop model, where millimolar concentrations of NE were used, to the activities of *E. coli* in a stressed host is unclear. Short-term starvation prior to surgery and manipulation of the intestines to construct ligated segments does not stimulate extensive adherence of *E. coli* O157:H7 per se (Fig. 6.2a), and one may therefore question the relevance of the observations. However, the concentration of NE in the intestinal tract of normal and surgically manipulated calves is unknown. Quantification of tissue-associated and free NE in the intestines is difficult for several reasons (Grassi and Esler 1999; Hjemdahl 1993); (1) microdialysis probes used to sample NE would not be suitable for use in the gut owing to blockage of the dialysis membrane and the fact that NE levels in the dialysate may not reach equilibrium with the surroundings over time, (2) high-pressure liquid chromatography is required to quantify NE in gut contents and recovery through purification steps cannot be estimated, (3) breakdown products of NE can be detected in the intestines, and it is not feasible to calculate the loss of NE through the activity of host and/or bacterial enzymes or other processes. Despite these limitations, the model offers the advantage that at least ten strains or treatments can be evaluated in triplicate in the same animal relative to internal positive and negative controls, thereby minimising the impact of inter-animal variation.

Analysis of muscle-stripped intestinal explants clamped between half-tubes in Ussing chambers provides a further tractable model to probe the role of bacterial and host factors in colonization of the mucosal surface (reviewed in Chap. 5). In this system, NE promotes adherence of *E. coli* O157:H7 to murine cecal mucosa (Chen et al. 2003) and porcine colonic mucosa (Green et al. 2004). NE also promoted uptake of *E. coli* O157:H7, but not a rodent commensal *E. coli* strain, into porcine jejunal Peyer's Patch tissue. EHEC are not widely considered to be invasive pathogens, and uptake by such tissue may reflect the antigen-sampling activity of M cells in follicle-associated epithelium (FAE). It is therefore noteworthy that FAE derived from rats subjected to chronic psychological (water avoidance) stress was found to take up higher levels of fixed fluorophore-conjugated *E. coli* K-12 as compared to epithelium from control rats (Velin et al. 2004). As inactivated bacteria were used and explants were not treated with exogenous neurochemicals, the data imply that stimulation of host tissues under stress may modulate subsequent interactions with enteric bacteria. Increased bacterial uptake was not observed in villus epithelium from stressed rats, indicating that the effect may be specific to FAE.

It is important to note that NE-stimulated adherence of *E. coli* O157:H7 to the luminal aspect of explants clamped in Ussing chambers occurred following the application of NE to the contraluminal aspect. Though it is not possible to preclude the possibility that some NE diffused to the luminal side, the amount of diffused NE and time of exposure is considered unlikely to have been adequate to promote a substantial increase in the number of bacteria available for adherence. Measurements of short circuit current and tissue electrical conductance indicated that the mucosa was viable and intact. In addition, pre-treatment of murine cecal explants with either the non-selective α-adrenergic receptor antagonist phentolamine or the β-adrenergic receptor antagonist propranolol prevented the action of NE, indicating that NE-promoted bacterial adherence at least partially reflects alterations in the host tissue (Chen et al. 2003). A similar effect of phentolamine on adherence and entry of *E. coli* O157:H7 into porcine jejunal Peyer's patch mucosa has been reported (Green et al. 2003), however in this system, propanolol did not inhibit the effect of NE (Green et al. 2004; Chen et al. 2006). Evidence exists that the ability of NE to stimulate adherence of EHEC to porcine intestinal explants may be more sensitive to the α_2-adrenergic receptor antagonist yohimbine than the α_1-adrenergic antagonist prazosin (Green et al. 2004; Chen et al. 2006). The importance of release of NE from endogenous stores has also been suggested by the use of sympathomimetic drugs. For example, the NE reuptake inhibitor cocaine and α_2-adrenergic receptor agonist UK-14,304 both stimulate adherence of *E. coli* O157:H7 to porcine colon explants (Chen et al. 2006), as does a combination of the NE reuptake blocker desipramine and pargyline, an inhibitor of monoamine oxidase that influences catecholamine metabolism (Green et al. 2004).

Though most studies with explants have focussed on NE, it has also been shown that DA increases adherence of *E. coli* O157:H7 to murine cecal mucosa when applied to the contraluminal aspect in a manner sensitive to the DA antagonist haloperidol (Chen et al. 2003). The stress-related peptide adrenocorticotropic hormone (ACTH) also augments adherence of *E. coli* O157:H7, but not a pig-adapted non-pathogenic *E. coli*, to porcine colonic mucosa (Schreiber and Brown 2005). Tyramine, which is structurally related to dopamine and abundant in certain dairy products owing to the tyrosine decarboxylase activity of *Enterococcus faecalis* starter cultures, can promote adherence of *E. coli* O157:H7 to murine cecal explants (Lyte 2004). In addition, extracts of banana that are rich in NE and serotonin (Waalkes et al. 1958) augment growth of Gram-negative bacteria (Lyte 1997), as do a number of non-catecholamine dietary catechols (Freestone et al. 2007c; reviewed in Chap. 4). Further studies are required to determine if neurochemicals consumed in the diet are able to modulate the outcome of enteric bacterial infections.

6.2.1 Possible Mechanisms of Action of Norepinephrine During E. coli Infection

Though evidence exists that catecholamines partly exert their effects by acting on host tissues (above), it is clear that during culture in vitro they promote the growth of Gram-negative bacteria of several genera and expression of their virulence factors.

E. coli O157:H7, *Salmonella enterica* and *Yersinia enterocolitica* vary markedly in their ability to grow in response to catecholamines (Freestone et al. 2007b). The response of *Y. enterocolitica* was limited to NE and DA, and these proved to be more potent inducers of growth than epinephrine in *E. coli* and *Salmonella*. The impact of catecholamines on bacterial growth is strictly dependent on inoculum density and media composition, being most prominent in a minimal salts medium supplemented with 30% (v/v) adult bovine serum (serum-SAPI) from inocula of 10^2–10^3 colony-forming units (CFU; reviewed in Chap. 3). From low inocula such media are bacteriostatic in the absence of catecholamines and are proposed to mimic the nutrient poor and iron-limited conditions encountered in vivo. The effect of catecholamines on growth is ameliorated during culture in rich media or at high inoculum densities. Growth stimulation is unlikely to be due to provision of a metabolite, as supplementation with the NE derivative normetanephrine, which contains one more methyl group than NE, failed to produce the effect (Lyte et al. 1996a, 1997a, b). Remarkably, α- (but not β-) adrenergic receptor antagonists are able to block NE- and epinephrine-induced growth and dopaminergic receptor antagonists inhibit the growth response to DA in the absence of effects on cell viability (Freestone et al. 2007a). These data imply that Gram-negative bacterial pathogens may possess elements that specifically interact with catecholamines and/or that antagonists can interfere with the ability of catecholamines to liberate factors required for bacterial growth.

Two major hypotheses have been put forward to explain the ability of NE to promote growth and virulence gene expression, and these are not mutually exclusive. The first posits that NE and related catecholamines facilitate the supply of iron to Gram-negative bacteria under iron-limiting conditions. As many virulence genes in Gram-negative bacteria are growth-phase regulated, this may explain downstream effects. The second hypothesis posits that virulence gene expression may result from microbial detection of catecholamines via specific receptors that initiate a signal transduction cascade on receipt of the signal leading to altered gene expression (Fig. 6.3). These hypotheses are considered in more detail subsequently.

The ability to liberate ferric iron from host storage proteins such as lactoferrin (Lf) and transferrin (Tf) is a key requirement in bacterial pathogenesis. Many Gram-negative bacteria secrete low molecular weight catecholate or hydroxamate siderophores to acquire iron, which are then imported via specific receptors. Significantly, NE, epinephrine and DA appear to facilitate iron supply from Lf and Tf, likely via the ability of the catechol moiety to complex ferric ion thereby lowering its affinity with Lf and Tf and releasing iron for siderophores to capture (Freestone et al. 2000, 2002, 2003). Recent data indicate that the formation of NE complexes with Lf and Tf leads to reduction of Fe(III) to Fe(II), for which Lf/Tf have a lower affinity (Sandrini et al. 2010). Siderophore synthesis and transport are required for NE-stimulated growth of *E. coli*, as strains with mutations affecting enterobactin synthesis (*entA*) or ferric-enterobactin transport (*fepA* or *tonB*) do not respond to NE in iron-limited serum-rich medium from a low inoculum (Burton et al. 2002; Freestone et al. 2003). This implies that NE does not act as a siderophore per se, although tritiated NE is taken into *E. coli* cells (Freestone et al. 2000;

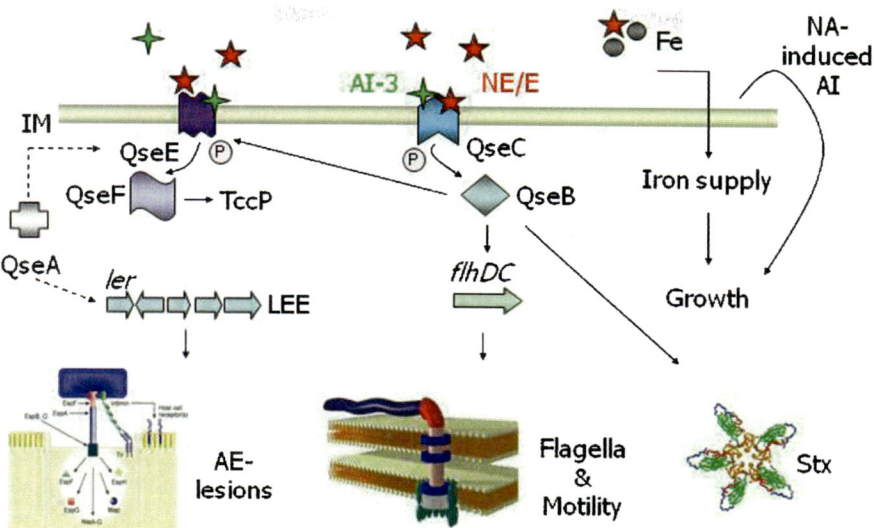

Fig. 6.3 Catecholamine sensing and signal transduction in *E. coli* O157:H7. QseC is an adrenergic sensor kinase that autophosphorylates on detection of epinephrine/NE/AI-3 and transfers the phosphate moiety to its cognate response regulator QseB, thereby activating transcription of the flagellar regulon (Clarke et al. 2006). The signalling cascade downstream of QseC in *E. coli* O157:H7 is known in some detail and is discussed elsewhere (Hughes et al. 2009). Transcription of genes encoding a second two-component system (QseEF) is sensitive to epinephrine and QseC (Reading et al. 2007). QseEF influences AE-lesion formation via activation of genes encoding a Type III secretion system and the Tir-cytoskeleton coupling protein (Reading et al. 2007), and recent evidence indicates that it directly senses catecholamines (Reading et al. 2009). Catecholamines also mediate iron supply by an ill-defined mechanism and promote bacterial replication in serum-rich iron-limited media from low inoculum densities by a mechanism that may involve induction of a heat-stable autoinducer (AI). NE stimulates production of Shiga toxins in *E. coli* O157:H7 however the signal transduction cascade leading to this event is unclear at the time of writing. Adapted from Reading et al. (2007)

Kinney et al. 2000) and the role played by NE-ferric ion complexes in growth induction and the mechanism of NE import remain unclear. NE has also been found to induce the production of the ferric-enterobactin receptor (FepA), indicating that it may facilitate iron acquisition by multiple mechanisms (Burton et al. 2002).

Iron supply may not be the only mechanism by which catecholamines induce bacterial growth. Supernatants of *E. coli* cultures collected after induction of growth by NE in iron-limited serum-rich medium from low inocula contain a heat-stable autoinducer of growth that stimulates replication of naïve bacteria to a comparable extent as catecholamines (Lyte et al. 1996b; Freestone et al. 1999). The autoinducer is produced rapidly after exposure to NE and is able to activate growth of Gram-negative bacteria of several genera (Lyte et al. 1996b; Freestone et al. 1999). It has also been implicated in the ability of NE to resuscitate viable but non-culturable *E. coli* and *Salmonella* and appears to act in manner independent of iron-supply from Lf and Tf (Reissbrodt et al. 2002).

The ability of catecholamines to promote interactions between pathogenic *E. coli* and intestinal mucosa may reflect not only bacterial outgrowth, but also the induction of virulence factors. In ETEC O9:K30:H-, NE promotes expression of the K99 pilus adhesin (Lyte et al. 1997a, b). K99 pili are important in the pathogenesis of ETEC-induced diarrhoea in pigs, and such regulation may partially explain the effect of stress on the outcome of ETEC infection in pigs (Jones et al. 2001). Increased expression of Type I pili has also been described in commensal *E. coli* following stress induced by partial hepatectomy or short-term starvation in mice (Hendrickson et al. 1999). Such fimbriae are vital in the pathogenesis of ascending urinary tract infections by *E. coli* and were proposed to mediate important interactions with intestinal epithelia following catabolic stress (Hendrickson et al. 1999). However, this does not offer an explanation for the increased adherence of *E. coli* O157:H7 to bovine intestinal mucosa in the presence of NE (Vlisidou et al. 2004; Fig. 6.2b), as *E. coli* O157:H7 fail to elaborate functional Type I fimbriae owing to mutations in the *fimA* promoter (Roe et al. 2001) and FimH adhesin (Bouckaert et al. 2006). Indeed, expression of Type I pili is negatively selected during colonization of the bovine intestines by EHEC O26:H- (van Diemen et al. 2005).

Norepinephrine has also been reported to stimulate the production of Shiga toxins by *E. coli* O157:H7 during growth from a low inoculum in serum-containing medium (Lyte et al. 1996a, 1997a). Toxin production could also be stimulated by the catecholamine-induced autoinducer of growth, indicating that NE may partially act via this secreted intermediary (Voigt et al. 2006). Enhanced production of Shiga toxins is significant in the context of EHEC pathogenesis in humans, as the toxins may cause acute renal and neurological sequelae via damage to microvascular endothelial cells. Shiga toxins have also been reported to influence persistence of *E. coli* O157:H7 in the intestines of mice (Robinson et al. 2006) and rabbits (Ritchie et al. 2003), and to deplete a subset of intraepithelial lymphocytes in the bovine intestines (Menge et al. 2004). However, it is unlikely that this would explain the ability of NE to augment adherence and *E. coli* O157:H7-induced enteritis in bovine ligated ileal loops as a non-toxigenic strain was used and Shiga toxin 1 does not play a significant role in EHEC-induced enteritis in this model (Stevens et al. 2002).

A more plausible explanation for the ability of NE to stimulate adherence of *E. coli* O157:H7 to mucosa surfaces was afforded by the finding that it increases the production of several factors required for the formation of AE lesions (Sperandio et al. 2003; reviewed in Chap. 12). AE lesions are characterised by intimate bacterial attachment to enterocytes and localised destruction of microvilli (Fig. 6.2c), and their formation relies on a Type III protein secretion system (T3SS) encoded by the locus of enterocyte effacement (LEE). This apparatus injects a set of bacterial proteins into enterocytes, one of which (the translocated intimin receptor, Tir) becomes localised in the apical leaflet of the host cell plasma membrane where it serves as a receptor for the bacterial outer membrane protein intimin. Both intimin, Tir, and the T3SS play pivotal roles in colonization of the bovine intestines by *E. coli* O157:H7 (Dziva et al. 2004; van Diemen et al. 2005; Vlisidou et al. 2006). A mutant lacking both intimin and Tir did not adhere to the surface of bovine ileal loops in response to NE 12 h after loop inoculation (Vlisidou et al. 2004). However, studies in Ussing chambers have indicated that intimin and the T3SS component EspA are not required for NE-stimulated early non-intimate adherence to porcine intestinal explants (Chen et al. 2006). Production of LEE-encoded proteins is regulated by the growth phase, however induction of the expression and secretion of Type III secreted proteins, as well as induction of transcription of LEE operons, was reported to occur in the absence of effects on growth (Walters and Sperandio 2006). Epinephrine and NE also promote motility of *E. coli* O157:H7 (Sperandio et al. 2003), by a mechanism that appears to involve induction of the flagella regulon via the master regulators FlhDC (Clarke et al. 2006). In a bovine model, an *E. coli* O157:H7 *flhC* mutant was impaired in its ability to persist in the intestines, but a mutant lacking the flagellin subunit FliC was not (Dobbin et al. 2006), indicating that if NE acted via this circuit to promote adherence in the bovine intestines, it may have required *flhDC*-regulated genes other than flagella genes for the effect.

A key contribution to our understanding of host-microbe communication was the identification of an adrenergic receptor in *E. coli* O157:H7 (Clarke et al. 2006; reviewed in Hughes and Sperandio 2008). This arose from the finding that the *qseBC* genes, which regulate flagella-mediated motility (Sperandio et al. 2002), are required for the ability of epinephrine to stimulate motility (Sperandio et al. 2003). QseBC exhibit homology to two-component systems, which typically comprise a sensor kinase (SK) that autophosphorylates at a conserved histidine on receipt of a specific signal and a response regulator (RR) to which the phosphate moiety is transferred from the cognate SK. Phosphorylation of the RR alters its activity, leading to altered expression of genes under its direct or indirect control. It was subsequently proven that QseC binds tritiated NE and that it autophosphorylates in response to epinephrine and NE when reconstituted in lipid micelles (Clarke et al. 2006). Transfer of the phosphate moiety to QseB after epinephrine stimulation of QseC could be detected, and this is believed to modulate transcription of genes under the control of QseB, which include *flhDC* (Clarke et al. 2006). The signalling cascade downstream of QseC is now known is some detail (Hughes et al. 2009). Catecholamine binding to QseC and autophosphorylation could be blocked by the α-adrenergic receptor antagonist phentolamine, but not by the β-adrenergic antagonist propanolol (Clarke et al. 2006), implying that QseC may structurally mimic eukaryotic adrenergic receptors. QseC also senses a bacterial autoinducer (AI-3) that provides a measure of population density (Sperandio et al. 2003). The interplay between such 'quorum sensing' systems and detection of host-derived catecholamines is reviewed in detail in Chap. 12.

The precise role of the QseBC system in activation of LEE genes remains to be determined. Interestingly, NE-stimulated adherence, production of LEE-encoded proteins and AE-lesion formation can be inhibited by the β-adrenergic receptor antagonist propanolol (Chen et al. 2003; Sperandio et al. 2003), even though this treatment does not impair QseC autophosphorylation in the presence of NE (Clarke et al. 2006). This finding, taken together with the fact that some enteric pathogens respond to NE but lack a QseC homologue (e.g. *Campylobacter jejuni* and *Yersinia enterocolitica*), may reflect the presence of other bacterial catecholamine receptors or alternative modes of action. In the case of *E. coli* O157:H7, it has recently been reported that a second two-component sensory system (QseEF), that is regulated by QseBC and required for AE-lesion formation (Reading et al. 2007), acts as a secondary receptor for epinephrine (Reading et al. 2009). It has been established that QseC is required for the induction of enteritis by rabbit enteropathogenic *E. coli* in infant rabbits (Clarke et al. 2006). However, mutations in two-component systems often have pleiotropic effects, and it remains to be shown that QseC is required for NE-induced phenotypes of *E. coli* O157:H7 in vivo. A specific inhibitor of QseC signalling (LED209) has recently been reported to interfere with actin nucleation by *E. coli* O157:H7 and the transcription of *flhDC* and *stx2A*; however, studies of its ability to control infection in infant rabbits were complicated by rapid adsorption of the inhibitor from the gut (Rasko et al. 2008).

In an attempt to define the global transcriptional response of enteric pathogens to catecholamines, microarray studies have been undertaken by several laboratories. Bansal et al. defined the transcriptome of *E. coli* O157:H7 during biofilm formation on glass wool during 7 h culture in Luria Bertani (LB)-glucose medium supplemented with 50 µM epinephrine or 50 µM NE. A total of 938 and 970 genes were differentially transcribed in response to epinephrine and NE, respectively, with 411 genes exhibiting the same pattern of transcription in response to both catecholamines (Bansal et al. 2007). The fact that epinephrine and NE exert different effects at the level of transcription is consistent with the observation that they do not stimulate the growth of *E. coli* O157:H7 to the same extent (Freestone et al. 2007a, b). Several genes involved in iron acquisition were upregulated in response to epinephrine and NE, however Shiga toxin and LEE genes were not found to be activated under the conditions used (Bansal et al. 2007), in contrast with observations at the protein level. Following growth in serum-SAPI medium in the presence of 50 mM NE from an inoculum of c. 10^6 CFU, Dowd detected differential regulation of 101 genes, including induction of genes involved in iron acquisition and the genes for intimin, T3SS components and Shiga toxins (Dowd 2007). Such changes mirror those seen at the protein level (Lyte et al. 1996a, 1997a; Voigt et al. 2006; Sperandio et al. 2003); however, other genes in the same operon as *eae* and *espAB* were not observed to be differentially regulated. The *fepA* gene was found to be repressed, in contrast with other studies (Burton et al. 2002), whereas induction of *feoB* and *fhuD* was also reported by Bansal et al. Culture of an *E. coli* O157:H7 *luxS* mutant (unable to synthesize autoinducer) in Dulbecco's modified Eagle's medium with 50 µM epinephrine has also been reported to induce transcription of LEE, *stx2* and flagella genes (Kendall et al. 2007). A key consideration with studies of this nature is the effect of exogenous catecholamines on bacterial growth. If control and treated cultures are collected at different points in the growth cycle, it is difficult to interpret whether altered virulence gene expression is a direct consequence of catecholamine-mediated regulation (e.g. via QseBC and QseEF) or associated with the change of growth phase. A further consideration is that at high concentrations catecholamines may sequester iron leading to indirect activation of iron acquisition genes. Such aspects are reviewed in detail elsewhere (Freestone et al. 2008).

6.3   Impact of Catecholamines on the Interaction of *Salmonella* with Intestinal Mucosa

In addition to the finding that pigs subject to social stress or transport exhibit increased faecal excretion of *S. enterica* serovar Typhimurium (Isaacson et al. 1999; Callaway et al. 2006), several studies have addressed the impact of pre-treating *Salmonella* or the host with catecholamines on the outcome of infection. Culture of *S.* Typhimurium in serum-SAPI medium containing 2 mM NE increased colonisation of selected tissues following oral inoculation of piglets compared to bacteria grown in Luria Bertani medium in the absence of NE (Toscano et al. 2007). However, it is likely that the physiological status of bacteria grown in serum-rich minimal salts medium is different to that of bacteria cultured in rich media and, though comparable numbers of bacteria were administered, they may have differed in growth phase and gene expression. Indeed, studies in the author's laboratory failed to reproduce the effect of precultivation in the presence of NE when *S.* Typhimurium strain 4/74 was grown in LB$\pm$5 mM NE prior to inoculation of 6-week-old Large White pigs (Stevens MP, unpublished observations).

In relation to the role of stress-related catecholamines during systemic salmonellosis, it has been reported that NE promotes *Salmonella* encephalopathy in calves challenged orally with clinical isolates of serovars Enteritidis, Montevideo and Saintpaul associated with neurological disease. Calves given 45 µg/kg NE daily via the intramuscular route developed neurological signs following inoculation with these isolates and had evidence of bacterial replication in the brain, whereas control animals given a placebo were only positive for bacteria by faecal culture (McCuddin et al. 2008). In the same study however, daily intramuscular administration of NE did not enhance faecal excretion of *S.* Typhimurium DT104 and even decreased excretion of *S.* Dublin (McCuddin et al. 2008), albeit that the relative ability of NE deposited in muscle to act on bacteria at enteric and systemic sites in calves at this dose is ill-defined. Intra-gastric administration of NE to mice the day before inoculation with *S.* Typhimurium increased cecal colonisation and translocation to the liver in a dose-dependent manner (Williams et al. 2006). Similarly, chicks given twice daily NE by crop instillation exhibited elevated levels of *S. enterica* serovar Enteritidis in the ceca and liver compared to controls (Methner et al. 2008).

Research in the author's laboratory has indicated that *S.* Typhimurium-induced fluid accumulation and ^{111}In-labelled neutrophil recruitment in bovine ligated ileal loops is increased by mixing the bacteria with NE immediately prior to inoculation (Pullinger et al. 2010). However, as with *E. coli* O157:H7, relatively high concentrations of NE (5 mM) were required to produce a statistically robust phenotype in this model. Using explants of porcine jejunal Peyer's patch mucosa clamped in Ussing chambers, addition of NE to the contraluminal aspect increased uptake of *S.* Choleraesuis in a phentolamine-sensitive manner (Green et al. 2003). Whilst this implies that NE may facilitate *S.* Choleraesuis invasion by a mechanism dependent on host adrenergic receptors, NE did not significantly enhance internalisation of *S.* Typhimurium DT104 by porcine jejunal Peyer's patch explants or ileal non-follicular mucosa in Ussing chambers (Schreiber et al. 2007). Indeed treatment of such explants with sympathomimetic drugs even decreased recoveries of *S.* Typhimurium DT104 (Brown and Price 2008). *S.* Typhimurium invades porcine intestinal mucosa more efficiently than *S.* Choleraesuis in Ussing chambers (Brown and Price 2008) and ileal loops (Paulin et al. 2007), and the effect of exogenous catecholamine on invasion may therefore be less pronounced if a finite level of uptake is possible under the assay conditions. Further studies are required to determine if there are strain- or serovar-specific mechanisms by which catecholamines may modulate the outcome of *S. enterica* infections.

Studies on the importance of release of catecholamines from endogenous stores on the virulence of *Salmonella* are lacking. A glucose analogue, 2-deoxy-D-glucose (2DG), previously shown to induce many of the hallmark parameters of physiological stress in pigs (Stabel 1999), failed to reactivate *S.* Choleraesuis at enteric or systemic sites when given to carrier animals (Stabel and Fedorka-Cray 2004), however catecholamine levels in the intestines or circulation of treated and control animals were not measured. Studies in the author's laboratory have indicated that administration of 40 mg/kg 6-hydroxydopamine intravenously to pigs 8 or 16 days after oral inoculation with *S.* Typhimurium caused a transient but statistically significant increase in the number of excreted *Salmonella* relative to infected pigs given diluent (Stevens MP, unpublished observations). This is consistent with the outgrowth of commensal coliforms described in 6-hydroxydopamine treated mice (Lyte and Bailey 1997).

6.3.1 Possible Mechanisms of Action of Norepinephrine During Salmonella Infection

As with *E. coli*, catecholamines have been proposed to supply iron to *Salmonella* and to act via an adrenergic receptor(s). In relation to iron, NE-induced growth of *S.* Typhimurium from low inocula in serum-SAPI medium requires the synthesis of enterobactin (Methner et al. 2008). However, mutants lacking the catecholate receptors IroN or FepA that mediate uptake of enterobactin, salmochelin S4 and 2,3-dihydroxybenzoylserine (DHBS) still grow in serum-SAPI in response to NE (Williams et al. 2006), as do those defective in enterobactin secretion (EntS) or conversion of enterobactin to salmochelin S4 (IroB; Methner et al. 2008). This implies that neither secretion nor uptake of enterobactin or its derivative salmochelin S4 is required for NE-mediated iron supply and suggests the possibility that enterobactin precursors or degradation products may be required for the effect. A mutant lacking Fes, which hydrolyzes enterobactin to DHBS failed to grow in serum-SAPI in response to NE (Methner et al. 2008) as did a triple mutant lacking IroN, FepA and Cir, all of which act as receptors for DHBS (Williams et al. 2006). The reduced ability of a *S.* Typhimurium *fes* mutant and an *iroN fepA cir* triple mutant to grow in response to NE has been independently reported (Bearson et al. 2008). A mutant lacking IroD, which hydrolyses salmochelin S4 to the linear glycosylated monomeric form SX, also blocked the growth response to NE (Methner et al. 2008). These data support the notion that degradation products of enterobactin and salmochelin S4 may be important for NE-mediated iron supply to *Salmonella*. The relevance of such events in vivo requires further study as whilst inoculation of mice pre-treated with NE did not result in elevated growth of a triple *iroN fepA cir* mutant (Williams et al. 2006), a mutant lacking TonB, which is required for uptake via these receptors, behaved in a manner comparable to the parent strain (Methner et al. 2008). Additionally, an *iroN fepA cir* triple mutant has been reported to show no defect in colonisation of the porcine intestines (Bearson et al. 2008), in contrast to observations in mice (Williams et al. 2006).

S. Typhimurium encodes orthologues of the *E. coli* O157:H7 QseC adrenergic sensor kinase (also known as PreB, YgiY or STM3178 in *S.* Typhimurium), and QseB response regulator (PreA, YgiX, STM3177). Although the ability of these orthologues to sense and respond to epinephrine and NE has yet to be proven, evidence exists that they encode a functional two-component system (TCS) that upregulates transcription of the *pmrCAB* genes encoding a further TCS that in turn regulates virulence gene expression (Merighi et al. 2006). The *E. coli* O157:H7 *qseBC* genes were able to functionally replace *preAB* as regulators of *pmrCAB* transcription in *S.* Typhimurium (Merighi et al. 2006), however reciprocal studies in *E. coli* have yet to be reported. Recent studies have indicated that the QseC orthologue in *S.* Typhimurium is required for full systemic virulence in mice (Merighi et al. 2009; Moreira et al. 2010) and influences NE-induced motility, invasion of IPEC-J2 porcine jejunal cells and intestinal colonisation of swine (Bearson and Bearson 2008). Further, the inhibitor of *E. coli* O157:H7 QseC signalling LED209 aids control of systemic salmonellosis in a murine model (Rasko et al. 2008). However, in calves only modest attenuation of a *S.* Typhimurium *qseC* mutant was observed (Pullinger et al. 2010). Two-component sensory systems are known to integrate multiple signals (indeed QseBC detects epinephrine, NE and AI-3), and evidence that QseC is required for full virulence does not provide evidence that catecholamine-sensing is required for full virulence. Indeed, recent studies have indicated that norepinephrine augments *S.* Typhimurium-induced enteritis in a bovine ligated ileal loop model of infection independently of QseC, QseE or both putative sensor kinases (Pullinger et al. 2010). The authors of this report also examined net replication of the bacteria using a plasmid partitioning system previously used to study in vivo growth of *S. enterica* serovars (Paulin et al. 2007). NE treatment enhanced net replication of *S.* Typhimurium in the same intestinal segments as it stimulated enteritis, consistent with NE-induced effects on growth in vitro (Pullinger et al. 2010).

NE has been reported to activate expression of the *S.* Typhimurium Type III secretion-related genes in a manner sensitive to LED209 and mutation of *qseC* (Rasko et al. 2008; Moreira et al. 2010). Type III secretion systems-1 and -2 are vital for persistence in the mammalian intestines and induction of enteritis (reviewed in Stevens et al. 2009). However, recent microarray studies of the QseBC regulon (Merighi et al. 2009) and the response of *S.* Typhimurium to epinephrine (Karavolos et al. 2008) do not support effects on Type III secretion loci. Stimulation of *S.* Typhimurium with NE also failed to induce the production or secretion of T3SS-1 or T3SS-2 proteins as detected by western blotting and use of translational fusions (Pullinger et al. 2010). T3SS-1 expression is sensitive to the availability of exogenous iron, being decreased in the presence of the iron chelator dipyridyl (Ellermeier and Slauch 2008). Thus the ability of NE to augment *Salmonella*-induced enteritis in ileal loops is unlikely to be due to T3SS-1 induction as a consequence of sequestering free iron by use of high concentrations of NE. It remains possible that differences in the role or mode of action of adrenergic sensor kinases may exist between *E. coli* and *Salmonella*, and functional redundancy may mask the effect of mutations studied to date. *S.* Typhimurium genes regulated by NE or epinephrine have recently been identified by a genome-wide transposon mutagenesis screen (Spencer et al. 2010), and further studies are required to understand the basis of NE-induced phenotypes in *Salmonella* and the basis of discrepancies between laboratories.

6.4 Modulation of the Activities of Other Enteric Pathogens by Catecholamines

It is becoming clear that other enteric bacterial pathogens possess the ability to respond to norepinephrine. NE promotes growth of *Campylobacter jejuni* in iron-limited media and bacterial uptake of ^{55}Fe (Cogan et al. 2007), as well as growth of the periodontal isolate *C. gracilis* (Roberts et al. 2002). *C. jejuni* lacks known siderophores but encodes homologues of siderophore receptors and it has been proposed that NE may bind iron and supply it to *Campylobacter* directly. In contrast to observations in *E. coli* (Freestone et al. 2000), no uptake of tritiated NE into *C. jejuni* cells could be detected (Cogan et al. 2007). Growth of *C. jejuni* in NE-supplemented medium markedly enhanced its ability to invade Caco-2 cell monolayers and disrupt tight junctions, as detected by a decrease in transepithelial electrical resistance and redistribution of occludin (Cogan et al. 2007). This may partially be explained by the ability of NE to promote motility of *C. jejuni* (Cogan et al. 2007), which mediates invasion of epithelial cells in vitro and in vivo. Stimulation of *C. jejuni* growth by NE is insensitive to the adrenergic receptor antagonists phenoxybenzamine, propanolol and metoprolol bitartrate (Cogan et al. 2007) and this, taken together with the fact that *C. jejuni* lacks known orthologues of the QseC and QseE adrenergic sensor kinases (Clarke et al. 2006), implies that NE may stimulate growth and gene expression in *Campylobacter* by distinct mechanisms. The impact of stress, whether induced during commercial practices or experimentally, on the carriage and virulence of *Campylobacter* requires further study. If correlated, analysis of bacterial gene expression in response to mediators of the host stress response and studies on the involvement of catecholate receptors in catecholamine-promoted phenotypes would be valuable.

In *Vibrio* species, NE, epinephrine and dopamine were found to promote growth of *V. parahaemolyticus* and *V. mimicus* in serum-SAPI medium, but not *V. cholerae* or *V. vulnificus* (Nakano et al. 2007b). *V. vulnificus* could be stimulated to grow in serum-SAPI with high concentrations of epinephrine, indicating that the specificity and magnitude of the response of catecholamines varies among *Vibrio* species (Nakano et al. 2007b), as described for *S. enterica*, *E. coli* and *Y. enterocolitica*. Stimulation of the growth of *V. parahaemolyticus* appears not to be due to provision of energy, as NE metabolites failed to elicit the effect (Nakano et al. 2007b). Although the effect of catecholamines on growth of the aquatic pathogen *V. splendidus* was not examined in this study, stress associated with physical shaking of oysters (*Crassostrea gigas*) challenged with *V. splendidus* increased both bacterial loads and mortality (Lacoste et al. 2001). Furthermore, physical stress was associated with elevated levels of circulating NE and injection of NE or adrenocorticotropic hormone, a key component of the oyster stress response, caused significantly higher mortality and increased accumulation of *V. splendidus* in challenged oysters (Lacoste et al. 2001). Although the molecular basis of this phenomenon is ill-defined, NE has been shown to increase the transcription of Type III secretion genes in *V. parahaemolyticus* and to augment cytotoxicity against Caco-2 cells and fluid accumulation in rat ligated ileal loops (Nakano et al. 2007a). NE-promoted fluid accumulation could be inhibited by phentolamine and the α_1-specific antagonist prazosin (Nakano et al. 2007a). As with *E. coli* and *Salmonella*, it remains unclear to what extent NE and adrenergic receptor antagonists exert their effect by acting on the bacteria or host.

6.5 Concluding Remarks

Enteric bacterial pathogens of many genera respond to catecholamine mediators of the host stress response. Whilst, the molecular mechanisms by which *S. enterica* and *E. coli* respond to catecholamines are rapidly being unravelled, the modes of iron supply and catecholamine sensing and signal transduction in other NE-responsive genera such as *Campylobacter*, *Citrobacter*, *Enterobacter*, *Enterococcus*, *Hafnia*, *Listeria*, *Shigella*, *Vibrio* and *Yersinia* are unknown. Further, the impact of stress-related catecholamines on gene expression in such genera and its relationship to the outcome of infection in stressed hosts is ill-defined. Whilst, the involvement of selected catecholate receptors in NE-promoted virulence of *S.* Typhimurium in mice and chickens has been probed, it remains to be proven that adrenergic sensors act to promote bacterial virulence because they sense catecholamines, as opposed to detection of other cues or activation of unrelated regulons. Care also needs to be taken in the design of experimental animal studies in relation to the physiological status of catecholamine-treated and untreated bacteria and interpretation of pheno-types resulting from high exogenous catecholamine doses.

It is clear that stress-related catecholamines may act via a number of mechanisms, both in relation to bacteria, where they may promote iron-acquisition, auto-inducer production and signal transduction, and in relation to the host. Further research will be needed to evaluate the effects of stress-related catecholamines on epithelial and immune cells in the gut, and the ability of adrenergic receptor antago-nists and sympathomimetic drugs to modulate subsequent interactions between the mucosa and enteric bacteria. Molecules that interfere with the dialogue between pathogen and host may represent a novel class of anti-infective agents (Rasko et al. 2008). However, highly selective agents will likely be needed as the commensal microflora is also sensitive to stress-related catecholamines and plays a vital role in intestinal development and homeostasis. Analysis of microflora dynamics and the frequency of enteric infection in humans given adrenergic receptor antagonists for the treatment of hypertension, heart failure, migraine and other conditions may provide valuable data for the development of such agents. In addition, reporter systems that provide a measure of adrenergic stimulation of bacterial growth or virulence gene expression will be useful for screening chemical libraries. Although many genera respond to catecholamines, the specificity and magnitude of the response may vary and thus solutions may need to be tailored to the pathogen. Nevertheless, work in this area is given impetus by the decline in discovery of new antibiotics, rise in antibiotic resistance and the fact that antibiotic treatment of some enteric infections is contraindicated.

Whilst microbial detection of stress-related catecholamines enjoys a high profile at present, the full extent of microbial cross-talk with the host remains unclear. *Ps. aeruginosa* responds to the endogenous opioid dynorphin (Zaborina et al. 2007), and it is noteworthy that the *E. coli envY* gene reportedly encodes a functional high-affinity opioid receptor (Cabon et al. 1993). Endogenous opioids are among the first signals to be released by tissues under stress and the relevance of EnvY during enteric bacterial infection requires further study. In addition, evidence is emerging that bacterial pathogens may sense the activation of host innate immunity. For example the pro-inflammatory cytokine interferon-γ binds the *Ps. aeruginosa* outer membrane porin OprF and activates PA-I lectin expression (Wu et al. 2005), tumour necrosis factor-α increases invasion by *Shigella flexneri* (Luo et al. 1993) and inter-leukin-1 stimulates growth of pathogenic *E. coli* (Porat et al. 1991). Such findings suggest that the interplay between bacterial pathogens and their hosts may be far more complex than at first thought and have far reaching implications in human and veterinary medicine.

Acknowledgements The author gratefully acknowledges the support of the Biotechnology and Biological Sciences Research Council U.K. (grant ref. BB/C518022/1).

References

Alverdy, J., Holbrook, C., Rocha, F., Seiden, L., Wu, R. L., Musch, M., Chang, E., Ohman, D., and Suh, S. 2000. Gut-derived sepsis occurs when the right pathogen with the right virulence genes meets the right host: evidence for *in vivo* virulence expression in *Pseudomonas aerugi-nosa*. Ann. Surg. 232:480–489.

Aneman, A., Eisenhofer, G., Olbe, L., Dalenback, J., Nitescu, P., Fandriks, L., and Friberg, P. 1996. Sympathetic discharge to mesenteric organs and the liver. Evidence for substantial mesenteric organ norepinephrine spillover. J. Clin. Invest. 97:1640–1646.

Bailey, M. T., Engler, H., and Sheridan, J. F. 2006. Stress induces the translocation of cutaneous and gastrointestinal microflora to secondary lymphoid organs of C57BL/6 mice. J. Neuroimmunol. 171:29–37.

Bansal, T., Englert, D., Lee, J., Hegde, M., Wood, T. K., and Jayaraman, A. 2007. Differential effects of epinephrine, norepinephrine, and indole on *Escherichia coli* O157:H7 chemotaxis, colonization, and gene expression. Infect. Immun. 75:4597–4607.

Barham, A. R., Barham, B. L., Johnson, A. K., Allen, D. M., Blanton, J. R. Jr., and Miller, M. F. 2002. Effects of the transportation of beef cattle from the feedyard to the packing plant on prevalence levels of *Escherichia coli* O157 and *Salmonella* spp. J. Food Prot. 65:280–283.

Bearson, B. L., and Bearson, S. M. 2008. The role of the QseC quorum-sensing sensor kinase in colonization and norepinephrine-enhanced motility of *Salmonella enterica* serovar Typhimurium. Microb. Pathog. 44:271–278.

Bearson, B. L., Bearson, S. M., Uthe, J. J., Dowd, S. E., Houghton, J. O., Lee, I., Toscano, M. J., and Lay, D. C. Jr. 2008. Iron regulated genes of *Salmonella enterica* serovar Typhimurium in response to norepinephrine and the requirement of *fepDGC* for norepinephrine-enhanced growth. Microbes. Infect. 10:807–816.

Bouckaert, J., Mackenzie, J., de Paz, J. L., Chipwaza, B., Choudhury, D., Zavialov, A., Mannerstedt, K., Anderson, J., Piérard, D., Wyns, L., Seeberger, P. H., Oscarson, S., De Greve, H., and Knight, S. D. 2006. The affinity of the FimH fimbrial adhesin is receptor-driven and quasi-independent of *Escherichia coli* pathotypes. Mol. Microbiol. 61:1556–1568.

Brown, D. R., and Price, L. D. 2008. Catecholamines and sympathomimetic drugs decrease early *Salmonella* Typhimurium uptake into porcine Peyer's patches. FEMS Immunol. Med. Microbiol. 52:29–35.

Burton, C. L., Chhabra, S. R., Swift, S., Baldwin, T. J., Withers, H., Hill, S. J., and Williams, P. 2002. The growth response of *Escherichia coli* to neurotransmitters and related catecholamine drugs requires a functional enterobactin biosynthesis and uptake system. Infect. Immun. 70:5913–5923.

Cabon, F., Morser, J., Parmantier, E., Solly, S. K., Pham-Dinh, D., and Zalc, B. 1993. The *E. coli envY* gene encodes a high affinity opioid binding site. Neurochem. Res. 18:795–800.

Callaway, T. R., Morrow, J. L., Edrington, T. S., Genovese, K. J., Dowd, S., Carroll, J., Dailey, J. W., Harvey, R. B., Poole, T. L., Anderson, R. C., and Nisbet, D. J. 2006. Social stress increases fecal shedding of *Salmonella typhimurium* by early weaned piglets. Curr. Issues Intest. Microbiol. 7:65–71.

Chen, C., Brown, D. R., Xie, Y., Green, B. T., and Lyte, M.. 2003. Catecholamines modulate *Escherichia coli* O157:H7 adherence to murine cecal mucosa. Shock 20:183–188.

Chen, C., Lyte, M., Steven, M. P., Vulchanova, L., and Brown, D. R. 2006. Mucosally-directed adrenergic nerves and sympathomimetic drugs enhance non-intimate adherence of *Escherichia coli* O157:H7 to porcine cecum and colon. Eur. J. Pharmacol. 539:116–124.

Clarke, M. B., Hughes, D. T., Zhu, C., Boedeker, E. C., and Sperandio, V. 2006. The QseC sensor kinase: a bacterial adrenergic receptor. Proc. Natl. Acad. Sci. USA103:10420–10425.

Cogan, T. A., Thomas, A. O., Rees, L. E., Taylor, A. H., Jepson, M. A., Williams, P. H., Ketley, J., and Humphrey. T. J. 2007. Norepinephrine increases the pathogenic potential of *Campylobacter jejuni*. Gut 56:1060–1065.

Costa, M., Brookes, S. J., and Hennig, G. W. 2000. Anatomy and physiology of the enteric nervous system. Gut 47(Suppl 4):iv15–iv19.

Dobbin, H. S., Hovde, C. J., Williams, C. J., and Minnich, S. A. 2006. The *Escherichia coli* O157 flagellar regulatory gene *flhC* and not the flagellin gene *fliC* impacts colonization of cattle. Infect. Immun. 74:2894–2905.

Dowd, S. E. 2007. *Escherichia coli* O157:H7 gene expression in the presence of catecholamine norepinephrine. FEMS Microbiol. Lett. 273:214–223.

Dowd, S. E., Callaway, T. R., and Morrow-Tesch, J. 2007. Handling may cause increased shedding of *Escherichia coli* and total coliforms in pigs. Foodborne Pathog. Dis. 4:99–102.

Dréau, D., Sonnenfeld, G., Fowler, N., Morton, D. S., and Lyte, M. 1999. Effects of social conflict on immune responses and *E. coli* growth within closed chambers in mice. Physiol. Behav. 67:133–140.

Dziva, F., van Diemen, P. M., Stevens, M. P., Smith, A. J., and Wallis, T. S. 2004. Identification of *Escherichia coli* O157:H7 genes influencing colonization of the bovine gastrointestinal tract using signature-tagged mutagenesis. Microbiology 150:3631–3645.

Eisenhofer, G., Aneman, A., Friberg, P., Hooper, D., Fåndriks, L., Lonroth, H., Hunyady, B., Mezey, E. 1997. Substantial production of dopamine in the human gastrointestinal tract. J. Clin. Endocrinol. Metab. 82:3864–3871.

Ellermeier, J. R., and Slauch, J. M. 2008. Fur regulates expression of the *Salmonella* pathogenicity island 1 type III secretion system through HilD. J. Bacteriol. 190:476–486.

Freestone, P. P., Haigh, R. D., Williams, P. H., and Lyte, M. 1999. Stimulation of bacterial growth by heat-stable, norepinephrine-induced autoinducers. FEMS Microbiol. Lett. 172:53–60.

Freestone, P. P., Lyte, M., Neal, C. P., Maggs, A. F., Haigh, R. D., and Williams, P. H. 2000. The mammalian neuroendocrine hormone norepinephrine supplies iron for bacterial growth in the presence of transferrin or lactoferrin. J. Bacteriol. 182:6091–6098.

Freestone, P. P., Williams, P. H., Haigh, R. D., Maggs, A. F., Neal, C. P., and Lyte, M. 2002. Growth stimulation of intestinal commensal *Escherichia coli* by catecholamines: a possible contributory factor in trauma-induced sepsis. Shock 18:465–470.

Freestone, P. P., Haigh, R. D., Williams, P. H., and Lyte, M. 2003. Involvement of enterobactin in norepinephrine-mediated iron supply from transferrin to enterohaemorrhagic *Escherichia coli*. FEMS Microbiol. Lett. 222:39–43.

Freestone, P. P., Haigh, R. D., and Lyte, M. 2007a. Blockade of catecholamine-induced growth by adrenergic and dopaminergic receptor antagonists in *Escherichia coli* O157:H7, *Salmonella enterica* and *Yersinia enterocolitica*. BMC Microbiol. 7:8.

Freestone, P. P., Haigh, R. D., and Lyte, M. 2007b. Specificity of catecholamine-induced growth in *Escherichia coli* O157:H7, *Salmonella enterica* and *Yersinia enterocolitica*. FEMS Microbiol. Lett. 269:221–228.

Freestone, P. P., Walton, N. J., Haigh, R. D., and Lyte, M. 2007c. Influence of dietary catechols on the growth of enteropathogenic bacteria. Int. J. Food Microbiol. 119:159–169.

Freestone, P. P., Sandrini, S. M., Haigh, R. D., and Lyte, M. 2008. Microbial endocrinology: how stress influences susceptibility to infection. Trends Microbiol. 16:55–64.

Furness, J. B. 2006. The Enteric Nervous System. Blackwell, Oxford.

Grassi, G., and Esler, M. 1999. How to assess sympathetic activity in humans. J. Hypertens. 17:719–734.

Green, B. T., Lyte, M., Kulkarni-Narla, A., and Brown, D. R. 2003. Neuromodulation of enteropathogen internalization in Peyer's patches from porcine jejunum. J. Neuroimmunol. 141:74–82.

Green, B. T., Lyte, M., Chen, C., Xie, Y., Casey, M. A., Kulkarni-Narla, A., Vulchanova, L., and Brown, D. R. 2004. Adrenergic modulation of *Escherichia coli* O157:H7 adherence to the colonic mucosa. Am. J. Physiol. Gastrointest. Liver Physiol. 287:G1238–G1246.

Hendrickson, B. A., Guo, J., Laughlin, R., Chen, Y., and Alverdy, J. C. 1999. Increased type 1 fimbrial expression among commensal *Escherichia coli* isolates in the murine cecum following catabolic stress. Infect. Immun. 67:745–753.

Hjemdahl, P. 1993. Plasma catecholamines-analytical challenges and physiological limitations. Baillieres Clin. Endocrinol. Metab.7:307–353.

Hughes, D. T., and Sperandio, V. 2008. Inter-kingdom signalling: communication between bacteria and their hosts. Nat. Rev. Microbiol. 6:111–120.

Hughes, D. T., Clarke, M. B., Yamamoto, K., Rasko, D. A., and Sperandio, V. 2009. The QseC adrenergic signaling cascade in enterohemorrhagic *E. coli* (EHEC). PLoS Pathog. 5:e1000553.

Humphrey, T. 2006. Are happy chickens safer chickens? Poultry welfare and disease susceptibility. Br. Poult. Sci. 47:379–391.

Isaacson, R. E., Firkins, L. D., Weigel, R. M., Zuckermann, F. A., and DiPietro, J. A. 1999. Effect of transportation and feed withdrawal on shedding of *Salmonella typhimurium* among experimentally infected pigs. Am. J. Vet. Res. 60:1155–1158.

Jones, P. H., Roe, J. M., and Miller, B. G. 2001. Effects of stressors on immune parameters and on the faecal shedding of enterotoxigenic *Escherichia coli* in piglets following experimental inoculation. Res. Vet. Sci. 70:9–17.

Karavolos, M. H., Spencer, H., Bulmer, D. M., Thompson, A., Winzer, K., Williams, P., Hinton J. C., and Khan, C. M. 2008. Adrenaline modulates the global transcriptional profile of *Salmonella* revealing a role in the antimicrobial peptide and oxidative stress resistance responses. BMC Genomics 9:458.

Kendall, M. M., Rasko, D. A., and Sperandio, V. 2007. Global effects of the cell-to-cell signaling molecules autoinducer-2, autoinducer-3, and epinephrine in a *luxS* mutant of enterohemorrhagic *Escherichia coli*. Infect. Immun. 75:4875–4884.

Kinney, K. S., Austin, C. E., Morton, D. S., and Sonnenfeld, G. 2000. Norepinephrine as a growth stimulating factor in bacteria-mechanistic studies. Life Sci. 67:3075–3085.

Kulkarni-Narla, A., Beitz, A. J., and Brown, D. R. 1999. Catecholaminergic, cholinergic and peptidergic innervation of gut-associated lymphoid tissue in porcine jejunum and ileum. Cell Tissue Res. 298:275–286.

Lacoste, A., Jalabert, F., Malham, S. K., Cueff, A., and Poulet, S. A. 2001. Stress and stress-induced neuroendocrine changes increase the susceptibility of juvenile oysters (*Crassostrea gigas*) to *Vibrio splendidus*. Appl. Environ. Microbiol. 67:2304–2309.

Laughlin, R. S., Musch, M. W., Hollbrook, C. J., Rocha, F. M., Chang, E. B., and Alverdy, J. C. 2000. The key role of *Pseudomonas aeruginosa* PA-I lectin on experimental gut-derived sepsis. Ann. Surg. 232:133–142.

Luo, G., Niesel, D. W., Shaban, R. A., Grimm, E. A., and Klimpel, G. R. 1993. Tumor necrosis factor alpha binding to bacteria: evidence for a high-affinity receptor and alteration of bacterial virulence properties. Infect. Immun. 61:830–835.

Lyte, M. 1997. Induction of gram-negative bacterial growth by neurochemical containing banana (Musa x paradisiaca) extracts. FEMS Microbiol. Lett. 154:245–250.

Lyte, M. 2004. The biogenic amine tyramine modulates the adherence of *Escherichia coli* O157:H7 to intestinal mucosa. J. Food Prot. 67:878–883.

Lyte, M., and Bailey, M. T. 1997. Neuroendocrine-bacterial interactions in a neurotoxin-induced model of trauma. J. Surg. Res. 70:195–201.

Lyte, M., and Ernst, S. 1992. Catecholamine-induced growth of Gram negative bacteria. Life Sci. 50:203–212.

Lyte, M., Arulanandam, B. P., and Frank, C. D. 1996a. Production of Shiga-like toxins by *Escherichia coli* O157:H7 can be influenced by the neuroendocrine hormone norepinephrine. J. Lab. Clin. Med. 128:392–398.

Lyte, M., Frank, C. D., and Green, B. T. 1996b. Production of an autoinducer of growth by norepinephrine cultured *Escherichia coli* O157:H7. FEMS Microbiol. Lett. 139:155–159.

Lyte, M., Arulanandam, B., Nguyen, K., Frank, C., Erickson, A., and Francis, D. 1997a. Norepinephrine induced growth and expression of virulence associated factors in enterotoxigenic and enterohemorrhagic strains of *Escherichia coli*. Adv. Exp. Med. Biol. 412:331–339.

Lyte, M., Erickson, A. K., Arulanandam, B. P., Frank, C. D., Crawford, M. A., and Francis, D. H. 1997b. Norepinephrine-induced expression of the K99 pilus adhesin of enterotoxigenic *Escherichia coli*. Biochem. Biophys. Res. Commun. 232:682–686.

McCuddin, Z. P., Carlson, S. A., and Sharma, V. K. 2008. Experimental reproduction of bovine *Salmonella* encephalopathy using a norepinephrine-based stress model. Vet. J. 175:82–88.

Menge, C., Stamm, I., van Diemen, P. M., Sopp, P., Baljer, G., Wallis, T. S., and Stevens, M. P. 2004. Phenotypic and functional characterization of intraepithelial lymphocytes in a bovine ligated intestinal loop model of enterohaemorrhagic *Escherichia coli* infection. J. Med. Microbiol. 53:573–579.

Merighi, M., Carroll-Portillo, A., Septer, A. N., Bhatiya, A., and Gunn, J. S. 2006. Role of *Salmonella enterica* serovar Typhimurium two-component system PreA/PreB in modulating PmrA-regulated gene transcription. J. Bacteriol. 188:141–149.

Merighi, M., Septer, A. N., Carroll-Portillo, A., Bhatiya, A., Porwollik, S., McClelland, M., and Gunn, J. S. 2009. Genome-wide analysis of the PreA/PreB (QseB/QseC) regulon of *Salmonella enterica* serovar Typhimurium. BMC Microbiol. 9:42.

Methner, U., Rabsch, W., Reissbrodt, R., and Williams, P. H. 2008. Effect of norepinephrine on colonisation and systemic spread of *Salmonella enterica* in infected animals: role of catecholate siderophore precursors and degradation products. Int. J. Med. Microbiol. 298:429–439.

Moreira, C. G., Weinshenker, D., and Sperandio, V. 2010. QseC mediates *Salmonella enterica* serovar Typhimurium virulence *in vitro* and *in vivo*. Infect. Immun. In press.

Nakano, M., Takahashi, A., Sakai, Y., and Nakaya, Y. 2007a. Modulation of pathogenicity with norepinephrine related to the type III secretion system of *Vibrio parahaemolyticus*. J. Infect. Dis. 195:1353–1360.

Nakano, M., Takahashi, A., Sakai, Y., Kawano, M., Harada, N., Mawatari, K., and Nakaya, Y. 2007b. Catecholamine-induced stimulation of growth in *Vibrio* species. Lett. Appl. Microbiol. 44:649–653.

Nance, D. M., and Sanders, V. M. 2007. Autonomic innervation and regulation of the immune system (1987–2007). Brain Behav. Immun. 21:736–745.

Nieuwenhuijzen, G. A., and Goris R. J. 1999. The gut: the 'motor' of multiple organ dysfunction syndrome? Curr. Opin. Clin. Nutr. Metab. Care 5:399–404.

O'Donnell, P. M., Aviles, H., Lyte, M., and Sonnenfeld, G. 2006. Enhancement of in vitro growth of pathogenic bacteria by norepinephrine: importance of inoculum density and role of transferrin. Appl. Environ. Microbiol. 72:5097–5099.

Paulin, S. M., Jagannathan, A., Campbell, J., Wallis, T. S., and Stevens, M. P. 2007. Net replication of *Salmonella enterica* serovars Typhimurium and Choleraesuis in porcine intestinal mucosa and nodes is associated with their differential virulence. Infect. Immun. 75:3950–3960.

Pullinger, G. D., Carnell, S. C., Sharaff, F. F., van Diemen, P. M., Dziva, F., Morgan, E., Lyte, M., Freestone, P. P., and Stevens, M. P. 2010. Norepinephrine augments *Salmonella enterica*-induced enteritis in a manner associated with increased net replication but independent of the putative adrenergic sensor kinases QseC and QseE. Infect. Immun. 78:372–380.

Porat, R., Clark, B. D., Wolff, S. M., and Dinarello, C. A. 1991. Enhancement of growth of virulent strains of *Escherichia coli* by interleukin-1. Science 254:430–432.

Rasko, D. A., Moreira, C. G., Li, de R., Reading, N. C., Ritchie, J. M., Waldor, M. K., Williams, N., Taussig, R., Wei, S., Roth, M., Hughes, D. T., Huntley, J. F., Fina, M. W., Falck, J. R., and Sperandio, V. 2008. Targeting QseC signaling and virulence for antibiotic development. Science 321:1078–1080.

Reading, N. C., Rasko, D. A., Torres, A. G., and Sperandio, V. 2009. The two-component system QseEF and the membrane protein QseG link adrenergic and stress sensing to bacterial pathogenesis. Proc. Natl. Acad. Sci. U. S. A. 106:5889–5894.

Reading, N. C., Torres, A. G., Kendall, M. M., Hughes, D. T., Yamamoto, K., and Sperandio, V. 2007. A novel two-component signaling system that activates transcription of an enterohemorrhagic *Escherichia coli* effector involved in remodeling of host actin. J. Bacteriol. 189:2468–2476.

Reissbrodt, R., Rienaecker, I., Romanova, J. M., Freestone, P. P., Haigh, R. D., Lyte, M., Tschäpe, H., and Williams, P. H. 2002. Resuscitation of *Salmonella enterica* serovar Typhimurium and enterohemorrhagic *Escherichia coli* from the viable but nonculturable state by heat-stable enterobacterial autoinducer. Appl. Environ. Microbiol. 68:4788–4794.

Ritchie, J. M., Thorpe, C. M., Rogers, A. B., and Waldor, M. K. 2003. Critical roles for *stx2*, *eae*, and *tir* in enterohemorrhagic *Escherichia coli*-induced diarrhea and intestinal inflammation in infant rabbits. Infect. Immun. 71:7129–7139.

Roberts, A., Matthews, J. B., Socransky, S. S., Freestone, P. P., Williams, P. H., and Chapple, I. L. 2002. Stress and the periodontal diseases: effects of catecholamines on the growth of periodontal bacteria in vitro. Oral Microbiol. Immunol. 17:296–303.

Robinson, C. M., Sinclair, J. F., Smith, M. J., and O'Brien, A. D. 2006. Shiga toxin of enterohemorrhagic *Escherichia coli* type O157:H7 promotes intestinal colonization. Proc. Natl. Acad. Sci. USA 103:9667–9672.

Roe, A. J., Currie, C., Smith, D. G., and Gally, D. L. 2001. Analysis of type 1 fimbriae expression in verotoxigenic *Escherichia coli*: a comparison between serotypes O157 and O26. Microbiology 147:145–152.

Sandrini, S. M., Shergill, R., Woodward, J., Muralikuttan, R., Haigh, R. D., Lyte, M., and Freestone, P. P. 2010. Elucidation of the mechanism by which catecholamine stress hormones liberate iron from the innate immune defense proteins transferrin and lactoferrin. J. Bacteriol. 192:587–594.

Schreiber, K. L., and Brown, D. R. 2005. Adrenocorticotrophic hormone modulates *Escherichia coli* O157:H7 adherence to porcine colonic mucosa. Stress 8:185–190.

Schreiber, K. L., Price, L. D., and Brown, D. R. 2007. Evidence for neuromodulation of enteropathogen invasion in the intestinal mucosa. J. Neuroimmune Pharmacol. 2:329–337.

Spencer, H., Karavolos, M. H., Bulmer, D. M., Aldridge, P., Chhabra, S. R., Winzer, K., Williams, P., Khan, C. M. 2010. Genome-wide transposon mutagenesis identifies a role for host neuroendocrine stress hormones in regulating the expression of virulence genes in *Salmonella*. J. Bacteriol. In press.

Sperandio, V., Torres, A. G., and Kaper, J. B. 2002. Quorum sensing *Escherichia coli* regulators B and C (QseBC): a novel two-component regulatory system involved in the regulation of flagella and motility by quorum sensing in *E. coli*. Mol. Microbiol. 43:809–821.

Sperandio, V., Torres, A. G., Jarvis, B., Nataro, J. P., and Kaper, J. B. 2003. Bacteria-host communication: the language of hormones. Proc. Natl. Acad. Sci. USA 100:8951–8956.

Stabel, T. J. 1999. Evaluation of 2-deoxy-D-glucose for induction of a stress response in pigs. Am. J. Vet. Res. 60:708–713.

Stabel, T. J., and Fedorka-Cray, P. J. 2004. Effect of 2-deoxy-D-glucose induced stress on *Salmonella choleraesuis* shedding and persistence in swine. Res. Vet. Sci. 76:187–194.

Stevens, M. P., Humphrey, T. J., and Maskell, D. J. 2009. Molecular insights into farm animal and zoonotic *Salmonella* infections. Philos. Trans. R Soc. Lond. B Biol. Sci. 364:2709–2723.

Stevens, M. P., Marchès, O., Campbell, J., Huter, V., Frankel, G., Phillips, A. D., Oswald, E., and Wallis, T. S. 2002. Intimin, Tir, and Shiga toxin 1 do not influence enteropathogenic responses to Shiga toxin-producing *Escherichia coli* in bovine ligated intestinal loops. Infect. Immun. 70:945–952.

Toscano, M. J., Stabel, T. J., Bearson, S. M. D., Bearson, B. L., and Lay, D. C. 2007. Cultivation of *Salmonella enterica* serovar Typhimurium in a norepinephrine-containing medium alters *in vivo* tissue prevalence in swine. J. Exp. Anim. Sci. 43:329–338.

van Diemen, P. M., Dziva, F., Stevens, M. P., and Wallis, T. S. 2005. Identification of enterohemorrhagic *Escherichia coli* O26:H-genes required for intestinal colonization in calves. Infect. Immun. 73:1735–1743.

Velin, A. K., Ericson, A. C., Braaf, Y., Wallon, C., and Söderholm, J. D. 2004. Increased antigen and bacterial uptake in follicle associated epithelium induced by chronic psychological stress in rats. Gut 53:494–500.

Vlisidou, I., Lyte, M., van Diemen, P. M., Hawes, P., Monaghan, P., Wallis, T. S., and Stevens, M. P. 2004. The neuroendocrine stress hormone norepinephrine augments *Escherichia coli* O157:H7-induced enteritis and adherence in a bovine ligated ileal loop model of infection. Infect. Immun. 72:5446–5451.

Vlisidou, I., Dziva, F., La Ragione, R. M., Best, A., Garmendia, J., Hawes, P., Monaghan, P., Cawthraw, S. A., Frankel, G., Woodward, M. J., and Stevens, M. P. 2006. Role of intimin-Tir interactions and the Tir-cytoskeleton coupling protein in the colonization of calves and lambs by *Escherichia coli* O157:H7. Infect. Immun. 74:758–764.

Voigt, W., Fruth, A., Tschäpe, H., Reissbrodt, R., and Williams, P. H. 2006. Enterobacterial autoinducer of growth enhances Shiga toxin production by enterohemorrhagic *Escherichia coli*. J. Clin. Microbiol. 44:2247–2249.

Waalkes, T. P., Sjoerdsma, A., Creveling, C. R., Weissbach, H., and Udenfriend, S. 1958. Serotonin, norepinephrine, and related compounds in bananas. Science 127:684–650.

Walters, M., and Sperandio, V. 2006. Autoinducer 3 and epinephrine signaling in the kinetics of locus of enterocyte effacement gene expression in enterohemorrhagic *Escherichia coli*. Infect. Immun. 74:5445–5455.

Williams, P. H., Rabsch, W., Methner, U., Voigt, W., Tschäpe, H., and Reissbrodt, R. 2006. Catecholate receptor proteins in *Salmonella enterica*: role in virulence and implications for vaccine development. Vaccine 24:3840–3844.

Wu, L., Holbrook, C., Zaborina, O., Ploplys, E., Rocha, F., Pelham, D., Chang, E., Musch, M., and Alverdy, J. 2003. *Pseudomonas aeruginosa* expresses a lethal virulence determinant, the PA-I lectin/adhesin, in the intestinal tract of a stressed host: the role of epithelia cell contact and molecules of the quorum sensing signaling system. Ann. Surg. 238:754–764.

Wu, L., Estrada, O., Zaborina, O., Bains, M., Shen, L., Kohler, J. E., Patel, N., Musch, M. W., Chang, E. B., Fu, Y. X., Jacobs, M. A., Nishimura, M. I., Hancock, R. E., Turner, J. R., and Alverdy, J. C. 2005. Recognition of host immune activation by *Pseudomonas aeruginosa*. Science 309:774–777.

Zaborina, O., Lepine, F., Xiao, G., Valuckaite, V., Chen, Y., Li, T., Ciancio, M., Zaborin, A., Petroff, E., Turner, J. R., Rahme, L. G., Chang, E., and Alverdy, J. C. 2007. Dynorphin activates quorum sensing quinolone signaling in *Pseudomonas aeruginosa*. PLoS Pathog. 3:e35.

Zhou, M., Hank Simms, H., and Wang, P. 2004. Increased gut-derived norepinephrine release in sepsis: up-regulation of intestinal tyrosine hydroxylase. Biochim. Biophys. Acta 1689:212–218.

Chapter 7
The Role of Microbial Endocrinology in Periodontal Disease

Anthony Roberts

7.1   Introduction to the Oral Cavity

Over 700 bacterial species have been identified as residents of the oral cavity (Moore and Moore, 1994; Kroes et al. 1999; Socransky and Haffajee 2000; Paster et al. 2001), and the number of clinical isolates recognised continues to increase with improved methods of isolation and identification. Within the estimated 215 cm^2 surface area of the oral cavity (Collins and Dawes 1987), there are several ecological niches, namely tongue, buccal mucosa, saliva and teeth where oral organisms may reside. On the tooth surface, bacteria are found within dental plaque, which is the primary aetiological agent responsible for dental caries and periodontal disease, which are the main causes of tooth loss throughout the world. Understanding dental plaque has therefore become a priority for oral microbiologists and those clinicians responsible for managing caries and periodontal diseases.

7.2   What Is Periodontal Disease?

The anatomical region where the tooth meets the gum (or gingiva) is particularly interesting as it is a junction between hard and soft tissue (Figs. 7.1a, b), where a nonshedding tooth surface may be colonised with dental plaque in close proximity to the supporting periodontal tissues. Within dental plaque, the pathogenic organisms and the various virulence factors they release, stimulate a reaction by the hosts' inflammatory and immune systems, indeed, a delicate equilibrium exists

A. Roberts (✉)
School of Dentistry, The University of Manchester, Higher Cambridge Street,
Manchester M15 6FH, UK
e-mail: anthony.roberts@manchester.ac.uk

M. Lyte and P.P.E. Freestone (eds.), *Microbial Endocrinology*,
Interkingdom Signaling in Infectious Disease and Health,
DOI 10.1007/978-1-4419-5576-0_6, © Springer Science+Business Media, LLC 2010

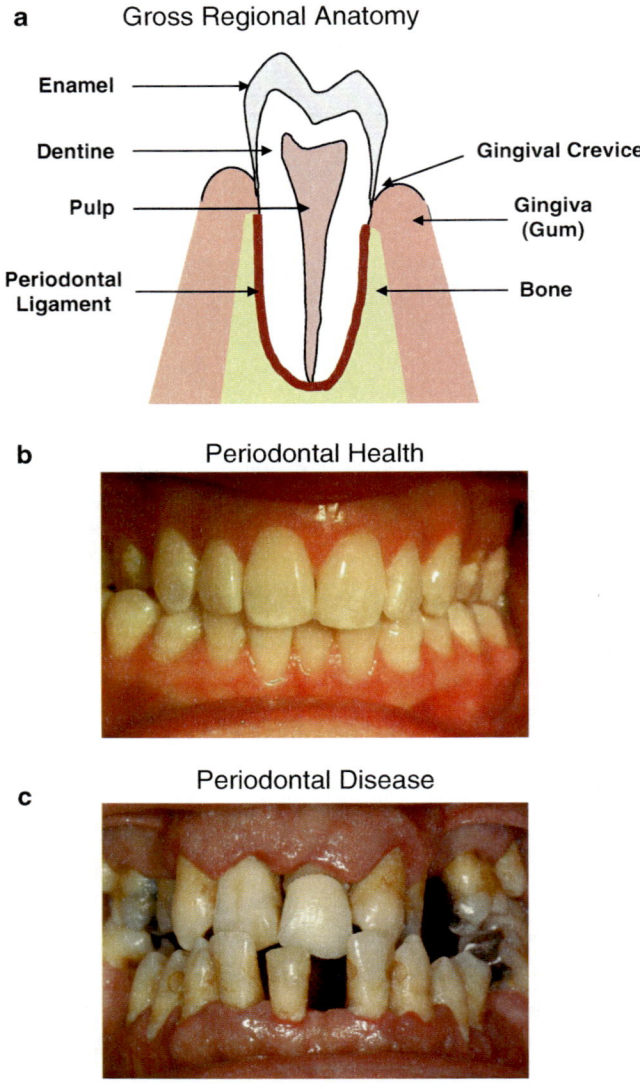

Fig. 7.1 (**a**) Gross regional anatomy. (**b**) Healthy periodontal tissues. (**c**) Periodontal disease

between the host and plaque microorganisms and changes in either component may result in periodontal tissue destruction (Fig. 7.1c).

The composition of dental plaque is highly complex and diverse containing bacteria imbedded in a polymer matrix derived from bacteria and saliva (Marsh and Martin, 1999). These components form a biofilm on the tooth surface within which are found areas of high and low bacterial populations. It is likely that the bacterial components of the plaque biofilm are able to communicate with one another as a prerequisite for the continued development of an ordered multi-species community observed in the plaque biofilm (Kolenbrander et al. 2002). This communication includes the receptors for mucins, agglutinins and alpha-amylases and the adhesins used for coaggregation and co-adherence between bacterial species, which are important in establishing the arrangement of organisms within the biofilm. The bacteria within the dental plaque biofilm in periodontal disease are primarily obligate anaerobes or capnophilic Gram-negative organisms, which represent a shift from the microflora within dental plaque from healthy subjects or those with gingivitis, which are primarily Streptococci and Actinomyces species.

The importance of Gram-negative bacterial species in the pathogenesis of periodontitis and in the production of autoinducers by organisms found within the gastrointestinal tract sparked the research effort into the microbial endocrinology of dental plaque.

7.3 Stress Hormone Research in Periodontal Disease

Early studies investigating the effects of epinephrine have shown an enhancement in infectivity of bacteria and in particular anaerobic pathogens (Evans et al. 1948). Saline suspensions of gingival crevice debris when injected subcutaneously into rabbits produced local abscesses, whose character changed when the injections were supplemented with epinephrine (Courant and Gibbons 1966). Addition of epinephrine resulted in marked haemorrhage and necrosis, with the size of the resultant lesions being greatest with smaller inoculae of gingival debris (Courant and Gibbons 1966). To determine whether these effects were due to epinephrine potentiating the effect of bacterial endotoxins present within the gingival debris, lipopolysaccharide was inoculated with and without epinephrine. No lesions developed and it was concluded that epinephrine was enhancing the virulence of gingival bacteria. These authors further postulated that any process that increased the local level of catecholamines in the gingiva, as hypothesised in necrotizing periodontal diseases (Goldhaber and Giddon 1964), could lead to the increased pathogenicity of bacteria found in the gingival crevice, and therefore contribute to the disease process (Courant and Gibbons 1966).

Epinephrine infusions (10 µg in 5 ml) in rabbits have produced vasoconstriction within the gingival vasculature, which was also observed with infusions of nicotine (0.081 mg in 5 ml). The combination of epinephrine and nicotine infusions demonstrated a synergistic effect on gingival blood flow with a more profound vasoconstriction observed than for either infusion alone (Clarke et al. 1981). The vessels delivering blood to the gingivae are end-arterioles with limited collateral vessels at the papillary tips (Shannon et al. 1974). This may provide a degree of vulnerability to gingival tissue viability during periods of ischaemia (epinephrine ± nicotine) and may provide a potential mechanism for the development of necrotising periodontal diseases. In humans, attempts to relate urinary catecholamines (norepinephrine, epinephrine and dopamine) to gingivitis in children ($n = 314$ aged 6–8 y) and to gingival bleeding index have failed to demonstrate any correlation (Vanderas et al. 1998).

Steroids (C_{21}, C_{19}, and C_{18}) are actively metabolised in the gingival tissues with their metabolism increasing with the degree of inflammation (Tilakaratne and Soory 1999). Indeed gingival tissue has been shown to convert cortisol into cortisone and 20α –dihydrocortisol (el-Attar 1968). Cortisol is one of the primary mediators of the stress response and is found within blood, both free and bound to proteins, at serum levels in the range of 200–800 nmol/l. It is the free form of cortisol which passively diffuses into the salivary acinar cells (Vining et al. 1983) and can be found in saliva (Aardal and Holm 1995). Another oral fluid is Gingival Crevicular Fluid (or GCF) which is found within the gingival crevice and is constantly replenished as it leaves the gingival crevice to become a component of saliva. GCF is a serum transudate when the gingival tissues are healthy but becomes a tissue exudate during disease. Using a modified radioimmunoassay technique, cortisol has been identified in GCF at levels approximately one tenth that of serum and substantially higher than salivary levels (Axtelius et al. 1998).

Investigations attempting to determine the influence of cortisone and epinephrine in the mouse, described clinical features consistent with periodontal disease (increased apical proliferation of the attachment apparatus, periodontal pocket formation, calculus deposition and inflammation and alveolar bone loss) following daily injections of cortisone. By contrast, no changes were observed in those mice receiving 0.05 mg of epinephrine daily (Glickman et al. 1960; Cohen et al. 1969). However, it is difficult to extrapolate to the human model because responses to catecholamines tend to vary between species (Taichman 1964).

7.4 Potential Implications of Stress Hormone Research Upon Periodontal Disease

There are many host related factors that may tip the microbe-host balance in favour of periodontal disease progression including smoking and psychological stress. The mechanisms whereby psychological stress effects the pathogenesis of periodontal diseases have yet to be fully elucidated, however the studies investigating the effects of stress hormones upon organisms found further along the gastrointestinal tract highlighted a number of potential avenues of research, which could be translated to the oral environment.

Previous investigations have demonstrated the ability of norepinephrine to increase the growth of *Escherichia coli* in a harsh serum-based medium (Lyte et al. 1996b). This non-nutritional increase in growth has been shown to be due to the production of an autoinducer of growth, which in addition to enhanced growth and proliferation causes virulence expression of many Gram-negative organisms. These Gram-negative strains both produce and respond to heat-stable norepinephrine-induced autoinducers of growth, most of which show a high degree of cross-species activity (Freestone et al. 1999). This suggests the existence of a novel family of

Gram-negative bacterial signalling molecules. Further, these investigations have shown the potential importance of norepinephrine-induced autoinducers in relation to infectious diseases caused by Gram-negative organisms. As a result, the Gram-negative organisms associated with periodontal diseases were an obvious target to commence a research effort investigating the catecholamine responses of oral organisms. Before a screen of dental plaque organisms for catecholamine responsiveness could commence, some key methodological issues needed to be addressed.

7.5   Modification of Research Methodology of Anaerobes

The use of an *E. coli* bacterial growth assay to assess catecholamine responsiveness has previously been restricted to organisms associated with the gastro-intestinal tract (see Chap. 3 of this book) and, in this respect, the methodology had not been developed to characterise the growth responses of organisms found within the oral cavity. In particular, the methodology required development to allow investigation of the catecholamine responses of putative periodontal pathogens requiring anaerobic conditions.

Previous investigations of use pour plate methodology for the investigation of catecholamine responsiveness of *E. coli* strains in aerobic culture conditions (Lyte and Ernst 1992; Freestone et al. 1999); however, this methodology would be technically demanding to perform for some of the strict anaerobes encountered in the dental plaque biofilm. Our research turned to optical density readings which produced comparative data to pour plate analysis, confirming them as a valid measure of bacterial growth assessment (Roberts et al. 2002). Indeed, in a previous study using the same model (Lyte and Ernst 1992), growth evaluation using spectrophotometry (optical density at 630 nm) has been utilised for determining the dose response of an *E. coli* strain (ATCC 23723) and the results are comparable to our findings. Optical density readings also had the advantage of being quicker, and were highly discriminatory as demonstrated by the small differences in the growth enhancement of the wild type *E. coli* when supplemented with norepinephrine and epinephrine which had not previously observed using pour plate analysis. Furthermore, spectrophotometry reduces the risk of contamination during any plating procedure and does not underestimate colony counts for strict anaerobes. In addition, the technique is cheaper (fewer consumables are required for the technique), however disadvantages of the technique are the lack of ability to identify contamination within the culture and the requirement to run a non-inoculated control to correct for any optical density changes caused by the supplement. Further, the comparison of optical density readings between cultures assumes that the size of the organism remains consistent with and without the addition of supplements. However, these data suggest that optical density readings may provide a suitable methodology for the evaluation of the catecholamine responses of oral organisms. In summary, a spectrophotometric approach to evaluate growth changes has enabled a large number of organisms to be screened for catecholamine responsiveness as well as eliminating the potential for colony death, and therefore underestimation of growth, during plating out procedures.

7.6 Screening of Periodontal Bacteria for Responses to Norepinephrine and Epinephrine

Once the model *E. coli* growth assay had been effectively adapted for use in assessing the growth of periodontal organisms, the purpose of subsequent experiments was to investigate the growth responses of 43 putative periodontal pathogens to norepinephrine and epinephrine under standard conditions of inoculum size, time course (24 or 48 h) and environmental growth conditions (anaerobic, 35°C) (Roberts et al. 2002)

This was the first study to demonstrate the direct role of stress hormones upon growth of those micro-organisms commonly found within sub-gingival plaque biofilms, although other studies have since demonstrated similar findings (Belay et al. 2003). A broad variation in the growth response of a large number of micro-organisms to norepinephrine and epinephrine was demonstrated with both positive and negative effects being identified. Of the oral bacteria studied roughly, half the species showed significant catecholamine-induced growth enhancement or inhibition indicating that stress hormones might directly modulate the growth and composition of the sub-gingival biofilm. As periodontal disease is a consequence of host-parasite interactions, the effects of stress hormones seen within the sub-gingival biofilm may well provide one potential mechanism linking stress to periodontal disease progression.

The microbial complexes (Socransky et al. 1998) were used to group sub-gingival species within this study as they provide a model of the associations and interrelations between different organisms (Fig. 7.2). These complexes relate closely to clinical measures of inflammation and periodontal disease activity and were used to provide a biological framework against which the diversity of catecholamine growth responses could be investigated. Briefly, the organisms most strongly associated with severe chronic periodontitis are found on the right-side of Fig. 7.2. Other organisms are found within the sub-gingival plaque biofilm are likely to be opportunistic pathogens capitalising upon local environmental changes as the disease advances.

Interestingly, there was considerable variation in growth response to norepinephrine and epinephrine both within and between sub-gingival microbial complexes as well as between sub-species of micro-organisms depending on the nature of the supplementation. It was obvious from the Roberts et al. (2002) study that a broad spectrum of growth responses to catecholamines exists, ranging from growth stimulation to growth inhibition. This indicates that the relationship between the growth of periodontal organisms and catecholamine hormones is likely to be a complex one, with the relative proportions of the resultant flora generated by the differential effects of these varied responses and interactions. It should be emphasised

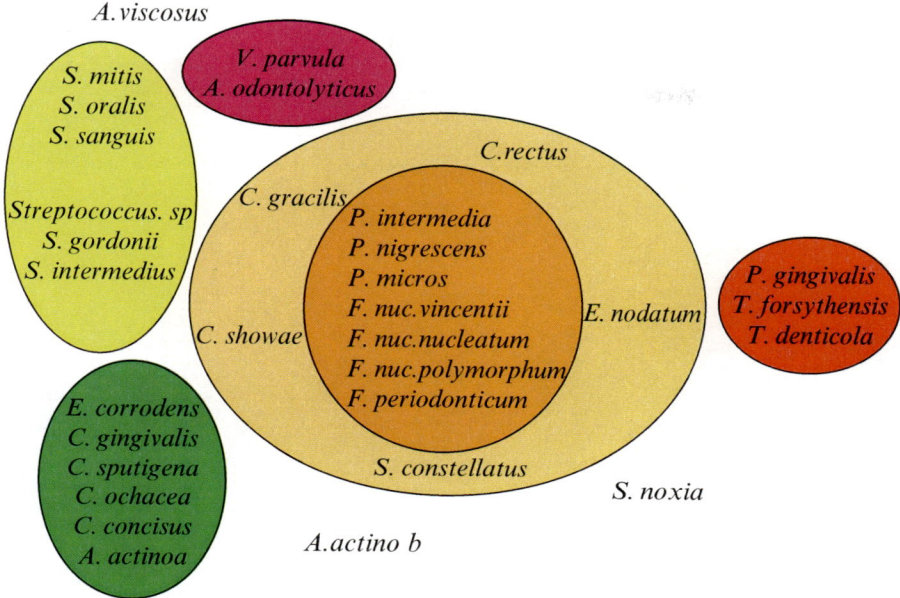

Fig. 7.2 Sub-gingival plaque microbial complexes

that negative growth effects are as significant as positive growth effects in determining the resultant flora within a sub-gingival biofilm. Inhibitory growth effects were observed for the "Red Complex" organisms (*P. gingivalis* and *T. forsythensis*), and it is possible that these organisms (and others showing negative growth responses) utilise catecholamines for up-regulation of virulence expression rather than growth. For example, *E. coli* strains increase the production of Shiga-like toxins (Lyte et al. 1997a) and K99 pili (Lyte et al. 1997b) in response to norepinephrine. It is therefore possible that catecholamines may have a dual effect on both virulence expression and growth response of periodontal organisms, which are entirely separate responses. In addition to the direct effect of catecholamine hormones on periodontal microorganisms, those organisms demonstrating increased growth may indirectly influence those organisms demonstrating growth inhibition either by production of growth factors or virulence factors within the biofilm. The experiments within the (Roberts et al. 2002) study dealt with single strain cultures of planktonic organisms, and it would be of interest in the future to consider the growth responses of periodontal organisms to catecholamines when in polymicrobial culture in a biofilm environment.

The study by (Roberts et al. 2002) investigated the growth effects of catecholamines on sub-gingival organisms. However, it is noticeable that many of the Actinomyces and Streptococci species demonstrated increases in growth to the catecholamines and are more commonly associated with the supra-gingival flora (Table 7.1). However, it has also been shown that the supra-gingival flora

Table 7.1 This table illustrates those oral organisms tested which responded with positive growth to norepinephine or epinephrine. + indicates a positive growth effect, – indicates a negative growth effect and those in within ( ) indicate a growth effect which was not statistically significant. No positive growth effects were observed from any organism found within the red complex

Socransky's microbial complex	ATCC strain	Species	Norepinephrine	Epinephrine
Actinomyces species	12104	A. naeslundii I	+	+
	23860	A. gerencseriae	+	+
Purple complex	17929	A. odontolyticus I	+	+
Yellow complex	10558	S. gordonii	(+)	+
	27335	S. intermedius	+	+
	35037	S. oralis	(−)	+
	49456	S. mitis	+	+
Green complex	23834	E. corrodens	+	+
	33612	C. sputigena	+	+
	33624	C. gingivalis	+	+
Orange complex	27823	S. constellatus	+	+
	33236	C. gracilis	+	−
	33270	P. micros	(+)	+
	33693	F. periodonticum	(+)	+
	49256	F nucleatum sp. vinc	−	+
Others	19696	N. mucosa	+	(+)
	33271	E. saburreum	−	+
	33397	S. anginosus	+	(−)
	14201	L. buccalis	+	+
	25175	S. mutans	+	+
	27337	P. anaerobius	+	+
	35308	P. denticola	+	+

plays a significant role in determining the composition of the sub-gingival flora (Dahlen et al. 1992); this may be another important influence on progression of periodontal disease.

Of considerable interest is the fact that a number of organisms associated with suppurative infections in humans responded positively to catecholamine supplementation (Roberts et al. 2002). *Streptococcus constellatus*, *S. intermedius* and *S. anginosus* are organisms associated with pus formation as well as Peptostreptococcus species (commonly isolated from dento-alveolar abscesses) and the Gram-negative anaerobic rods (including *E. corrodens* and Capnocytophagia species) that have been reported to cause lateral periodontal abscesses (Marsh and Martin 1999). Anecdotal evidence has existed for an association between stressful episodes and abscess formation for some time, and the positive growth changes in these organisms to stress hormones observed in this study, could provide one potential mechanism.

Other important periodontal diseases of considerable interest are the necrotising periodontal diseases. Two important organisms associated with these diseases are *Fusobacterium nucleatum* subsp. *vincentii* which showed positive growth effects with epinephrine and *Prevotella intermedia* which showed a tendency for increased growth with epinephrine (Loesche et al. 1982). The positive growth effects observed in these organisms with epinephrine could also provide a potential mechanism linking stress hormones and necrotising periodontal diseases. The spirochaetes are an important group of organisms also associated with these diseases, however the experimental medium could not support their growth.

In summary, the results suggest that changes in catecholamine levels may trigger a shift in the balance of sub-gingival species within that environment, with certain organisms able to utilise host-derived hormones to influence their growth and possible virulence expression. In this manner, biofilm composition could change in relation to a stressed host, with one or two organisms within a "complex" orchestrating change in response to their local environment. The growth effects of stress hormones could therefore have a major influence on periodontal disease pathogenesis.

7.7 Mechanisms of Catecholamine Responses by Periodontal Pathogens

Two main avenues of research have been explored so far in relation to catecholamine responsiveness of periodontal bacteria. The first relates to the ability of periodontal pathogens to respond to the *E. coli* norepinephrine induced-autoinducer. The second relates to the proposal that catecholamines proposed act as siderophore-like molecules and essentially de-stabilised iron binding by host proteins such as transferrin and lactoferrin (Freestone et al. 2000), facilitating iron uptake by micro-organisms. Since transferrin and lactoferrin are both found within the oral environment, then this could provide a potential mechanism whereby sub-gingival organisms can survive in an iron restricted environment or gain a survival advantage.

7.8 Screening for Responses to Iron and Autoinducer

These studies (Roberts et al. 2005) demonstrated that 15 of the 43 oral bacterial strains tested (the same organisms previously screened for catecholamine responsiveness) gave >45% positive growth responses to a novel stress-associated autoinducer of growth produced by *E. coli*. The data supports and extends the previous work performed on organisms found within the gastrointestinal tract (Freestone et al. 2002) as both Gram-positive and Gram-negative organisms were able to respond to the *E. coli* autoinducer. Thus, the data suggest that stress-related autoinducers similar to that produced by *E. coli* may play a role in the growth and development of the sub-gingival biofilm and be implicated in the pathogenesis of those periodontal diseases that are associated with human stress responses (Lyte et al. 1996a).

A major difference in the *E. coli* norepinephrine induced-autoinducer-induced growth enhancement of sub-gingival plaque organisms and those derived from the gastrointestinal tract is the magnitude of the response. Gastrointestinal organisms cultured in serum-SAPI minimal under aerobic conditions invariably show substantial positive growth effects (2–4 log increases in 24 h) to the autoinducer (Freestone et al. 1999), whereas those elicited in sub-gingival organisms are smaller and typically less than a doubling of growth in 24–48 h of anaerobic culture. The finding of only small increases in growth yield for the oral bacterial species could indicate that the growth effects detected are due to the presence of a limiting nutrient within the spent medium rather than a true autoinducer response. This conclusion is unlikely as in the dose response experiments, using different amounts of autoinducer-containing spent medium on two of the responsive organisms (*S. constellatus* and *F. nucleatum* subsp. *polymorphum*), a typical autoinducer response was observed that plateaued at 0.3% supplementation Roberts et al. (2005).

While the *E. coli* autoinducer does cause statistically significant increases in the growth of sub-gingival organisms at 24–48 h of exposure (Roberts et al. 2005), it could be argued that such responses would not be biologically significant in respect of the pathogenesis of periodontal disease. However, the results have demonstrated that for sub-gingival organisms grown in serum-SAPI medium, 24–48 h incubation represent very early stages in the growth curve and indicate that over longer time courses (96–120 h), more prolonged responses of a greater magnitude are seen. Further experiments (Roberts, unpublished data) have investigated the longer term effects of *E. coli* autoinducer, Fe, norepinephrine and epinephrine upon periodontal bacteria to ensure that the magnitude of the responses reported does not underestimate the true longer term effects.

The results involving the same panel of 43 sub-gingival organisms studied previously Roberts et al. (2002), have demonstrated that the catecholamine hormones have significant effects upon growth *in vitro* with 12 species exhibiting similar levels of growth enhancement to those found with *E. coli* autoinducer (species showing greater than 25% catecholamine-induced growth enhancement) (Roberts et al. 2002). The mechanisms underlying these stress-hormone effects are unknown, and it has yet to be determined whether the levels of catecholamine hormones used in the *in vitro* studies ($50\,\mu M$) are reflective of those found within inflamed (plaque induced) periodontal tissues.

7.9   Host Iron-Binding Protein Experiments

A further mechanism that may explain the reported catecholamine-induced enhancement of growth in some sub-gingival organisms is the ability of catecholamines to act as siderophores and supply Fe derived from host iron binding proteins, such as transferrin, to *E. coli* (Freestone et al. 2000). The findings for dental plaque bacteria using the same experimental model as *E. coli*, have demonstrated that norepinephrine can play a similar role for oral organisms (Roberts et al. 2005).

In contrast to the findings in *E. coli*, enhanced incorporation of [55]Fe does not automatically ensure that growth enhancement of the organism(s) is observed (Roberts et al. 2005). For example, *F. nucleatum* polymorphum demonstrated a 1,139% increase in [55]Fe incorporation with the addition of norepinephrine and yet a growth inhibition of 9.5% was observed (Roberts et al. 2005). It is therefore likely that the enhanced Fe acquisition by *F. nucleatum* polymorphum is used for alternative purposes such as virulence factor production. Indeed, catecholamines enhance the production of virulence factors in *E. coli* such as Shiga-like toxins by over 150-fold (Lyte et al. 1996a), the synthesis of K99 pili by over 1,600-fold (Lyte et al. 1997b) and cause a two- to eightfold increase in levels of virulence associated proteins of *S. typhimurium* (Rahman et al. 2000). Furthermore, norepinephrine may increase the aerotolerance of micro-organisms (*Spirillum volutans* and *Campylobacter fetus* subsp. *jejuni*; Bowdre et al. 1976) which may have implications within the sub-gingival biofilm. An interesting finding was that "cold" Fe increased the [55]Fe incorporation of the oral organisms tested. This may reflect the "cold" Fe causing the dissociation of [55]Fe from transferrin as it competes with [55]Fe for binding sites allowing free [55]Fe in solution to become available to the oral organisms. This may allow the free [55]Fe to associate with norepinephrine for transportation into the organism. Alternatively, the free Fe could have acted to increase the iron-uptake mechanisms of the organisms in a positive feedback manner.

7.10   Catecholamine Levels in the Oral Cavity

The detection of catecholamines within the oral environment has previously been restricted to studies specifically investigating the levels of norepinephrine and epinephrine within saliva and dental pulp (Schachman et al. 1995; Mitome et al. 1997). Putative periodontal pathogens are likely to be exposed to catecholamines within saliva; however, this exposure is likely to exert an effect only on the supra-gingival microflora. The purpose of the next set of experiments was to determine the levels of sub-gingival catecholamines. GCF constantly bathes the sub-gingival environment, including any periodontal pathogens within dental plaque and therefore was an ideal fluid to begin the search for catecholamines (Fig. 7.3). Attempts have been made to measure catecholamines locally in GCF and plaque using HPLC with electrochemical detection and MALDI-ToF Electro-Spray Mass Spectrometry, however the small volumes of GCF and plaque made analysis difficult.

Previous studies have investigated biological fluids which are readily obtained in millilitre quantities (Mitome et al. 1997; Forster and Macdonald 1999) however despite attempts to pool the sub-microlitre volume GCF samples, this provided only sporadic identification of catecholamine content. Specifically, norepinephrine was identified in two pooled samples of GCF using the Dionex equipment using a 30 min elution into sterile water followed by acidification with formic acid. Allowing a 30 min elution into sterile water improved the catecholamine recovery from Periopaper™ strips (Fig. 7.3) which was an unexpected result as

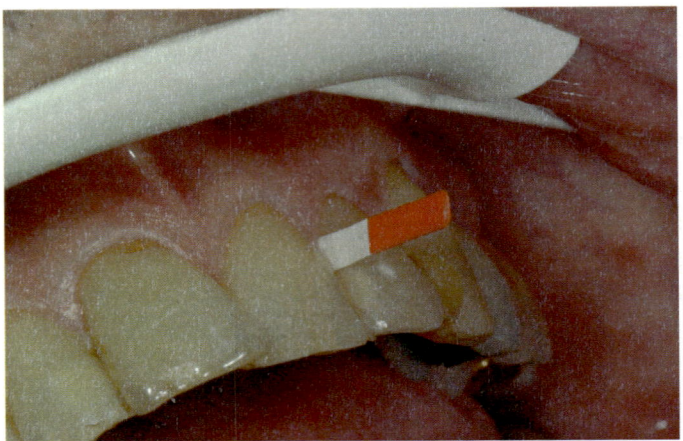

Fig. 7.3 GCF sampling being performed using Periopaper strips

catecholamines are generally unstable at neutral or basic pH as their catechol rings are vulnerable to oxidation in the presence of air and light (Callingham and Barrand 1979; Weir et al. 1986). Catecholamines are stable at low pH and/or in the presence of antioxidants such as cysteine, sodium metabisulphite, or ascorbic acid (Weir et al. 1986) and as a result, acid preservatives and/or stabilizing agents are added to urine collections for subsequent catecholamine studies (Moerman and de Schaepdryver 1984; Davidson and Fitzpatrick 1985). The recovery of norepinephrine from the Periopaper™ strips was improved by having a 30 min elution period in sterile water before acidification. Further, it is also possible that immediate acidification with formic acid was likely to have caused precipitation of the protein content within the GCF sample preventing the elution of catecholamines from the Periopaper™ strips.

Calculations based upon the results from the 30 min elution – delayed acidification method, indicated that the recovery rate for this technique was in the range of 83.4–96.5% and that GCF concentrations in positive samples were in the order of 0.58–3.08 µmol/l. Typical plasma levels of norepinephrine are 300 pg/ml (1.8 nmol/l), and salivary levels are 7.8–9.9 pg/ml (46.1–58.5 pmol/l) and therefore the GCF levels detected are approximately 300–1,700 times the level found within plasma. GCF is thought to be a serum transudate in health and inflammatory exudate in disease. Indeed, the levels of drugs within GCF have been shown to be several fold greater than circulating levels (Thomason et al. 1997; Lavda et al. 2004). The crude calculations performed here, indicate that this may also be true for norepinephrine and epinephrine, but further work is needed to establish more reproducible methods of detection.

In a further attempt to determine the presence or absence of catecholamines in GCF samples, a MALDI-ToF ESMS methodology was used which has the advantage of requiring small sample volumes. The minimum sample volume for this technique was 2 μl and masses could be determined on 50–500 fmol quantities of low molecular weight (<5,000) peptides with an average mass accuracy of ±0.1% (Williams 2006). Whilst norepinephrine and epinephrine could not be identified in a pool of three GCF samples (total volume 0.974 μl), epinephrine was identified in two healthy and one chronic periodontitis plaque sample (Roberts, unpublished data). Further, norepinephrine was also identified in the chronic periodontitis plaque sample as well as a saliva sample from a periodontally healthy subject (Roberts, unpublished data). These data support the HPLC method findings that catecholamines within GCF were not consistently found, however, it is interesting that MALDI-ToF ESMS confirmed the presence of catecholamines within plaque and saliva samples. Catecholamines present within saliva are likely to originate from the sympathetic innervation of the salivary glands, and therefore the salivary norepinephrine is likely to bathe the supra-gingival environment, including dental plaque. The presence of norepinephrine within a chronic periodontitis plaque sample and absence within healthy plaque samples, may reflect the quantity of plaque collected in each case or that the levels of norepinephrine within plaque correlate with the extent of periodontal disease. Epinephrine was identified in all plaque samples, and therefore it is possible that catecholamine levels within plaque influence the microbial flora, which in turn may influence the pathogenesis of periodontal disease. Overall, peaks or signals which appear to correspond to norepinephrine or epinephrine were found in GCF, plaque and saliva however, further attempts to quantify catecholamine levels in small volumes of biological samples are required as no published data currently exists.

7.11   Future Areas of Research

It would be interesting to investigate any changes in growth effects which may be seen when culturing the micro-organisms in a poly-microbial environment, where a multitude of organisms co-exist and individual responses within the biofilm mass may change. Further, it will be necessary to confirm whether recently collected clinical isolates behave in a similar manner to the laboratory strains used in this investigation and to determine at what time interval the growth responses are maximal.

Studies are also planned to elucidate the effects of catecholamines and autoinducer on virulence factor expression within periodontal organisms as it has been shown that both autoinducer and catecholamines can affect virulence expression in gastro-intestinal tract organisms (Lyte et al. 1996a; Lyte et al. 1997b). For example, norepinephrine may increase the aerotolerance of micro-organisms (*Spirillum volutans* and *Campylobacter fetus* subsp. *jejuni* (Bowdre et al. 1976), which again may have implications within the oral plaque biofilm.

In summary, the pathogenesis of the periodontal diseases is complex and involves multiple interactions between Gram-positive and Gram-negative organisms within the sub-gingival biofilm and between such organisms and the host (Socransky et al. 1998) which sits very well with the findings from the gastro-intestinal tract.

References

Aardal, E., Holm, A.C. 1995. Cortisol in saliva–reference ranges and relation to cortisol in serum. European Journal of Clinical Chemistry & Clinical Biochemistry 33, 927–932.

Axtelius, B., Edwardsson, S., Theodorsson, E., Svensater, G., Attstrom, R. 1998. Presence of cortisol in gingival crevicular fluid. A pilot study. Journal of Clinical Periodontology 25, 929–932.

Belay, T., Aviles, H., Vance, M., Fountain, K., Sonnenfeld G. 2003. Catecholamines and in vitro growth of pathogenic bacteria: enhancement of growth varies greatly among bacterial species. Life Sci. 73, 1527–35.

Bowdre, J.H., Krieg, N.R., Hoffman, P.S., Smibert, R.M., 1976. Stimulatory effect of dihydroxy-phenyl compounds on the aerotolerance of *Spirillum volutans* and *Campylobacter fetus* subspecies *jejuni*. Applied & Environmental Microbiology. 31, 127–133.

Callingham, B.A., Barrand, M.A. 1979. The catecholamines. Adrenaline; noradrenaline; dopamine. London: Academic Press.

Clarke, N.G., Shephard, B.C., Hirsch, R.S. 1981. The effects of intra-arterial epinephrine and nicotine on gingival circulation. Oral Surgery, Oral Medicine, Oral Pathology. 52, 577–582.

Cohen, M.M., Shusterman, S., Shklar, G. 1969. The effect of stressor agents on the grey lethal mouse strain periodontium. Journal of Periodontology 40, 462–466.

Collins, L.M., Dawes, C. 1987. The surface area of the adult human mouth and thickness of the salivary film covering the teeth and oral mucosa. Journal of Dental Research. 66, 1300–1302.

Courant, P.R., Gibbons, R.J. 1966. Epinephrine potentiation of response to gingival crevice bacteria. Archives of Oral Biology 11, 737–740.

Dahlen, G., Lindhe, J., Sato, K., Hanamura, H., Okamoto, H. 1992. The effect of supragingival plaque control on the subgingival microbiota in subjects with periodontal disease. Journal of Clinical Periodontology. 19, 802–809.

Davidson, D.F., Fitzpatrick, J. 1985. A simple, optimised and rapid assay for urinary free catecholamines by HPLC with electrochemical detection. Annals of Clinical Biochemistry. 22, 297–303.

El-Attar, T.M., 1968. Study of the metabolism of cortisol. Arthritis & Rheumatism 11.

Evans, D.G., Miles, A.A., Niven, J.S.F., 1948. The enhancement of bacterial infections by adrenaline. British Journal of Experimental Pathology 29, 20–39.

Forster, C.D., Macdonald, I.A., 1999. The assay of the catecholamine content of small volumes of human plasma. Biomedical Chromatography. 13, 209–215.

Freestone, P.P., Haigh, R.D., Williams, P.H., Lyte, M. 1999. Stimulation of bacterial growth by heat-stable, norepinephrine-induced autoinducers. FEMS Microbiology Letters. 172, 53–60.

Freestone, P.P., Lyte, M., Neal, C.P., Maggs, A.F., Haigh, R.D., Williams, P.H. 2000. The mammalian neuroendocrine hormone norepinephrine supplies iron for bacterial growth in the presence of transferrin or lactoferrin. Journal of Bacteriology. 182, 6091–6098.

Freestone, P.P., Williams, P.H., Haigh, R.D., Maggs, A.F., Neal, C.P. Lyte, M. 2002. Growth stimulation of intestinal commensal *Escherichia coli* by catecholamines: a possible contributory factor in trauma-induced sepsis. Shock 18, 465–470.

Glickman, T., Stone, I.C., Chawla, T.N. 1960. Effect of systemic administration of cortisone upon the periodontium. Journal of Periodontology. 31, 161–166.

Goldhaber, P., Giddon, D.B. 1964. Present concepts concerning the etiology and treatment of acute necrotizing ulcerative gingivitis. International Dental Journal. 14, 468–496.

Kolenbrander, P.E., Andersen, R.N., Blehert, D.S., Egland, P.G., Foster, J.S., Palmer, R.J., Jr. 2002. Communication among oral bacteria. Microbiology & Molecular Biology Reviews. 66, 486–505.

Kroes, I., Lepp, P.W., & Relman, D. A. 1999. Bacterial diversity within the human subgingival crevice. Proceedings of the National Academy of Sciences of the United States of America 96, 14547–14552.

Lavda, M., Clausnitzer, C.E., Walters, J.D. 2004. Distribution of systemic ciprofloxacin and doxycycline to gingiva and gingival crevicular fluid. Journal of Periodontology. 75, 1663–1667.

Loesche, W.J., Syed, S.A., Laughon, B.E., Stoll, J. 1982. The bacteriology of acute necrotizing ulcerative gingivitis. Journal of Periodontology. 53, 223–230.

Lyte, M., Arulanandam, B.P., Frank, C.D. 1996a. Production of Shiga-like toxins by *Escherichia coli* O157:H7 can be influenced by the neuroendocrine hormone norepinephrine. Journal of Laboratory & Clinical Medicine. 128, 392–398.

Lyte, M., Arulanandam, B., Nguyen, K., Frank, C., Erickson, A. & Francis, D. 1997a. Norepinephrine induced growth and expression of virulence associated factors in enterotoxigenic and enterohemorrhagic strains of *Escherichia coli*. Advances in Experimental Medicine & Biology. 412, 331–339.

Lyte, M., Erickson, A.K., Arulanandam, B.P., Frank, C.D., Crawford, M.A., Francis, D.H. 1997b. Norepinephrine-induced expression of the K99 pilus adhesin of enterotoxigenic *Escherichia coli*. Biochemical & Biophysical Research Communications. 232, 682–686.

Lyte, M. and Ernst, S. 1992. Catecholamine induced growth of gram negative bacteria. Life Sciences. 50, 203–212.

Lyte, M., Frank, C. D. and Green, B. T. 1996b. Production of an autoinducer of growth by norepinephrine cultured *Escherichia coli* O157:H7. FEMS Microbiology Letters. 139, 155–159.

Marsh, P., Martin, M. 1999. Oral Microbiology: Oxford: Wright.

Mitome, M., Shirakawa, T., Kikuiri, T., Oguchi, H. 1997. Salivary catecholamine assay for assessing anxiety in pediatric dental patients. Journal of Clinical Pediatric Dentistry. 21, 255–259.

Moerman, E.J., De Schaepdryver, A.F. 1984. Quantitation of catecholamines in urine and in plasma. Clinica Chimica Acta. 139, 321–333.

Moore, W.E., Moore, L.V. 1994. The bacteria of periodontal diseases. Periodontology 2000 5, 66–77.

Paster, B.J., Boches, S.K., Galvin, J.L., Ericson, R.E., Lau, C.N., Levanos, V.A., Sahasrabudhe, A., Dewhirst, F.E. 2001. Bacterial diversity in human subgingival plaque. Journal of Bacteriology 183, 3770–3783.

Rahman, H., Reissbrodt, R. & Tschape, H. 2000. Effect of norepinephrine on growth of *Salmonella* and its enterotoxin production. Indian Journal of Experimental Biology. 38, 285–286.

Roberts A., Matthews JB., Socransky SS., Freestone PPE., Williams PH., Chapple ILC., 2002. Stress and the periodontal diseases: Effects of catecholamines on the growth of periodontal bacteria in vitro. Oral Microbiology & Immunology 17, 296–303.

Roberts A., Matthews JB., Socransky SS., Freestone PPE., Williams PH., Chapple ILC. 2005. Stress and the periodontal diseases: Growth responses of periodontal bacteria to *Escherichia coli* stress associated autoinducer and exogenous Fe. Oral Microbiology & Immunology. 20: 147–153.

Schachman, M. A., Rosenberg, P. A. and Linke, H. A. 1995. Quantitation of catecholamines in uninflamed human dental pulp tissues by high-performance liquid chromatography. Oral Surgery Oral Medicine Oral Pathology Oral Radiology & Endodontics 80, 83–86.

Shannon, I., Kilgore, W.G., T.J., O. L. 1974. Aetiology of acute necrotizing ulcerative gingivitis: a hypothetical explanation. Journal of Periodontology 45, 830–832.

Socransky, S.S., Haffajee, A.D. 2000. Evidence of bacterial etiology: a historical perspective. Periodontology 5, 7–25.

Socransky, S.S., Haffajee, A.D., Cugini, M.A., Smith,C., Kent, R.L., Jr. 1998. Microbial complexes in subgingival plaque. Journal of Clinical Periodontology. 25, 134–144.

Taichman, N.S. 1964. The production of hemorrhagic necrosis by epinephrine and endotoxin in the hamster cheek pouch. Journal of Dental Research suppl 43, 795–796.

Thomason, J.M., Ellis, J.S., Kelly, P.J., Seymour, R.A. 1997. Nifedipine pharmacological variables as risk factors for gingival overgrowth in organ-transplant patients. Clinical Oral Investigations 1, 35–39.

Tilakaratne, A., Soory, M. 1999. Androgen metabolism in response to oestradiol-17beta and progesterone in human gingival fibroblasts (HGF) in culture. Journal of Clinical Periodontology 26, 723–731.

Vanderas, A.P., Kavvadia, K., Papagiannoulis, L. 1998. Urinary catecholamine levels and gingivitis in children. Journal of Periodontology 69, 554–560.

Vining, R.F., Mcginley, R.A., Symons, R.G. 1983. Hormones in saliva: mode of entry and consequent implications for clinical interpretation. Clinical Chemistry. 29, 1752–1756.

Weir, T.B., Smith, C.C.T., Round, J.M., Betteridge, D.J. 1986. Stability of catecholamines in whole blood, plasma, and platelets. Clinical Chemistry. 35, 882–883.
Williams, K. 2006. http://info.med.yale.edu/ycc/research/proteomics.html. Yale-New Haven Hospital, Yale University. (Vol. 2006): Yale University, School of Medicine, New Haven, CT. 2006-02-02.
Wooldridge, K.G., Williams, P.H. 1993. Iron uptake mechanisms of pathogenic bacteria. FEMS Microbiology Reviews. 12, 325–348.

Chapter 8
Staphylococci, Catecholamine Inotropes and Hospital-Acquired Infections

Primrose P.E. Freestone, Noura Al-Dayan, and Mark Lyte

8.1 Introduction: Nosocomial Infections

Patients in hospital intensive care units (ICU) are at particular risk of developing infections from bacteria, fungi, and viruses acquired from within the hospital environment; such infections are also often multispecies in their presentation (Vincent et al. 1995). Surveys on the incidence of ICU patients developing nosocomial infections during their hospital stay have produced varying figures, but it is generally agreed that over 1:5 patients will pick up a hospital-acquired infection, resulting in extended stays in intensive care beds, increased overall patient morbidity and, of course, avoidable and greater economic outlays (Vincent et al. 1995). The main patient-associated risk factors for the development of nosocomial infections are recognized as coming from lowered physical fitness due to severe prior illness, accidental or intentional tissue trauma such as surgery, reduction in immune competence (which can be medication-induced), or colonization by infectious microbes. Other factors recognized as leading to increased risk of infection involve invasive treatment procedures, particularly use of endotracheal tubes (with mechanical ventilation), urinary catheters, surgical drains, and intravascular catheters.

An additional factor that might also predispose acutely ill patients to development of potentially life-threatening infections has recently been recognized as coming from the medications they are given. Surveys of drug usage within ICUs indicate patients may receive up to 20 medications during their stay, with up to half of these patients receiving catecholamine inotrope (dopamine, epinephrine, norepinephrine, dobutamine, isoprenaline) support to maintain renal and cardiac function

P.P.E. Freestone (✉) and N. Al-Dayan
Department of Infection, Immunity and Inflammation, School of Medicine,
University of Leicester, University Road, Leicester, LE1 9HN, UK
e-mail: ppef1@le.ac.uk

M. Lyte
Department of Pharmacy Practice, School of Pharmacy, Texas Tech University
Health Sciences Center, 3601 4th Street, MS 8162, Lubbock, TX 79430, USA

M. Lyte and P.P.E. Freestone (eds.), *Microbial Endocrinology*,
Interkingdom Signaling in Infectious Disease and Health,
DOI 10.1007/978-1-4419-5576-0_8, © Springer Science+Business Media, LLC 2010

(Smythe et al. 1993). Work from our laboratories has shown that all of these drugs are able to markedly increase the growth and infectivity of pathogenic and commensal bacteria (see Chap. 3) by rendering blood and serum markedly less bacteriostatic, and by direct effects upon the bacteria themselves, such as enhanced expression of virulence factors (Freestone et al. 2008). Surveys of the types of bacteria causing the majority of nosocomial infections have identified the following as most prevalent: *Staphylococcus aureus* (30%), *Pseudomonas aeruginosa* (29%), coagulase-negative staphylococci such as *Staphylococcus epidermidis* (19%), yeasts such as *Candida* (17%), endogeneous enterics such as *Escherichia coli* (13%), Gram-positive enterococci (12%), *Acinetobacter* (9%), and *Klebsiella* (8%) (Spencer 1996, Vincent et al. 1995). Interestingly, all of the bacterial species shown are responsive to the catecholamine inotropes (Freestone et al. 1999, 2000, 2002, 2008; Freestone and Lyte 2008; Neal et al. 2001). Other chapters in this book examine catecholamine interactions with Gram-negative bacteria in some detail, and since the staphylococci are statistically the most prevalent pathogens in the ICU, they will form the focus of the remainder of this chapter.

8.2 Historical Evidence Suggesting a Role for Microbial Endocrinology in Infectious Diseases of the Acutely Ill

Catecholamine inotropes are routinely employed in the critical care setting to support heart function through their ability to increase cardiac contractility (see Chap. 3 for the structures of the catecholamine inotropes). For example, the inotropic agent, dobutamine is used in the treatment of congestive heart failure, epinephrine for the treatment of anaphylactic shock and dopamine to support renal function. Clinical evidence of the role of catecholamines, both endogenous and administered in the infectious disease process, has only been appreciated in the last decade or so, even though evidence of its existence has been around for nearly 80 years (see Chap. 1 for a fuller history of the field of microbial endocrinology and its relevance to medical practice). The ability of neuroendocrine hormones to influence the in vivo growth of pathogenic bacteria was first observed in 1930 (Renaud and Miget 1930). Prior to the advent of disposable syringes, metal needles and glass syringes were reused constantly between patients with only a cursory cleaning in alcohol. Patient to patient transmission of infectious disease was frequently encountered due to the inadequate alcohol treatment of syringe needles, which could only marginally kill actively growing (vegetative) bacterial cells, but not bacterial spores. As is well understood today, certain vegetative bacteria such as *Clostridium perfringes*, the causative agent of gas gangrene, can undergo sporulation. Such spores, which are formed from the vegetative cells under conditions of nutritional deprivation, are totally resistant to alcohol treatment and can only be killed by autoclaving.

From the 1930s onward, reports associating the use of contaminated needles with administration of epinephrine solutions in the development of rapidly disseminating infections began to increase in number (Brocard 1940; Cooper 1946; Evans et al. 1948). A previously used syringe needle to treat a gas gangrene patient was then used to administer epinephrine to a patient for urticaria. Within 6 h, a fatal fulminating gas gangrene infection developed. These reports noting the rapidity of infectious spread in patients receiving epinephrine injections with contaminated needles led A.A. Miles in 1948 (Evans et al. 1948) to begin a series of experiments examining the role of catecholamines in bacterial pathogenesis. In these experiments, the ability of epinephrine to modulate the in vivo growth of both Gram-positive and Gram-negative bacteria in a guinea pig model was conclusively demonstrated in tissue slices with enhancement of growth of bacteria coinjected with epinephrine that was log orders greater than that for control slices coinjected with saline. It should be noted that norepinephrine was not investigated. The authors concluded that the ability of epinephrine to dramatically enhance bacterial growth was due to some protective coating of the bacteria by epinephrine or an epinephrine-induced inhibition of immune cell function. The testing of each of these possible mechanisms, however, met with failure (Evans et al. 1948). Significantly, at no time did these authors or others suggest that the action of epinephrine on bacterial growth was due to a direct, nonimmune effect as is currently proposed (Lyte 1992, 1993). These references had not been previously recognized since they occurred before 1966 and as such are not referenced by Medlars or other computer-based bibliographic retrieval services. Interestingly, one of the frequently used techniques by microbiologists to enable gas gangrene infections to "take" in mice has been the coinjection of epinephrine along with *C. perfringes*. It can reasonably be inferred that this practice dates back to the decades old reports described earlier.

A clearer realization that the catecholamine might be directly interacting with the infectious agents rather than aspects of host immunity were not made, and it took more than 20 years more before even an indirect association between plasma levels of catecholamine stress hormones and infectious disease episodes was realized. This came about when Gruchow (1979) observed that infectious disease episodes in patients had a tendency to occur after medical procedures that led to increased levels of catecholamines, while Groves et al. (1973) reported that postoperative patients who developed acute septic states had higher levels of epinephrine and norepinephrine in their blood than patients who experienced straightforward non-infected postoperative recoveries.

Catecholamine levels within the human body are tightly regulated and plasma clearance is usually rapid in the healthy, resulting in normal circulatory levels in the nanomolar range (Goldstein et al. 2003). However, catecholamine concentrations in patients receiving inotropic support can be several orders of magnitude higher. For example, dopamine is typically infused intravenously into acutely ill patients over a concentration range of 1–15 µg/kg/min (British National Formulary 2009). Dopamine has a half life of several minutes (Goldstein et al. 2003), and steady state levels in plasma vary according to infusion levels and general metabolic fitness, with acutely ill patients showing slower elimination rates. There is therefore a wide variation in dopamine plasma concentrations in patients receiving inotrope supplementation, ranging from ~50 nM (Johnston et al. 2004) to nearly 5,000 nM (5 µM)

(Girbes et al. 2000). We have used in vitro analyses of bacteria-catecholamine interactions to show that μM concentrations of dopamine and norepinephrine are high enough to induce bacterial growth in serum and blood based media. Once within the circulation, dopamine also undergoes enzymatic conversion to norepinephrine, and metabolism of both leads to a range of compounds including dihydroxyphenylacetic acid (which can exist in plasma at 50 times the level of dopamine), dihydroxymandelic acid, and dihydroxyphenylglycol (Goldstein et al. 2003). In analyses of bacterial interactions with inotropic agents, we have found that dihydroxyl-containing intermediates of dopamine and norepinephrine (which include dihydroxymandelic acid, and dihydroxyphenylglycol), though pharmacologically inactive, still retain their ability to induce bacterial growth to a level comparable with the original catecholamine (Neal et al. 2001; Freestone et al. 2002). Catecholamine and metabolite effects can be additive (our unpublished data), and consideration of catecholamine-levels alone may not fully reveal the possible effects of metabolized inotropic agents on any bacteria that come into contact with them. The combinational effects of endogeneous catecholamines plus administered catecholamines plus their metabolites could result in changes in the blood that lead to the proliferation of infectious bacteria.

8.3   Staphylococcal Infections in the ICU

Most infections caused by the staphylococci are due to *S. aureus*, an often antibiotic resistant microbe that possesses a diverse array of virulence factors. However, in recent years and in correlating with increase usage of invasive medical procedures, the more opportunistic pathogen coagulase-negative staphylococci (C-NS), and in particular *S. epidermidis*, have become one of the most important causes of nosocomial infections (Huebner and Goldman 1999; Geffers et al. 2003). The C-NS constitute a major component of the skin microflora and were for many years regarded as saprophytes, or at least as organisms with no or low virulence. However recently, the C-NS, in particular *S. epidermidis*, have become recognized as serious nosocomial pathogens associated with indwelling medical devices such as catheters and prosthetic joints Surveys have shown that of the 5 million intravascular catheters employed each year, approximately 250,000 CVC-related bloodstream infections are reported with an attributable mortality of up to 25%. The microbes most commonly isolated from indwelling medicals devices are the C-NS. The increasing incidence of infections caused by these normally commensal bacteria is largely due to their affinity for the biomaterials of the invasive technologies integral to modern medicine. In association with appropriate biomaterial surfaces, such as those ranging from polysilicone in catheters to steel in hip replacements, *S. epidermidis* and other C-NS adhere and proliferate to form biofilms, highly complex structures that represent functional communities of microbes. The site of insertion through the skin is one of the most common sources of bacterial contamination of intravenous catheters. Skin commensals are thought to migrate along the external surface of the catheter and colonize the intravascular tip eventually forming a biofilm. Bloodstream infections can occur if sections of the biofilm shear off and bacteria are flushed into the circulation. In a biofilm mode of existence, bacteria are protected against attack from both the immune system and antibiotic treatment, and infections by such organisms are particularly difficult to eradicate.

8.4 Catecholamine Inotropes Induce Staphylococcal Growth and Biofilm Formation on Intravascular Catheters

The requirement for iron in growth is recognized for the vast majority of pathogenic and commensal bacteria including the staphylococci (Ratledge and Dover 2000). Under normal conditions, the human body seeks to severely limit the availability of free iron to approximately 10^{-18} M through the production of the iron sequestering transferrin (blood) and lactoferrin (mucosal secretions and the gastrointestinal tract), to a concentration of iron that is below the level required to support the growth of bacteria. In an effort to obtain iron from host iron-binding proteins such as transferrin, bacteria have developed a variety of mechanisms including ferric reductase, transferrin and lactoferrin binding proteins, and ferric iron sequestering siderophores (Ratledge and Dover 2000). The importance of iron as a determinant of *S. epidermidis* biofilm formation has been demonstrated in studies that have employed iron restriction. For example, staphylococcal strains of *S. epidermidis* which were initially biofilm negative, were induced to form biofilms when grown in iron-starved medium (Deighton and Borland 1993), and iron limitation generally induces slime production by the staphylococci (Baldassarri et al. 2001).

Work from our laboratories has shown that catecholamine inotropic drugs may have additional side effects to those recognized pharmacologically. We have shown that they can serve as an etiological factor in the staphylococcal colonization of indwelling medical devices due to their ability to stimulate *S. epidermidis* growth and biofilm formation (Freestone et al. 1999; Neal et al. 2001; Lyte et al. 2003). The exposure of less than 100 CFU per ml of *S. epidermidis* to pharmacologically relevant concentrations of dopamine, dobutamine, and norepinephrine resulted in only 24 h of a greater than 10,000-fold increased growth in serum and blood-based media. The mechanism of this growth enhancement involved the drugs enabling the staphylococci access to normally inaccessible transferrin-sequestered iron (Freestone et al. 2000; Neal et al. 2001; Lyte et al. 2003; Freestone et al. 2008) (see also Chap. 3). Importantly, as well as inducing growth in normally bacteriostatic host tissue fluids, inotropes can also enhance other aspects of C-NS physiology highly relevant to their presentation in the ICU, specifically biofilm formation. This is demonstrated in Fig. 8.1, which shows a series of scanning electron micrographs of the biofilm formation that occurs when less than 100 *S. epidermidis* cells were seeded onto intravenous catheter grade polystyrene in the presence of tissue fluids they will encounter in vivo (plasma) and incubated without (a) and with (b–f) concentrations of norepinephrine that would be administered via the intravenous catheter. As is

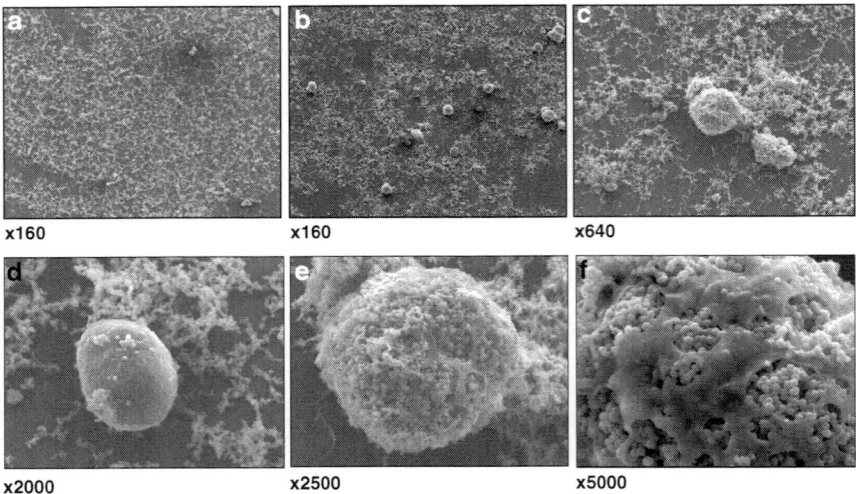

Fig. 8.1 Catecholamine-induced biofilm formation in *Staphylococcus epidermidis*. The image panels show scanning electron micrographs of biofilms of *S. epidermidis* adhering to polystyrene after overnight growth in freshly prepared plasma-SAPI medium in the absence (**a**) or presence (**b**–**f**) of 0.1 mM norepinephrine as described in Lyte et al. (2003). Initial inocula for both *S. epidermidis* cultures were approximately 10^2 CFU per ml. Higher magnification scanning electron micrographs showing details of the bacteria-exopolysaccharide clusters are shown in panels (**b**)–(**f**). This figure was taken with permission from Freestone and Lyte (2008)

clearly visible, exposure of *S. epidermidis* to the inotrope led to massive increases in biofilm formation and production of exopolysaccharide (Lyte et al. 2003). The increasing magnification shows the mushroom like structures of bacteria and exopolysaccharide, all of which became evident in less than 48 h.

The findings demonstrated in Fig. 8.1 are clinically significant, as bacterial colonization of intravenous catheters, predominantly by the C-NS, is recognized as the most common infection encountered in the intensive care setting. There are also a number of methodological issues in the data shown in Fig. 8.1 that are worth noting. In demonstrating catecholamine inotrope induction of *S. epidermidis* biofilm, we attempted to make the analytical conditions close to those likely to be occurring in the clinical scenario. We therefore used two methodological aspects that were markedly different from the previous studies of bacterial biofilm formation. Only 10–100 *S. epidermidis* cells were used to seed the polystyrene plastic section on which the biofilm was to be established, a low inoculum chosen to reflect the number of bacteria likely to be encountered at the beginning of the catheter colonisation. This is in contrast to many prior biofilm studies which have used several log orders higher levels of bacteria to establish biofilms. The second important difference was the use of culture conditions that employed a plasma supplemented minimal media. This again differs from studies which have used rich microbiological media to study biofilm formation, and in so doing, it may be argued, do not realistically reflect in vivo conditions in which host factors present in plasma, which are recognized to play a role in initial bacterial adhesion, are not present.

8.5 Catecholamine Inotropes Can Resuscitate Antibiotic Damaged Staphylococci

Intravenous catheter-related bloodstream infections are invariably associated with increases in length of time in intensive care units and of course hospital costs (Crnich and Maki 2001). In an effort to combat bacterial colonization of catheter lines and any possible subsequent progression to catheter-related bloodstream infections, the use of antimicrobial impregnated catheters, particularly those incorporating antibiotics such as rifampin and minocycline, is becoming increasingly adopted in the clinical care setting, However, there is a measure of doubt concerning the real efficacy of antibiotic-impregnated plastics in the prevention of catheter-related blood stream infections (Crnich and Maki 2004; McConnell et al. 2003). Concern exists as to whether they may also contribute to the emergence of antibiotic-resistant nosocomial pathogens (Sampath et al. 2001). A recent study from our laboratory might explain why antibiotics impregnation of the polymers used to construct intravenous catheter lines has not proven to be as effective as anticipated.

In this investigation (Freestone et al. 2008), we undertook two methodological approaches using norepinephrine and dopamine to investigate whether these inotropes were capable of facilitating the recovery and growth of antibiotic-damaged staphylococci (S. epidermidis, S. haemolyticus and two S. aureus strains, Newman and 832-4). We employed the antibiotics rifampin and minocycline as our test antimicrobials, as they are the two most frequently employed antibiotics used to coat catheters, and carried out the analyses in a serum-based culture medium (serum-SAPI) that more closely approaches the environment the bacteria would have experienced in vivo. The first experimental approach used a minimum inhibitory concentration (MIC) assay, in which the staphylococci were incubated with norepinephrine and dopamine in the presence of increasing concentrations of rifampin or minocycline (up to 100 times the MIC for each antibiotic). Representative data for rifampin is shown in Fig. 8.2 for S. epidermidis (a) and S. aureus strain Newman (b). The responses for S. haemolyticus and S. aureus strain 832-4 were very similar to those shown for S. epidermidis and S. aureus Newman and so are not shown. As can be seen, for both strains simultaneous administration of inotrope and rifampin allowed more than 2 logs greater growth in the presence of rifampin concentrations around the MIC (determined to be ~0.1 µg per ml), although the catecholamines did not significantly protect the staphylococci from the inhibitory effects of higher rifampin concentrations. A similar response profile of greater growth in the presence of the inotropes was also obtained when the staphylococci were exposed to minocycline in the presence/absence of catecholamine inotropes (data not shown).

Our second experimental approach involved analyzing whether catecholamine inotropes could rescue the growth of staphylococci pretreated with antibiotics,

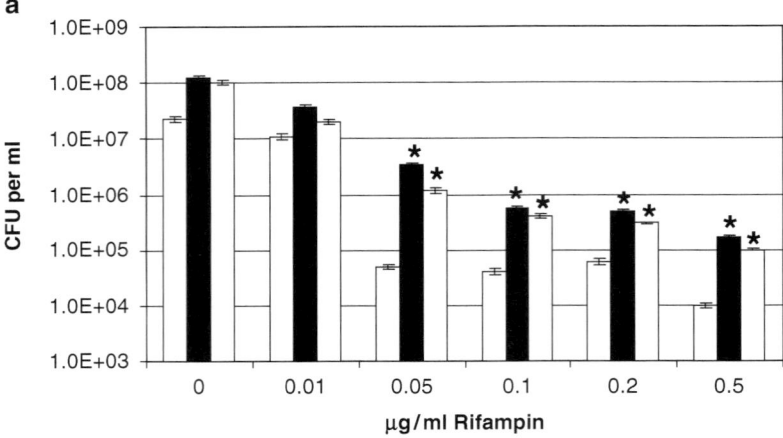

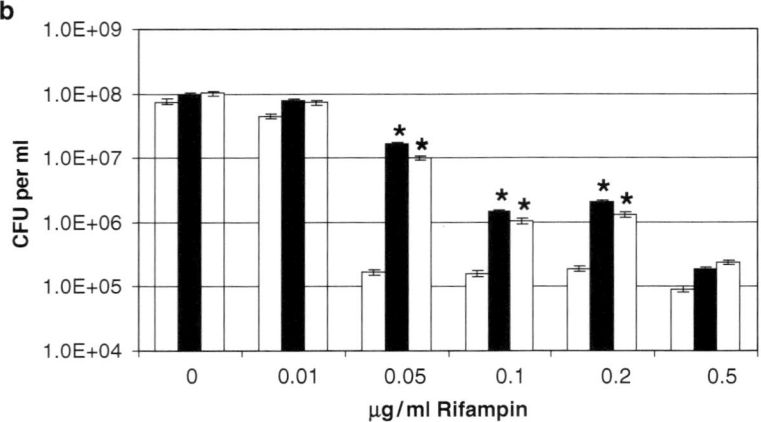

Fig. 8.2 Staphylococcal response to increasing rifampin concentrations in the presence of cate-cholamine inotropes. Analyses of simultaneous exposure of *S. epidermidis* and *S. aureus* to anti-biotics and catecholamine inotropes were performed in serum-SAPI medium. Bacteria were incubated for 24 h 37°C in a 5% CO_2 humidified incubator and measured for growth and measured for growth using plate analysis. The results shown represent the mean of triplicate cultures; standard error of the mean was not greater than 7% for all cultures shown. The asterisk, *, indicates a statistically significant increase in growth level over the corresponding non-inotrope supplemented control culture ($P < 0.0001$). (**a**) Growth response of *S. epidermidis* (inoculum size 1×10^6 CFU per ml); (**b**) Growth response of *S. aureus* Newman (inoculum size 0.92×10^6 CFU per ml). *White bar*, no additions (control); *black bar*, 100 μM norepinephrine; *gray bar*, 100 μM dopamine. This figure was taken with permission from Freestone et al. (2008)

as it is well known that following antibiotic treatment, some acute bacterial infections can become subclinical, re-activating in response to not always understood changes in the host environment. Figure 8.3 shows the ability of norepinephrine and dopamine to rescue growth of *S. epidermidis* pre-treated for 4 h with a rifampin concentration

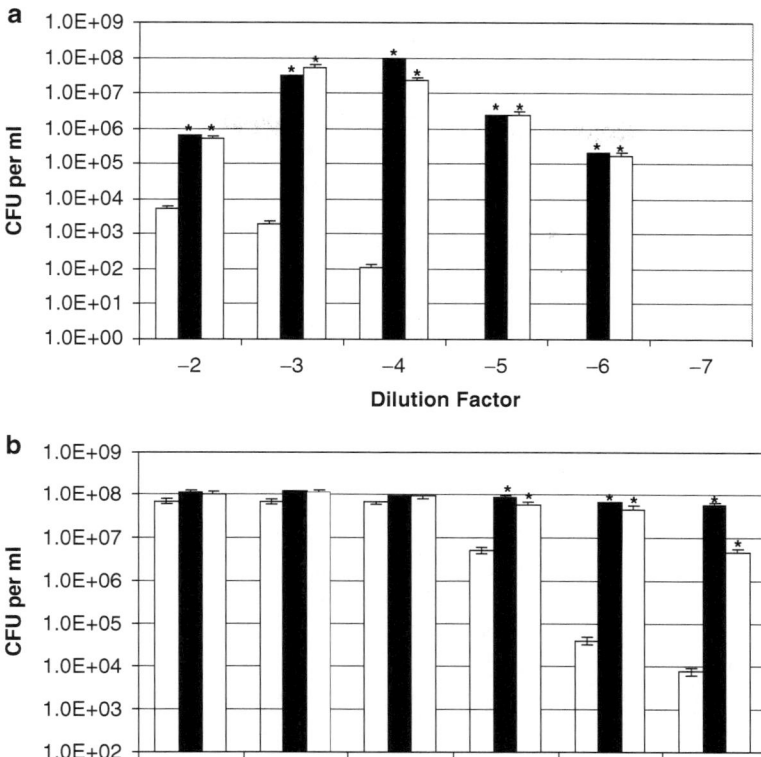

Fig. 8.3 Effect of catecholamine inotropes on coagulase-negative staphylococci pre-exposed to rifampin. Replicates of exponentially growing *S. epidermidis* cultures grown in serum-SAPI medium were incubated for 4 h at 37°C with 5 μg per ml rifampin (100 times MIC). The antibiotic-treated (inoculum size 5.40×10^4 CFU per ml) (**a**) and control cultures (inoculum size 1.05×10^8 CFU per ml) (**b**) were each then serially diluted in tenfold dilution steps into fresh serum-SAPI with no additions (*white bar*) or supplemented with norepinephrine (*black bar*) or dopamine (*grey bar*) each at a concentration of 100 μM. Test and control were incubated at 37°C for 24 h and enumerated for viable cell levels using pour-plate analysis. The asterisk, *, indicates a statistically significant increase in growth level over the corresponding non-catecholamine inotrope supplemented control culture ($P < 0.0001$). This figure was taken with permission from Freestone et al. (2008)

of 5 μg per ml (approximately 100 times the MIC) and then serially diluted into serum-SAPI medium containing no additions or the catecholamine inotropes and grown on for 24 h. Figure 8.3a, b show the growth profiles of control and antibiotic treated *S. epidermidis,* and reveal that the rifampin had reduced the viable count of the treated bacteria by nearly 4 log orders. However, the presence of the inotropes significantly increased the growth of *S. epidermidis* preexposed to rifampin, even when the antibiotic carryover was near to the MIC (the 10^{-2} dilution) ($P < 0.0001$).

Earlier work from our laboratories had shown that *S. aureus* and other coagulase-positive staphylococcal strains showed little growth responsiveness to catecholamines in serum-based medium (Freestone et al. 1999; Neal et al. 2001). This observation was confirmed by the responses of nonantibiotic-treated *S. aureus* strain Newman to norepinephrine and dopamine (Fig. 8.4b), which showed that growth enhancement by the inotropes was evident at only very low cell densities (~10 CFU per ml). However, antibiotic-treatment of the *S. aureus* cultures (Fig. 8.4a) caused the bacteria to become significantly more responsive to the inotropes ($P<0.0001$). This being so, the growth profile observed over the *S. aureus* culture dilutions is different to that obtained for *S. epidermidis* (compare b of Figs. 8.3 and 8.4). This is most likely because unlike the C-NS, the coagulase positive staphylococci possess highly efficient acquisition systems for the scavenging of host sequestered iron stores, which then enables them to overcome the iron-limitation of tissue fluids such as serum and blood without aid of the catecholamines. Indeed, when we analyzed the ability of *S. epidermidis* and *S. aureus* to remove iron from ^{55}Fe-labeled transferrin, we found that the coagulase positive staphylococci were able to extract around 10 times more iron than the C-NS. Adding the inotrope enabled the coagulase positive staphylococci to obtain around 2.5-3 times more iron from the transferrin while for the C-NS, the inotrope allowed much more Fe uptake from transferrin – around 30 times more than the noninotrope supplemented controls, such that when the inotrope were present, the final levels of uptake of transferrin-complexed iron was similar between the staphylococci. Figure 8.5 shows the ability of inotropes to rescue growth of staphylococci pretreated with 2 µg per ml minocycline (~100 times MIC). Although both *S. epidermidis* and *S. aureus* recovered better from the minocycline assault, than they did from the rifampin, inclusion of norepinephrine and dopamine still induced a much greater rate of recovery.

8.6   Fighting Back: Blockade of Staphylococcal Catecholamine Responsiveness

Previously we had shown that blockade of catecholamine growth responsiveness in enteric bacteria was possible using drugs employed therapeutically as catecholamine receptor antagonists (Freestone et al. 2007). To determine if we could similarly inhibit staphylococcal responses to the inotropes, we examined the ability of a range of α and β-adrenergic and dopaminergic antagonists to block *S. epidermidis* responses to norepinephrine and dopamine. As can be seen in Table 8.1, the β-adrenergic receptor antagonist propranolol had no effect on the ability of either catecholamine to stimulate growth of *S. epidermidis*. Other β-adrenergic antagonists such as labetalol, atenolol, and yohimbine also had no effect (data not shown). However, the α-adrenergic antagonists phentolamine and prazosin (data not shown) were able to inhibit growth induction by norepinephrine by over three log-orders.

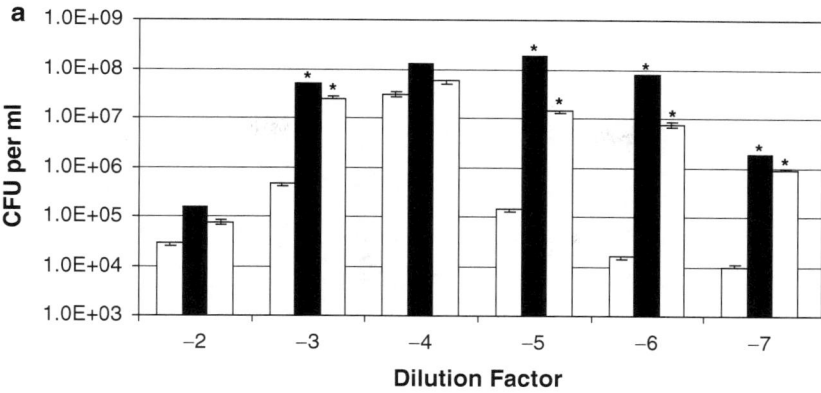

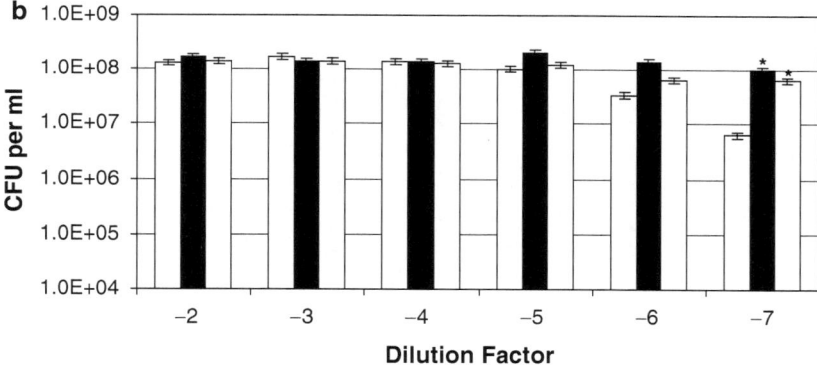

Fig. 8.4 Effect of catecholamine inotropes on coagulase-positive staphylococci pre-exposed to rifampin. Replicates of an exponentially growing *S. aureus* strain Newman cultures grown in serum-SAPI medium were incubated for 4 h at 37°C with 5 μg per ml rifampin (100 times MIC). The antibiotic-treated (inoculum size 2.21×10^5 CFU per ml) (**a**) and control cultures (inoculum size 2.22×10^8 CFU per ml) (**b**) were each then serially diluted in tenfold dilution steps into fresh serum-SAPI with no additions (*white bar*) or supplemented with norepinephrine (*black bar*) or dopamine (*grey bar*) each at a concentration of 100 μM. Test and control were incubated at 37°C for 24 h and enumerated for viable cell levels using pour-plate analysis. The asterisk, *, indicates a statistically significant increase in growth level over the corresponding non-catecholamine inotrope supplemented control culture ($P < 0.0001$). This figure was taken with permission from Freestone et al. (2008)

None of the α- or β-antagonists when tested alone induced growth of *S. epidermidis*, even at 500 μM. Furthermore, addition of Fe overcame the antagonist blockade of growth induction (Table 8.1), indicating that growth inhibition by the α-adrenergic receptor antagonists was not due to any cellular toxicity of the antagonist, but instead represents a specific antagonism of the staphylococcal response to the catecholamines.

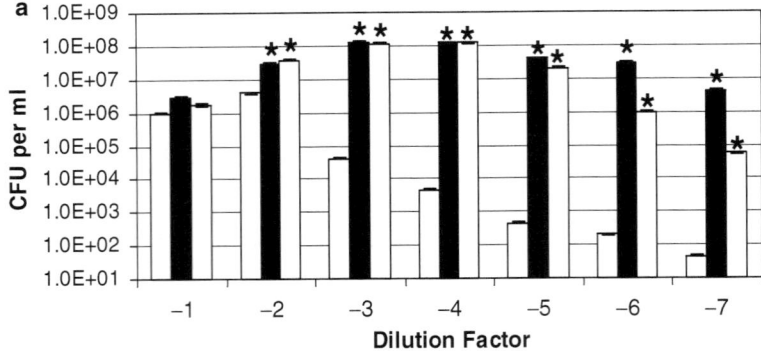

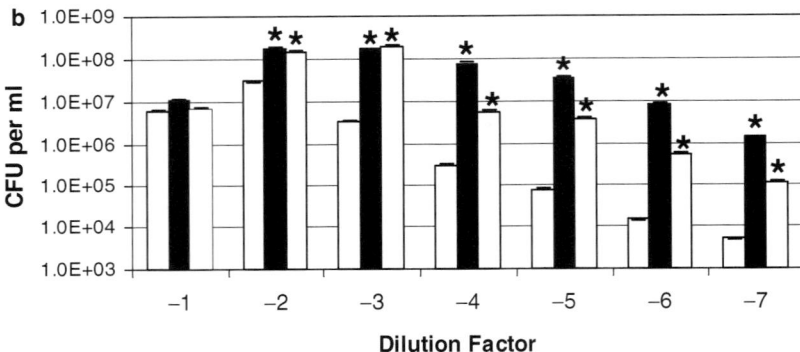

Fig. 8.5 Effect of catecholamine inotropes on staphylococci pre-exposed to minocycline. Replicates of an exponentially growing *S. epidermidis* and *S. aureus* strain Newman cultures grown in serum-SAPI medium were incubated for 4 h at 37°C with 2 μg per ml minocycline (100 times MIC). The antibiotic treated *S. epidermidis* culture (inoculum size 5.40×10^4 CFU per ml) (**a**) and *S. aureus* Newman culture (inoculum size 8.90×10^4 CFU per ml) (**b**) were each then serially diluted in tenfold dilution steps into fresh serum-SAPI with no additions (*white bar*) or supplemented with norepinephrine (*black bar*) or dopamine (*grey bar*) each at a concentration of 100 μM. The two sets of cultures were then incubated at 37°C for 24 h and enumerated for viable cell levels using pour-plate analysis. The asterisk, *, indicates a statistically significant increase in growth level over the corresponding non-catecholamine inotrope supplemented control culture ($P < 0.0001$). This figure was taken with permission from Freestone et al. (2008)

The next question we asked was whether we could use phentolamine and chlorpromazine to prevent the inotrope from rescuing the growth of antibiotic-treated staphylococci. Figure 8.6 shows the effect of including the antagonists in the catecholamine resuscitation experiments described in Figs. 8.3 and 8.4. As seen in Fig. 8.6, the α-adrenergic and dopaminergic antagonists were able to specifically block norepinephrine and dopamine resuscitation of rifampin-damaged *S. epidermidis* and *S. aureus*. Further work showed that the inotrope blocking activity of a single addition of the antagonists was retained for at least 3 days for both strains

Table 8.1 Blockade of catecholamine-induced staphylococcal growth responsiveness

Antagonist	CA	Antagonist concentration (µM)										
		0	0.1	1	10	20	50	75	100	200	300	N/A[b]
Phentolamine	NE	8.01[a]	7.94	7.86	6.90	6.08	5.15	4.90	4.60	3.75	3.75	3.90
	NE+Fe[c]	8.21	8.13	8.13	8.13	8.06	8.06	8.04	8.15	8.09	7.99	
	Dop	8.01									7.96	
Propranolol	NE	8.02	8.02	7.99	8.03	7.68	7.76	7.91	7.90	7.81	7.60	4.00
	Dop	8.01									7.95	
Chlorpromazine	Dop	8.00	7.98	7.98	7.81	7.34	6.92	6.66	6.35	5.27	4.81	4.03
	Dop+Fe[c]	8.06	8.08	8.08	8.16	8.11	8.08	8.08	8.08	8.11	8.06	
	NE	8.03									7.81	

S. epidermidis was inoculated at 10^2 CFU per ml into serum–SAPI medium containing the catecholamine (CA) plus the concentrations of antagonists shown in the table, incubated statically for 24 h 37°C in a 5% CO_2 humidified incubator and measured for growth ([a]expressed as $\log^{-10}$ CFU per ml) using pour plate counts. Growth levels of non-catecholamine supplemented cultures ([b]N/A) are shown for comparison purposes. Norepinephrine (NE) and dopamine (Dop) were both used at 50 µM and Fe ($Fe(NO_3)_3$) at 100 µM. Results shown are representative data from at least three separate experiments; all data points showed variation of no more than 5% Data table was taken with permission from Freestone et al. (2008) [c]Note that Fe was included to show that the effects of the antagonists were not due to toxicity of the compounds.

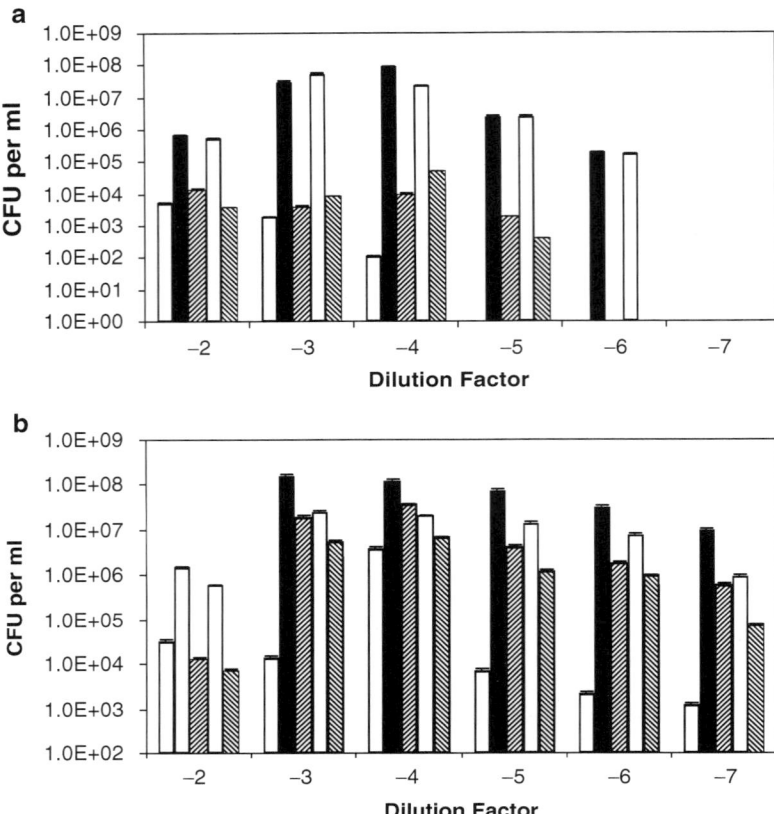

Fig. 8.6 Effects of adrenergic and dopaminergic antagonists on the ability of catecholamines to rescue of growth of antibiotic-stressed staphylococci. Replicates of exponentially growing cultures of the bacteria shown were cultured in serum-SAPI medium and then pre-incubated with 5 μg per ml rifampin as described in the legends to Fig. 8.3, and similarly further processed except that additional catecholamine assays containing norepinephrine were supplemented with 200 μM phentolamine and dopamine supplemented with 200 μM chlorpromazine. Test and non-catecholamine inotrope supplemented controls were incubated at 37°C for 24 h, and enumerated for viable cell levels using pour-plate analysis. Viable counts of the inoculating antibiotic treated *S. epidermidis*, and *S. aureus* Newman cultures were 2.21×10^5 and 7.8×10^5 CFU per ml respectively. The values shown represent means of triplicate plate counts; standard error of the mean was not greater than 7% for all cultures shown. (**a**) *S. epidermidis*, (**b**) *S. aureus White bar* no additions (control), *black bar* 100 μM norepinephrine, *right diagonal hatch* 100 μM norepinephrine plus 200 μM phentolamine, *gray bar* 100 μM dopamine, *left diagonal hatch* 100 μM dopamine plus 200 μM chlorpromazine. This figure was taken with permission from Freestone et al. (2008)

(data not shown). For *S. aureus* (Fig. 8.6b), the antagonists were less potent, and at best phentolamine and chlorpromazine were able to reduce norepinephrine and dopamine resuscitation by at most a log order in magnitude.

8.7   Future Thoughts: Side Effects and Covert Side Effects

Until recently, side effects of drugs used in clinical practice have been evaluated in a somewhat limited way when considering their effects upon microorganisms. This perspective may have to change, as the evidence increases that certain nonantibiotic compounds, the catecholamine inotropes principally but also other administered drugs such as steroids and opioids may directly increase bacterial infectivity (Neal et al. 2001; Zaborina et al. 2007). Humans are the residence of up to 10^{14} highly diverse bacteria, fungi, viruses, and protozoa, and since the advent of microbial endocrinology, it is perhaps no longer surprising that changes in the hormonal milieu of the host do not go un-remarked by these prokaryotic coinhabitants. Singer (2007) has suggested that inotrope effects on bacteria such be viewed as 'covert side effects'. Given the potentially devastating consequences of the covert side effects of the catecholamine inotropes class of drugs, there is a need for further research into whether any of the other widely used pharmaceuticals can also affect infectivity of microbes through neuro-homormonal-mediated interactions that may play a part in nosocomial infections. A final thought is that it might become necessary is to extend the patient side effects information on drug labels to also include effects on the microbes inhabiting the patient.

References

Baldassarri, L., Bertuccini, L., Ammendolia, M. G., Arciola, C. R., and Montanaro, L. 2001. Effect of iron limitation on slime production by *Staphylococcus aureus*. Eur J Clin Microbiol Infect Dis 20, 343–345.

Brocard, H. 1940. Production, chez le gobaye, de phlegmons gazeux mortels par l'injection simultanee d' adrenaline et de cultures de *Bacillus perfringes* avirulent. C R Seances Soc Biol Fil 134, 567–569.

British National Formulary, 2009. http://www.bnf.org.

Cooper, E. 1946. Gas gangrene following injection of adrenaline. Lancet i, 459–461.

Crnich, C. J., and Maki, D. G. 2001. The role of intravascular devices in sepsis. Curr Infect Dis Rep 3, 496–506.

Crnich ,C. J., and Maki, D. G. 2004. Are antimicrobial-impregnated catheters effective? Don't throw out the baby with the bathwater. Clin Infect Dis 38, 1287–1292.

Deighton, M., and Borland, R. 1993. Regulation of slime production in *Staphylococcus epidermidis* by iron limitation. Infect Immun 61, 4473–4479.

Evans, D. G., Miles, A. A., and Niven J. S. F. 1948. The enhancement of bacterial infections by adrenaline. Br J Exp Pathol 29, 20–39.

Freestone, P. P. E., Haigh, R. D., Williams, P. H., Lyte, M. 1999. Stimulation of bacterial growth by heat-stable, nor-induced autoinducers. FEMS Microbiol Lett 172, 53–60.

Freestone, P. P., Williams, P. H., Haigh, R. D., Maggs, A. F., Neal, C. P., and Lyte, M. 2002. Growth stimulation of intestinal commensal *Escherichia coli* by catecholamines: a possible contributory factor in trauma-induced sepsis. Shock 18, 465–470.

Freestone, P. P., Haigh, R. D., and Lyte, M. 2007. Blockade of catecholamine-induced growth by adrenergic and dopaminergic receptor antagonists in *Escherichia coli* O157:H7, *Salmonella enterica* and *Yersinia enterocolitica*. BMC Microbiol 7, 8.

Freestone, P. P., and Lyte, M. 2008. Microbial endocrinology: experimental design issues in the study of interkingdom signalling in infectious disease. Adv Appl Microbiol 64, 75–105.

Freestone, P. P. E., Haigh, R. D., and Lyte, M. 2008. Catecholamine inotrope resuscitation of antibiotic-damaged staphylococci and its blockade by specific receptor antagonists. J Infect Dis 197, 2044–1052.

Geffers, C., Zuschneid, I., Eckmanns, T., Ruden, H., and Gastmeier, P. 2003. The relationship between methodological trial quality and the effects of impregnated central venous catheters. Intensive Care Med 29, 403–409.

Girbes, A., Pattern, M., McCloskey, B., Groeneveld, A., and Hoogenberg, K. 2000. The renal and neurohumoral effects of the addition of low-dose dopamine in septic critically ill patients. Intensive Care Med 26, 1685–1689.

Goldstein, D. S., Eisenhofer, G., and Kopin, I. J. 2003. Sources and significance of plasma levels of catechols and their metabolites in humans. J Pharmacol Exp Ther 305, 800–811.

Groves, A. C. et al. 1973. Plasma catecholamines in patients with serious postoperative infection. Ann Surg 178, 102–107.

Gruchow, H. W. 1979. Catecholamine activity and infectious disease episodes. J. Human Stress 5, 11–17.

Huebner, J., and Goldman, D. A. 1999. Coagulase-negative staphylococci: Role as pathogens. Ann Rev Med 50, 223.

Johnston, A. J., Steiner, L. A., O'Connell, M., Chatfield, D. A., Gupta, A. K., and Menon, D. K. 2004. Pharmacokinetics and pharmacodynamics of dopamine and norepinephrine in critically ill head-injured patients. Intensive Care Med 30, 45–50.

Lyte, M. 1992. The role of catecholamines in gram-negative sepsis. Med Hypotheses 37, 255–258.

Lyte, M. 1993. The role of microbial endocrinology in infectious disease. J Endocrinol 137, 343–345.

Lyte, M., Freestone, P. P., Neal, C. P., Olson, B. A., Haigh, R. D., Bayston, R., and Williams, P. H. 2003. Stimulation of *Staphylococcus epidermidis* growth and biofilm formation by catecholamine inotropes. Lancet 361, 130–135.

McConnell, S. A., Gubbins, P. O., and Anaissie, E. J. 2003. Do antimicrobial-impregnated central venous catheters prevent catheter-related bloodstream infection? Clin Infect Dis 37, 65–72.

Neal, C. P., Freestone, P. P., Maggs, A. F., Haigh, R. D., Williams, P. H., and Lyte, M. 2001. Catecholamine inotropes as growth factors for *Staphylococcus epidermidis* and other coagulase-negative staphylococci. FEMS Microbiol Lett 194, 163–169.

Ratge, D., Steegmuller, U., Mikus, G., and Kohse, K. P. 1990. Dopamine infusion in healthy subjects and critically ill patients. Clin Exp Pharmacol Physiol 17, 361–369.

Ratledge, C., and Dover, L. G. 2000. Iron metabolism in pathogenic bacteria. Ann Rev Micro 54, 881–941.

Renaud, M., and Miget, A. 1930. Role favorisant des perturbations locales causees par l' adrenaline sur le developpement des infections microbiennes. C R Seances Soc Biol Fil 103, 1052–1054.

Sampath, L.A., Tambe, S.M., and Modak, S.M. 2001. In vitro and in vivo efficacy of catheters impregnated with antiseptics or antibiotics: evaluation of the risk of bacterial resistance to the antimicrobials in the catheters. Infect Control Hosp Epidemiol 22, 640–646.

Singer, M. 2007. Catecholamine treatment for shock – equally good or bad? Lancet 370, 636–637.

Spencer, R.C. 1996. Predominant pathogens found in the European Prevalence of infection in Intensive Care Study. Eur J Clin Microbiol Infect Dis 15, 281–285.

Smythe, M. A., Melendy, S., Jahns, B., and Dmuchowski, C. 1993. An exploratory analysis of medication utilization in a medical intensive care unit. Crit Care Med 21, 1319–1323.

Vincent, J. L., Bihari, D., and Suter, P. M. 1995. The prevalence of nosocomial infection in intensive care units in Europe: the results of the EPIC study. JAMA 274, 639–644.

Zaborina, O., Lepine, F. Xiao, G., Valuckaite, V., Chen, Y., Li, T., Ciancio, M., Zaborin, A., Petrof, E. O., Turner, J. R., Rahme, L. G., Chang, E. and Alverdy, J. C. 2007. Dynorphin activates quorum sensing quinolone signaling in *Pseudomonas aeruginosa*. PLoS Pathog 3, e35.

Chapter 9
The Microbial Endocrinology of *Pseudomonas aeruginosa*

John C. Alverdy, Kathleen Romanowski, Olga Zaborina, and Alexander Zaborin

9.1 Epidemiology of *Pseudomonas aeruginosa*

Pseudomonas aeruginosa is a model pathogen with which to advance the notion that microbial endocrinology plays a central role in the pathogenesis of bacteria and other microbes. *P. aeruginosa* is a gram-negative opportunistic pathogen that can infect a variety of host species, including Arabidopsis, Drosophila, *Caenorhabditis elegans*, rodents, and man. Like many opportunistic pathogens, virulence expression in *P. aeruginosa* is not an invariant phenotype. Some investigators consider *P. aeruginosa* to be an accidental pathogen to man given that it does not appear to have co-evolved with the human immune system; as such it has been assumed to be rarely part of the normal commensal flora. Yet more comprehensive genome-based analyses of the human intestinal microflora suggest that *P. aeruginosa* is present in up to 20% of normal healthy individuals (Marshall 1991). Although primarily considered to be a nosocomial pathogen that infects the injured and immunocompromised host, *P. aeruginosa* appears to be the most common cause of infection-related deaths among patients with cystic fibrosis, a genetic disorder of the respiratory epithelium. In this latter host, *P. aeruginosa* is a chronic colonizer that can persist for many years where it often exerts only moderate virulence. In hospitalized patients, however, *P. aeruginosa* is most commonly isolated from the aero-digestive tract where it can colonize up to 50% of patients after as little as 3 days in hospital (Marshall 1991). Widespread and promiscuous use of antibiotics in the critically ill and injured appears to be among the various causes of the persistent prevalence of this pathogen in hospitalized patients. Attempts at predicting which colonizing pathogens are associated with the highest rates of virulence, and hence associated with the worst outcome has traditionally been assessed by genotyping and the use of antibiotic resistance profiles. Even attempts at predicting outcome from patients

J.C. Alverdy (✉), K. Romanowski, O. Zaborina, and A. Zaborin
University of Chicago, Chicago, IL, USA
e-mail: jalverdy@surgery.bsd.uchicago.edu

M. Lyte and P.P.E. Freestone (eds.), *Microbial Endocrinology*,
Interkingdom Signaling in Infectious Disease and Health,
DOI 10.1007/978-1-4419-5576-0_9, © Springer Science+Business Media, LLC 2010

who are infected versus colonized have yielded highly paradoxical results. In a recent study, patients with lung infection, i.e., pneumonia versus those with lung colonization (culture positive without pneumonia) demonstrated that the mortality rates were higher in colonized patients versus those that were clinically infected (Zhuo et al. 2008). Despite strict control measures, the prevalence and mortality rate of *P. aeruginosa* in hospitalized patients remain high and have not appreciably decreased in the last 10 years. The highly opportunistic nature of *P. aeruginosa*, its ability to colonize and remain clinically elusive in antibiotic resistant biofilms, and its highly lethal virulence repertoire make this pathogen particularly difficult to detect and treat. The emergence of strains that are multi-drug resistant poses a real and present danger to patients who suffer burn injury, solid organ and bone marrow transplantation, traumatic injury, major surgical intervention, or severe immuno-compromise such as HIV/AIDS. Many of these infections arise from endogenous sources, the most common of which is the digestive tract reservoir. *P. aeruginosa* continues to carry the highest case fatality rate (60%) among nosocomial pathogens and is responsible for a variety of clinical infections, including keratitis, otitis, pneumonia, bacteremia, catheter-related sepsis, echthema gangrenosa, and severe diarrhea (Neuhauser et al. 2003). The finding that *P. aeruginosa* can be cultured from the lung, urine, and feces of more than 50% of critically ill and immunocom-promised patients, yet often can remain indolent through the course of illness and recovery, suggest that, in addition to its genotype and antibiotic resistance profile, this organism undergoes major phenotype transition when present in a stressed host (Murono et al. 2003). Yet the precise signals that are unique to the environmental niches where *P. aeruginosa* colonizes remain to be elucidated. Many of the locally secreted signals that activate the virulence of *P. aeruginosa* can be classified as endocrine compounds released during stress.

9.2   Microbial Endocrinology of *P. aeruginosa* Virulence Activation in the Intestinal Tract Following Surgical Injury

Our laboratory has been most interested in how *P. aeruginosa* causes host injury from within the intestinal tract of a critically ill and injured host, because many of its subversive tactics and its lethal effect can be identified in animal models of intestinal infection with *P. aeruginosa*. Although not considered to be an intestinal pathogen in the classic sense, *P. aeruginosa* occupies a ubiquitous presence in the feces of critically ill patients where antibiotic penetration, and hence the ability of systemic antibiotics to eradicate this organism, is limited. As the intestinal tract epithelial surface occupies 200 m^2 of area from mouth to anus, the ability of this organism to remain undetected in this organ and mount a toxic offensive against its host, makes the intestinal tract a particularly interesting area of focus to model microbial endocrinology. Intriguingly, in severe life-threatening sepsis in immuno-compromised patients, the site of infection is identified in less than 50% of cases (Kropec et al. 1993). This condition has been termed as "culture negative sepsis" or "gut-derived sepsis" to reflect the pathogens such as *P. aeruginosa* that can activate a systemic pro-inflammatory effect without detection by conventional techniques such as blood culture. However, the term culture negative sepsis misrepresents the fact that this process is driven by a microbial process that can initiate and sustain systemic inflammation. Work from our laboratory has demonstrated that intestinal *P. aeruginosa* can mount a toxic offensive against the host and at the same time subvert immune clearance mechanisms. Virulence factors such as the type III secretion apparatus, the ability to disregulate epithelial tight junctional permeability to lethal cytotoxins, and the production of protective biofilms allow *P. aeruginosa* to induce host inflammation and lethality at arm's length from the immune system. Our work has focused on elucidation of key mechanisms by which intestinal *P. aeruginosa* is cued to transform its phenotype to a lethal pathogen using highly subversive tactics that elude clinical detection and immune elimination. How this occurs specifically during host injury and stress has been a major focus of our laboratory.

9.3   Host Derived Bacterial Signaling Compounds (HDBSC's): How Microbial Pathogens Sense Host Stress at the Intestinal Epithelial Surface

The discovery by Lyte that host epinephrine and other catecholamines provide major signals to *E. coli* and other pathogens to express a more adherent and hence virulent phenotype against the intestinal mucosal epithelium set the stage for the discovery of similar compounds released during host stress that might signal pathogenic organisms to express enhanced virulence (Chen et al. 2003). We began to work with *P. aeruginosa* in a unique animal model that we developed in our laboratory. To recreate operative trauma and the colonization of *P. aeruginosa* in the distal intestinal, we performed a 30% surgical hepatectomy in mice and simultaneously injected *P. aeruginosa* strain 27853 into the cecum. To mimic the clinical circumstances of a major surgical intervention, mice were fasted for 48 h where they drank only D5W in their drinking water. This model recapitulated a typical major surgical intervention by imposing surgical injury, intestinal colonization with a hospital pathogen, and a short period of partial starvation. In this model, mice undergoing hepatectomy and intestinal inoculation with *P. aeruginosa* develop a high mortality rate (75–100%) whereas mice without *P. aeruginosa* inoculation or without operative injury recover and remain healthy. We identified a key virulence determinant in *P. aeruginosa* in this model, the PA-I lectin, that becomes "in vivo expressed" when present in the intestine of mice following surgical injury (Laughlin et al. 2000). We further identified the key role of the PA-I lectin as a mediator of lethality in this model since mutant strains lacking PA-I were non-lethal in this model. Work with cultured human intestinal epithelial cells revealed that the PA-I lectin causes a permeability defect to exotoxin A and elastase, lethal cytotoxins of *P. aeruginosa*, causing mortality. This effect was mediated in part by the ability of the PA-I lectin to disrupt the tight junctional proteins occludin and ZO-1. Blockade of the PA-I lectin with specific sugars such as GalNaC and galactose prevented *P. aeruginosa* induced mortality in this model. In summary, the intestinal tract of surgically stressed mice appears to be a unique environmental niche in which *P. aeruginosa* virulence activation developed leading a lethal effect that was mediated by the PA-I lectin/adhesin (Laughlin et al. 2000).

We next performed a series of in vitro and in vivo experiments to determine the relative contributions of host factors that might be locally secreted into the intestinal tract during surgical injury that would serve as "host cues" actively sensed by *P. aeruginosa* to express a virulent and lethal phenotype. Analogous to the work of Lyte and others we identified three classes of host compounds released during physiologic stress that are actively sensed by *P. aeruginosa* leading to virulence activation including immune elements (interferon-gamma), opioids (morphine, dynorphin) and end-products of ischemia (inosine, adenosine) (Wu et al. 2005, Zaborina et al. 2007, Patel et al. 2007). Although technically not part of the endocrine system, many of these compounds function as paracrine signals that are cytoprotective to host cells during physiologic stress and host injury. It is intriguing that pathogens like *P. aeruginosa* have evolved to intercept these signals and use them to activate their virulence by incorporating them into specific elements of the quorum sensing system.

9.4 Mechanisms By Which Bacterial Membrane Proteins, Cytoplasmic Transcriptional Regulators, Microbial Enzymes, and Components of the Quorum Sensing Signaling System Gather, Process, and Transduce Host Compounds Released During Physiologic Stress

9.4.1 Interferon-γ

Using activated T-cells, we discovered that interferon-γ is released and binds to the *P. aeruginosa* membrane protein OprF activating the quorum sensing signaling system (QS) via its RhlRI branch (Wu et al. 2005). Using reporter constructs, mutants, and mass spectrometry, we demonstrated that the outer membrane porin OprF, is the putative binding site on *P. aeruginosa* for interferon gamma. Direct protein–protein binding experiments confirmed these observations. OprF, originally discovered by Dr. Robert Hancock, also functions as an adhesin of *P. aeruginosa* in respiratory epithelial cells and is one of the first examples of a bacterial membrane protein with bidirectional functionality (Azghani et al. 2002). Further work demonstrated that binding of interferon-γ to OprF increased the expression of the PA-I lectin and pyocyanin production in *P. aeruginosa* causing strains to adhere to and disrupt the intestinal epithelium in a manner that disrupted its barrier function (Wu et al. 2005). Taken together these studies provide compelling evidence that

P. aeruginosa can sense immune activation in its host through the release of soluble elements of the adaptive immune system and respond with enhanced virulence. Precisely how these findings are interpreted in the context of complex infection with *P. aeruginosa* remains to be determined. For example, interferon knockout mice are more susceptible to lung inoculation with *P. aeruginosa*, as interferon is a major mechanism of bacterial clearance. Paradoxically, interferon gamma administration during lung infection with *P. aeruginosa* does not reduce, and in some experimental models and clinical studies, worsens infection related outcome (Babalola et al. 2004). These studies demonstrate the complex interplay that exists between a pathogen and its host. This process often involves a fragile balance between the ability of a microbe to express virulence and subvert the immune system and the immune response of the host to clear the pathogen. In the case of opportunistic pathogens like *P. aeruginosa* that rarely cause infection in a healthy host, one would predict that targeted disruption of immune clearing mechanisms are likely to result in worse outcome. However, within the complex immune activated environment of a stressed or injured host, how highly clever and subversive pathogens like *P. aeruginosa* respond to immune clearing mechanisms might be a matter distinct from experimental models where all other variables are held constant against a background of targeted immune disruption. For these reasons, a more detailed understanding of microbial endocrinology in the context of physiologic perturbations in the host that directly affect bacterial behavior is likely to be more representative of in vivo conditions that lead to the development of clinically significant infection.

9.4.2   Adenosine/Inosine

Adenosine, a HIF-1α regulated cytoprotectant, is released during hypoxia onto the apical surface of intestinal epithelial and binds to the A2b receptor to enhance epithelial barrier function. Thus, hypoxia induces a cytoprotective response designed to protect the intestinal epithelium against microbial invasion (Zhou et al. 2009). To determine if pathogenic bacteria such as *P. aeruginosa* can sense intestinal epithelial hypoxia, we first observed that, as expected, hypoxic intestinal epithelial cells (Caco-2) or intestinal epithelial cells with forced expression of HIF-1α, secreted massive quantities (>10,000-fold increase) of adenosine into the apical media (Patel et al. 2007). We next determined that adenosine induced the expression of quorum sensing dependent virulence factor, the PA-I lectin/adhesin in *P. aeruginosa*, in a dose-dependent manner. Similarly, inosine, the immediate downstream metabolite of adenosine, also induced the expression of the PA-I lectin/adhesin; however, it was tenfold more potent than adenosine. Surprisingly, inosine was not detected in the media of hypoxic Caco-2 cells, most likely as a result of a lack of expression of adenosine deaminase during hypoxia. As shown by others, hypoxia induced down regulation of adenosine deaminase by epithelial cells allows for the concentration of adenosine at the apical surface where it can be available to exert its cytoproctective action (Kobayashi et al. 2000). Hypothesizing that *P. aeruginosa* colonizing the surface of hypoxic intestinal epithelial cells might be clever enough to sense and metabolize adenosine to inosine to generate a potent quorum sensing activation signal, we next tested whether *P. aeruginosa* could metabolize adenosine to inosine. Using various experimental manipulations and mutants of *P. aeruginosa* deficient in adenosine deaminase, we demonstrated that *P. aeruginosa* metabolizes adenosine to inosine in a manner that is accelerated by soluble products released by hypoxic intestinal epithelial cells (Caco-2) (Patel et al. 2007). Thus, hypoxic intestinal epithelial cells not only release adenosine that is converted to inosine by *P. aeruginosa*, but the conversion itself is accelerated by a yet un-identified compound (s) released into the apical media. Taken together these findings demonstrate that *P. aeruginosa* is capable of subverting the intestinal epithelial cytoprotective response by depriving the epithelium of a key cytoprotectant adenosine while at the same time enriching the local milieu with a QS mimic, inosine, to activate its virulence circuitry (Patel et al. 2007). These studies again confirm the highly subversive and clever tactics that opportunistic pathogens can deploy and establish how host compounds are sensed, intercepted, and transduced by colonizing pathogens. Thus understanding microbial endocrinology at a basic level can uncover previously unidentified mechanism of bacterial pathogenesis.

9.4.3   Dynorphin

Dynorphin, an endogenous κ-opioid receptor agonist is another example by which microbial endocrinology functions to alter the pathogenesis of *P. aeruginosa*. Dynorphin has been identified mostly in the brain as an important opioid neurotransmitter. However, its presence and function in the gut has been poorly described. To determine if dynorphin is released into the gut during physiologic stress, we subjected mice to a period of intestinal ischemia and reperfusion injury, as might occur during a major operation such as cardiac, vascular, or liver transplant surgery. Results of these studies demonstrated that (1) following intestinal ischemia/reperfusion (I/R), dynorphin positive cells densely expand to populate the intestinal mucosa migrating from the crypts to the villus tips, an effect which is further enhanced when *P. aeruginosa* is present in the gut; (2) Dynorphin is released into the intestinal lumen after I/R, diffuses into the cytoplasm of *P. aeruginosa*, and activates the QS system via the MvfR pathway to express enhanced virulence against epithelial cells as measured by alterations in transepithelial electrical resistance of cultured human intestinal epithelial monolayers (Zaborina et al. 2007). In addition, *P. aeruginosa* exposed to κ-opioids is enhanced in its killing effect against *C. elegans* (Zaborina et al. 2007). Most importantly, exposure of *P. aeruginosa* to κ-opioids enhances its killing effect against colonizing organism considered to be probiotic to the intestinal epithelium, such as lactobacillus species, via mechanisms that appear to involve the release of the toxic compound

HQNO. Taken together these studies demonstrate that dynorphin is yet another example of how physiologic stress is sensed by certain pathogenic bacteria and can be incorporated into their virulence circuitry to mount a toxic offensive against the host.

9.5   Phosphatonins and Phosphate Sensing by Pathogenic Bacteria: Microbial Endocrinology in Action

For decades it has been recognized that serum levels of inorganic phosphate [Pi] can precipitously fall to dangerous levels following severe catabolic stress such as occurs following major surgical injury (liver resection, cardiac surgery) and following inflammatory injury (burn injury, pancreatitis, multiple system trauma). Only recently however, has the hormone phosphatonin been identified to be a major mediator of this response as it is released in high concentration following many of these injuries (Salem and Tray 2005). Acute serum Pi depletion is especially dangerous as it can lead to cardiac conduction abnormalities, heart failure, and death. Elevated circulating phosphatonin seems to drive serum Pi depletion (hypophosphatemia) as a result of phosphatonin binding to renal epithelial cells that causes excess Pi excretion in urine. Stress induced hypophosphatemia was originally believed to be due to reprioritization of high energy phosphates to form ATP leading to serum phosphate depletion, however bioenergetic failure following injury and sepsis does not appear to be an early finding when phosphate depletion is severe, thus mitigating this hypothesis. Importantly two recent studies have demonstrated a strong correlation between hypophosphatemia among critically ill patients and infectious related mortality (Hoffmann et al. 2008; Shor et al. 2006). While Pi levels in blood can be easily re-established in critically ill patients with intravenous Pi administration, whether Pi becomes depleted at microbial colonization niches such as within the lung and gut, remain poorly understood.

Our laboratory recently demonstrated that when mice are subjected to a 30% liver resection and short term starvation (24 h), an otherwise recoverable injury, Pi becomes critically depleted with the intestinal mucus, the major site of bacterial colonization in the gut (Long et al. 2008). This effect is further enhanced when mice are intestinally inoculated with *P. aeruginosa*. Retrieval of intestinal *P. aeruginosa* from mice following 30% liver resection (hepatectomy) and intestinal inoculation with *P. aeruginosa* demonstrate that they display enhanced expression of PstS, an important component of the mechanism by which bacteria sense low environmental phosphate. In fact expression was 34-fold increased compared to controls. Further study of the PstS protein in *P. aeruginosa* by our lab demonstrated that expression is not only dependent on local Pi concentration, but also plays a key role in the ability of *P. aeruginosa* to disrupt the intestinal epithelial barrier. Importantly, when mice are subjected to a 30% hepatectomy and intestinally inoculated with *P. aeruginosa*, mortality rates are 60%, however, when intestinal mucus Pi levels are maintained with oral supplementation, mortality is eliminated

(Long et al. 2008). Intestinal mucus Pi levels cannot be maintained when Pi is provided intravenously and luminal levels of Pi do not correlate with PstS expression and mortality in mice as it is well established that for intestinal bacteria to successfully colonize and invade they must adhere to the mucus layer. Therefore, more information on the [Pi] at sites of microbial colonization such as mucus are needed to determine the role of Pi in microbial virulence activation. In higher mammals including humans, oral Pi provision increases phosphatonin levels resulting in phosphate excretion, therefore understanding how local levels of phosphate fall in response to sepsis and injury at colonization niches of important pathogenic bacteria will be necessary to determine how Pi supplementation can prevent infection.

9.5.1 Pi Sensing by P. aeruginosa and Its Interaction with the QS System: A Conserved Mechanism of Virulence Activation in Nosocomial Pathogens

Pi concentration [Pi] appears to be not only important to the activation of *P. aeruginosa* to express a lethal phenotype, but also to multiple pathogens relevant to gut-derived sepsis following surgical injury and stress. From the standpoint of the microbe, [Pi] could be viewed as a marker of relative energy abundance in eukaryotic cells and hence a surrogate marker for tissue homeostasis versus dysfunction. On this level, it seems logical that bacteria have evolved sophisticated sensory systems to detect [Pi]. A highly conserved two-component phosphor-sensor system consisting of an inner membrane histidine kinase PhoR and transcriptional regulator PhoB is present in various bacteria associated with gut-derived sepsis. This system controls the expression of the PhoB regulated genes via binding of PhoB to its specific DNA binding site *pho*. The *pho* box has been identified upstream of multiple genes including those of the QS regulon. As such Pi depletion can induce the expression of virulence through its effect on quorum sensing (Fig. 9.1c) (Zaborin et al. 2009). Although there is considerable information on the regulation of microbial metabolism and virulence during low Pi conditions in vitro, there is virtually no information on how Pi concentration affects microbial lethality in vivo. As mentioned above, remarkably the concentration of Pi within tissue sites of microbial colonization during health or disease is essentially unstudied. It is assumed that the intestinal tract does not represent a site of low [Pi] as luminal Pi is predicted to be abundant from oral nutrition. However, as we have shown, mucus [Pi] may be of more importance to intestinal pathogens as they attach to mucus during colonization and invasion. Given the importance of local [Pi] to be sensed by such a wide variety of intestinal bacteria, more detailed study of the concentration of phosphate is needed to understand its relevance in infection-related mortality during severe host stress (Lamarche et al. 2008).

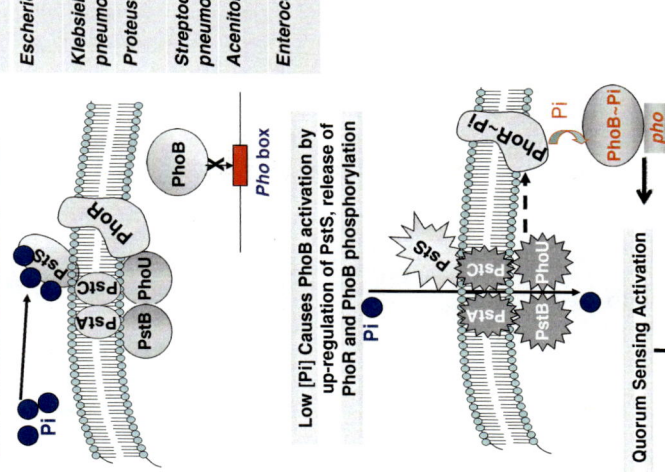

Fig. 9.1 The PstS–PhoB phosphate sensing system. (*Panel A*) High extracellular [Pi] binds PstS causing PhoR to remain bound to the multi-component membrane complex. (*Panel B*) When PstS is activated by low [Pi], PhoR release leads to PhoB phosphorylation (PhoB-Pi), which then activates QS leading to virulence expression. (*Panel C*) Various pathogens of clinical relevance activate their virulence in response to low [Pi] via PhoB (Van Dien and Keasling 1998)

9.5.2 Modeling In Vivo Phosphate Depletion in the C. elegans–P. aeruginosa System: Discovery of "Red Death"

In order to verify that acute Pi depletion alone was sufficient to shift the lethality of *P. aeruginosa* in vivo, we modeled Pi depletion in the *P. aeruginosa–C. elegans* system (Fig. 9.2) (Zaborin et al. 2009). *C. elegans* normally feed on lawns of *E. coli* OP50 that grow on nematode growth media (NGM), a high [Pi] media (25 mM). We used *P. aeruginosa* PAO1, known to be non-lethal to *C. elegans* on NGM media, and depleted the media of Pi to determine if low [Pi] would shift *P. aeruginosa* to express lethality. To accomplish this, *C. elegans* feeding on *E. coli* OP50 were transferred onto lawns of *P. aeruginosa* PAO1 growing on NGM media at high [Pi] (25 mM) or low [Pi] (~0.1 mM). Although initially we did not observe any mortality under low [Pi], *C. elegans* progeny formation, an important marker of viability, was significantly attenuated. In order to deplete the worm digestive tube of any residual Pi upon transfer to the lawn of *P. aeruginosa*, we subjected worms to a period of short-term starvation (18 h) and performed reiterative studies. Results demonstrated that following this approach, *C. elegans* survival dramatically decreased on low [Pi] lawns compared to those feeding on high [Pi] lawns (30% versus 100% at 48 h – Fig. 9.2). To determine if these results were due to the stress of starvation versus nutrient deprivation itself, we exposed non-starved worms to mild heat stress (35°C 2 h), performed reiterative studies, and obtained similar results (data not shown) (Note: worms normally thrive and feed at 20°C). Thus analogous to our mouse model, host stress is required for intestinal *P. aeruginosa* to express a lethal phenotype against its host. Importantly, the Pi concentration at which *P. aeruginosa* produced mortality in *C. elegans* (~0.1 mM) was similar to that which increased the virulence output of *P. aeruginosa* in vitro and to its concentration in the intestinal tract of mice following surgical injury.

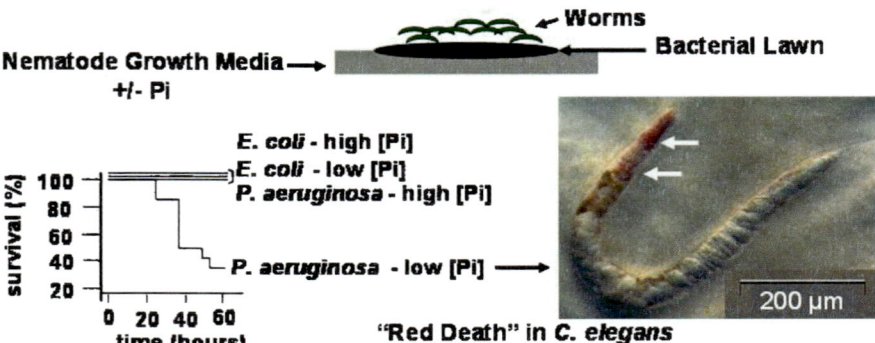

Fig. 9.2 *C. elegans* growing on lawns of *P. aeruginosa* on Nematode growth media (NGM) at high and low phosphate [Pi]. Worms are transferred onto lawns of *E. coli* or *P. aeruginosa* at high or low [Pi]. Only *P. aeruginosa* growing on low [Pi] NGM causes mortality with *red* material accumulating in the digestive tube (*arrows*). *Red* material was not observed in surviving worms

We unexpectedly observed red pigmented material in the digestive tube of worms and discovered that this was due to the release of a toxic PQS iron complex in response to low [Pi] which we termed red death (Zaborin et al. 2009). We also determined that genes encoding the byosynthesis of rhamnolipids, surfactants known to solubilize PQS, were also overexpressed during low [Pi] and that the triple mixture (PQS, Fe, rhamnolipids) induced severe epithelial apoptosis in the cecum of mice (Fig. 9.3) (Zaborin et al. 2009). Taken together, these data provide compelling evidence that low [Pi] is sufficient to shift *P. aeruginosa* to express a lethal phenotype from within the digestive tube of *C. elegans* and mice.

Lastly, we tested the synergistic effect of both low [Pi] and exposure to the various host compounds (i.e., opioids) and found that low [Pi] enhances the responsiveness of *P. aeruginosa* to host compounds released during stress (unpublished observations). Thus, it is likely that multiple aspects of the local microenvironment within which pathogens like *P. aeruginosa* colonize and feed are important for the activation of virulence and lethality. As these aspects are uncovered and pathways identified that are conserved across a variety of pathogens, compounds can be developed that might molecularly silence *P. aeruginosa* and other pathogens from expressing virulence as they course through the intestinal tract of critically ill and immunocompromised patients.

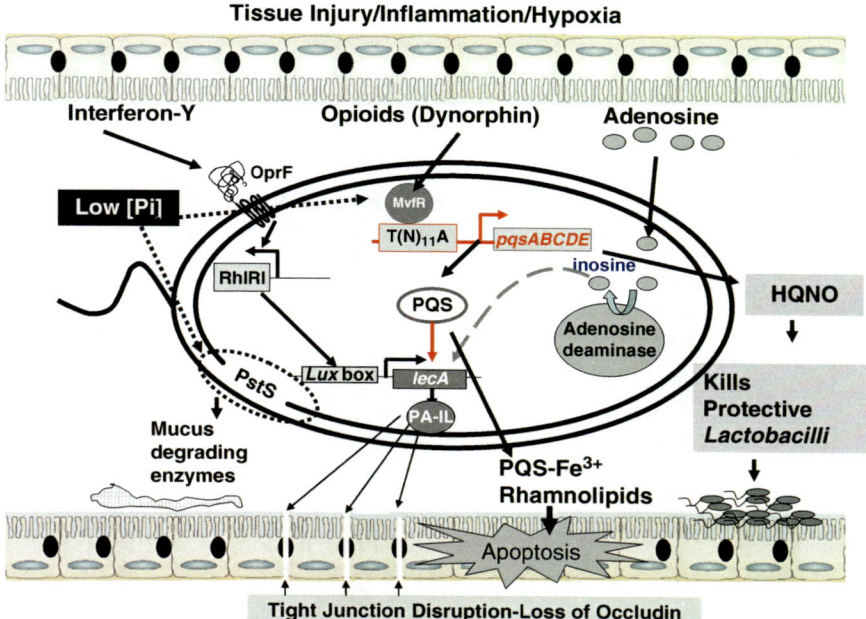

Fig. 9.3 Multi-pronged attack of *P. aeruginosa* against the intestinal epithelium in response to host stress tissue compounds and local environmental "cues" present in the gut during surgical injury. Host death is hypothesized to result from various toxins and barrier disrupting compounds released by *P. aeruginosa* in response to host stress compounds and low phosphate conditions

9.6   Summary and Conclusions

Pseudomonas aeruginosa is among various microbes capable of sensing and responding to multiple elements released by host tissues and the host environment that are important to its pathogenicity. As it is now widely recognized that virulence expression is not an invariant trait for most pathogens, understanding how and why bacteria make the fundamental tradeoff to mount a toxic offensive against their host during physiologic stress has the potential to uncover a multitude of novel pathways of virulence expression and metabolism present in highly problematic pathogens of concern to human and animal health. New tools are emerging that allow for high throughput screening of host and local environmental cues that activate microbial virulence and may lead to non-antibiotic-based therapies that can offer the promise to contain these pathogens without killing them.

References

Alverdy J, Holbrook C, Rocha F, Seiden L, Wu RL, Musch M, Chang E, Ohman D, Suh S. Gut-derived sepsis occurs when the right pathogen with the right virulence genes meets the right host: evidence for in vivo virulence expression in *Pseudomonas aeruginosa*. Ann Surg. 2000; 232:480–9.

Azghani AO, Idell S, Bains M, Hancock RE. *Pseudomonas aeruginosa* outer membrane protein F is an adhesin in bacterial binding to lung epithelial cells in culture. Microb Pathog. 2002; 33(3):109–14.

Babalola CP, Nightingale CH, Nicolau DP. Effect of adjunctive treatment with gamma interferon against *Pseudomonas aeruginosa* pneumonia in neutropenic and non-neutropenic hosts. Int J Antimicrob Agents. 2004; 24(3):219–25.

Chen C, Brown DR, Xie Y, Green BT, Lyte M. Catecholamines modulate *Escherichia coli* O157:H7 adherence to murine cecal mucosa. Shock. 2003; 20:183–8.

Hoffmann M, Zemlin AE, Meyer WP, Erasmus RT. Hypophosphataemia at a large academic hospital in South Africa. J Clin Pathol. 2008; 61(10):1104–7.

Kobayashi S, Zimmermann H, Milhorn DE. Chronic hypoxia enhances adenosine release in rat PC12 cells by altering adenosine metabolism and membrane transport. J. Neurochem. 2000; 74(2):621–632.

Kropec A, Huebner J, Riffel M, Bayer U, Benzing A, Geiger K, Daschner FD. Exogenous or endogenous reservoirs of nosocomial *Pseudomonas aeruginosa* and *Staphylococcus aureus* infections in a surgical intensive care unit. Intensive Care Med. 1993; 19:161–5.

Lamarche MG, Wanner BL, Crépin S, Harel J. The phosphate regulon and bacterial virulence: a regulatory network connecting phosphate homeostasis and pathogenesis. FEMS Microbiol Rev. 2008; 32(3):461–73.

Laughlin RS, Musch MW, Holbrook CJ, Rocha FM, Chang EB, Alverdy JC. The key role of *Pseudomonas aeruginosa* PA-I lectin on experimental gut-derived sepsis. Ann Surg. 2000; 232:133–42.

Long J, Zaborina O, Holbrook C, Zaborin A, Alverdy J. Depletion of intestinal phosphate after operative injury activates the virulence of *P. aeruginosa* causing lethal gut-derived sepsis. Surgery. 2008; 144(2):189–97.

Marshall JC. The ecology and immunology of the gastrointestinal tract in health and critical illness. J Hosp Infect. 1991; 19 Suppl C:7–17.

Murono K, Hirano Y, Koyano S, Ito K, Fujieda K. Molecular comparison of bacterial isolates from blood with strains colonizing pharynx and intestine in immunocompromised patients with sepsis. J Med Microbiol. 2003; 52:527–30.

Neuhauser MM, Weinstein RA, Rydman R, Danziger LH, Karam G, Quinn JP. Antibiotic resistance among gram-negative bacilli in US intensive care units: implications for fluoroquinolone use. JAMA. 2003; 289:885–8.

Patel NJ, Zaborina O, Wu L, Wang Y, Wolfgeher DJ, Valuckaite V, Ciancio MJ, Kohler JE, Shevchenko O, Colgan SP, Chang EB, Turner JR, Alverdy JC. Recognition of intestinal epithelial HIF-1alpha activation by *Pseudomonas aeruginosa*. Am J Physiol Gastrointest Liver Physiol. 2007; 292(1):G134–42.

Salem RR, Tray K. Hepatic resection-related hypophosphatemia is of renal origin as manifested by isolated hyperphosphaturia. Ann Surg. 2005; 241:343–8.

Shor R, Halabe A, Rishver S, Tilis Y, Matas Z, Fux A, Boaz M, Weinstein J. Severe hypophosphatemia in sepsis as a mortality predictor. J Ann Clin Lab Sci. 2006; 36(1):67–72.

Van Dien SJ, Keasling JD. A dynamic model of the *Escherichia coli* phosphate-starvation response. J. Theor. Biol. 1998; 190:37–49.

Wu L, Estrada O, Zaborina O, Bains M, Shen L, Kohler JE, Patel N, Musch MW, Chang EB, Fu YX, Jacobs MA, Nishimura MI, Hancock RE, Turner JR, Alverdy JC. Recognition of host immune activation by *Pseudomonas aeruginosa*. Science. 2005; 309(5735):774–7.

Zaborin A, Romanowski K, Gerdes S, Holbrook C, Lepine F, Long J, Poroyko V, Diggle SP, Wilke A, Righetti K, Morozova I, Babrowski T, Liu DC, Zaborina O, Alverdy JC. Red death in *Caenorhabditis elegans* caused by *Pseudomonas aeruginosa* PAO1. Proc Natl Acad Sci USA. 2009; 106(15):6327–32.

Zaborina O, Lepine F, Xiao G, Valuckaite V, Chen Y, Li T, Ciancio M, Zaborin A, Petrof EO, Turner JR, Rahme LG, Chang E, Alverdy JC. Dynorphin activates quorum sensing quinolone signaling in *Pseudomonas aeruginosa*. PLoS Pathog. 2007; 3:e35.

Zhou Y, Mohsenin A, Morschl E, Young HW, Molina JG, Ma W, Sun CX, Martinez-Valdez H, Blackburn MR. Enhanced airway inflammation and remodeling in adenosine deaminase-deficient mice lacking the A2B adenosine receptor. J Immunol. 2009; 182(12):8037–46.

Zhuo H, Yang K, Lynch SV, Dotson RH, Glidden DV, Singh G, Webb WR, Elicker BM, Garcia O, Brown R, Sawa Y, Misset B, Wiener-Kronish JP. Increased mortality of ventilated patients with endotracheal *Pseudomonas aeruginosa* without clinical signs of infection. Crit Care Med. 2008; 36(9):2495–503.

Chapter 10
Mechanisms of Stress-Mediated Modulation of Upper and Lower Respiratory Tract Infections

Cordula M. Stover

10.1 Respiratory Immunity

Immune defenses of the upper respiratory tract encompass both host and bacterial factors (Wilson et al. 1996). Common to all mucosal surfaces, the symbiosis of resident bacteria and epithelial cell layer with its mucous layer protects the healthy steady state of the tissue. Studies using mice reared in a pathogen-free environment reveal that on introduction of bacteria, those that will become commensals are instrumental in shaping the mucosal immune repertoire (Ichimiya et al. 1991). Below the mucus layer that entraps particles and bacteria, there is a so-called periciliary layer composed of epithelial surface-bound mucins and glycolipids the state of hydration of which is a factor in optimal clearance function of the mucus, thereby preventing biofilm growth and subsequent inflammation (Randell et al. 2006). The periciliary fluid layer contains a multitude of antimicrobial factors (Grubor et al. 2006). Commensals are able to adhere to the mucus, thereby occupying a niche where they essentially discourage the claims of pathogens to the same environment they are said to compete. Mucociliary clearance by ciliary beat of specialized epithelial cells is especially important in the upper airway tract where these cells are found – the lower respiratory tract (beyond the bronchial bifurcation) is essentially sterile and forms a significant innate immune barrier.

Respiratory epithelial cells are actively involved in the recognition of pathogens by their expression of pattern recognition receptors such as Toll-like receptors and secretion of antimicrobial peptides (Bals and Hiemstra 2004). Furthermore, epithelial cells are active players in instructing the respiratory immune response by physically anchoring intraepithelial dendritic cells, as well as synthesizing factors that influence T-cell mediated effector functions (Schleimer et al. 2007). Once the mucociliary escalator is impaired, and transit time for bacteria is prolonged,

C.M. Stover (✉)
Department of Infection, Immunity and Inflammation,
University of Leicester, University Road, Leicester, LE1 9HN, UK
e-mail: cms13@le.ac.uk

M. Lyte and P.P.E. Freestone (eds.), *Microbial Endocrinology*,
Interkingdom Signaling in Infectious Disease and Health,
DOI 10.1007/978-1-4419-5576-0_10, © Springer Science+Business Media, LLC 2010

adherence to epithelial cells ensues and a potential respiratory infection begins. Production of virulence factors by the adherent bacteria leads to an extensive local respiratory immune response that affects intactness of mucociliary clearance, epithelial cell layer, IgA and function of attracted leukocytes (Wilson et al. 1996).

Nasal associated lymphatic tissue is the mucosal inductive site for humoral immune response of the upper respiratory tract (Zuercher et al. 2002). IgA or IgG deficiency, complement C4 deficiencies, common variable immune deficiency, and atopy all predispose the development of respiratory disease. It is widely known that virally-caused epithelial cell damage in itself can lead to bacterial super-infection by exposing neoligands for pathogen adherence.

10.2   Respiratory Mucosal Immunity, Neuronal Innervation, and Its Stress-Related Perturbations

The innervation of the respiratory tree is summarized in Table 10.1. Nervous system effectors play major roles in regulating respiratory function. For instance, airway muscle tone generally, and physiological reflexes relevant to infection control such as coughing, bronchodilation, and mucociliary clearance, are all regulated by the sympathetic (adrenergic) and parasympathetic (cholinergic) branches of the autonomic nervous system (Nadel and Barnes 1984). Human tracheal gland cells respond to exogenous epinephrine and norepinephrine (Merten et al. 1993). Nebulized epinephrine is also used as a bronchodilator in intubated ventilated patients. Therefore, catecholamines will be present on host respiratory surfaces and so be components of the host secretions that will be detected by respiratory pathogens.

Table 10.1 Innervation of the respiratory tree and its physiologic effects

Designation of nerves	Bronchial smooth muscle	Mucociliary system	Bronchial vessels
Parasympathetic nerves[1]	Constriction	Secretion	Dilation
Sympathetic nerves[2]	a		
Inhibitory nonadrenergic noncholinergic nerves[3]	Dilation		
Excitatory nonadrenergic noncholinergic nerves[4]	Constriction	Secretion	Dilation
		Cough	Increase in permeability

The tone of the respiratory tree is under control of dominant, cholinergic-parasympathetic, adrenergic sympathetic mechanisms and non-adrenergic, non-cholinergic neural pathways (Joos, 2001). The effects are listed.
[a]No innervation, but expression of beta-adrenergic receptors, which trigger dilation on stimulation. The mediators are acetylcholine (1), norepinephrine (2), vasoactive intestinal peptide and nitric oxide (3), substance P and neurokinin A (4).

The connection between psychological status and airway health has long been recognized, and over the centuries realization of the need to reduce stress has incorporated in the treatment regime of both infectious and inflammatory respiratory disease, though usually without understanding the underlying biological mechanisms. For instance, in the pre-antibiotic era sanatoria practice, among others, environmental approaches to controlling tuberculosis (TB) involved low stress and emphasis on rest and relaxation techniques. Now, TB is less of a problem to the developed world due to usage of generally effective antibiotics. But, stress is still widely recognized as a predisposing factor to exacerbation of the disease, and re-activation of latent TB infections. In the modern context, asthma is a much more common human respiratory condition, and asthma sufferers worldwide are offered a multitude of supportive treatment methods beyond drugs only, aimed at influencing positively the psychological–endocrinological–immunological feedback loops now viewed as being important in the etiology of the disease.

Is the stress that can exacerbate TB or other respiratory conditions such as asthma causal or reactive in its effects? It is now widely believed that stress-related physiological reactions are contributory to the pathogenesis of a large number of clinical conditions. Importantly, stress may lead not only to a down-regulation of the humoral and cellular immune responses, but depending on the duration of stressor it may also impact the inflammatory threshold (Kemeny and Schedlowski 2007). Perception of stress in human normal and asthmatic subjects was found to positively correlate with the percentage of TNF-alpha producing T cells in the peripheral blood of the asthmatics, but not the control subjects (Joachim et al. 2007). Levels of TNF-alpha were also found to be elevated in chronic obstructive pulmonary disease (Barnes 2008), another condition affected by stress. It is important to understand that an increase in pro-inflammatory cytokines, when localized, is associated with tissue damage, which in itself increases the susceptibility to further injury (which can allergic, infectious, or fibrotic in origin). Unresolved, chronic inflammation predisposes respiratory disease patients to the development of the immunological phenomenon of epitope spreading, in which the adaptive immune response is widened, but para-doxically, remains specific. Such allergen cross-reactivity is thought to play a key role in the overall deterioration of health in patients suffering from asthma (Burastero 2006).

The physiological feedback loops, which are heightened in situations of stress, are the hypothalamus–pituitary–adrenal cortical axis and the sympathetic-adrenal-medullary axis (Chen and Miller 2007). These turn into pathophysiological feed-back loops when cognitive and emotional evaluations of a perceived external threat lead to an enhancement of inflammatory response. Cells of the immune system express receptors for catecholamine and glucocorticoid stress hormones, and other factors released during stress, such as neuropeptides (Reiche et al. 2004). Catecholamines and glucocorticoids have been shown to alter secretions or recep-tor densities of immune cells (Malarkey and Mills 2007). Airway macrophages have an activation profile that differs from other macrophages in the body, and possess receptors for epinephrine to which they respond in a way that increases severity of conditions such as asthma (James and Nijkamp 2000). Neuropeptides can also directly mediate stress-related inflammatory responses because they stimulate mast cells in local proximity to sensory nerve endings (Black 2002). Together, stress released hormones and peptides determine not only the duration of infectious disease, but also the sensitization of the respiratory tract, thereby changing the phenotype of initial airway disease. Stress hormones can also directly exacerbate inflammatory respiratory conditions. When mammals are exposed to repeat stress, alveolar macrophages can become pre-disposed to the development of a more inflammatory phenotype (Broug-Holub et al. 1998). This changes the respiratory milieu, making the organism more likely to show hyperreactivity to stimuli, which can in turn exacerbate the response to respiratory pathogens. Substance P is an example of a neuropeptide the release of which by bronchopulmonary nerves is increased on stress and in addition, is produced by inflammatory cells, which are increased in bronchoalveolar lavage in situations of stress (Joachim et al. 2006).

The lower respiratory tract in several respects represents an immunological entity in its own right because resident immune cells, in particular alveolar macrophages acting as antigen presenting cells and interacting T cells can both influence the type of immune response that favors antibody production, a so-called Th2 immune response. Furthermore, draining lymph nodes do not have a role in dealing with antigen, unlike other parts of the body, excepting immune privileged central nervous tissue. Rather, bronchus-associated lymphoid tissue, so-called BALT and, in the case of more chronic inflammation, tertiary lymphoid organs, assume a more predominant role in the immune response (Constant et al. 2002). Alveolar macrophages also contribute to the epinephrine content of broncho-alveolar lavage (Flierl et al. 2007) and are able to increase the production of epinephrine on activation.

Since stress is such a potent modulator of the immune response and its effector mechanisms (Elenkov et al. 2000), it follows that it may be a contributor to the success of immunization, a mainstay strategy in the prevention of respiratory infections such as influenza and pneumococcal disease. In a recent trial involving young adults receiving the hepatitis B vaccine, those subjects experiencing psychological distress had significantly lower specific antibody responsiveness (Marsland et al. 2006). Drummond and Hewson-Bower (1997) found a correlation between lower serum IgA/albumin ratio and stress in children with recurrent upper respiratory tract infections. Another study concluded that a healthy individual's increased sociability decreased the risk of viral upper respiratory tract infection (Cohen et al. 2003).

Relatively little is currently known about the impact of immune stressors on transmissibility and infectivity of bacteria colonizing the human respiratory mucosa. A recent murine study demonstrated that repeated stress led to prolonged, induced pathogen carriage, effecting delayed lethality on the first isogenic challenge but impaired protective immunity on second isogenic challenge (Gonzales et al. 2008). The implications of this chain of events for stress in the human work force may be considerable.

10.3 Stress and Its Influence on Susceptibility to Respiratory Infection via Modulation of Respiratory Pathogen Growth and Virulence

While the respiratory tract is usually seen to be in a "low state of alert" vis-à-vis the plethora of aerogenic stimuli (Holt et al. 2008), responding to an infectious organism brings with it the need to do this in a measured and balanced way, in order that the host may regain its steady state integrity, minimizing so-called bystander damage. This is a dual edged sword that the host is wielding and is exemplified by molecules like nitric oxide, which, on the one hand, is part of the respiratory tract's arsenal to respond to infectious organisms, which on the other hand, under excessive production and conversion to reactive metabolites can lead to tissue damage of the host (van der Vliet et al. 2000). Any stress-related tipping of this fine balance can lead to potentially deleterious local over-inflammation.

At the mucosal interface between host and microorganism, adhesion of bacteria to respiratory epithelial cells is influenced by very different stress-associated factors: bacterial adhesion is improved through expression of proteins that are induced by the paucity of iron which is effected by the host as a means of protection against iron-requiring pathogens (Perez Vidakovics et al. 2007). To the possible detriment of the host, bacterial binding to respiratory epithelia also tends to occur with greater efficiency in the hypersecretory conditions of inflamed airways, when the decoration of mucin glycoconjugates is altered (Davril et al. 1999).

Those microorganisms able to form biofilms are more likely to possess the ability to translocate the respiratory epithelial cell layer into deeper tissues than those that have lesser cohesive and adhesive organization (Yamazakki et al. 2006), and they thereby mediate the progression from inflammation to infection. It is therefore of interest that several of the immune effectors released during stress, particularly the catecholamines norepinephrine, epinephrine and dopamine, can induce changes in the bacterial phenotype relevant to biofilm formation such as adhesion to host epithelia (Vlisidou et al. 2004). In terms of the respiratory tract, catecholamines are naturally present in secretions due the role they play in regulating respiratory function (Table 10.1).

All patient groups share an increased risk of respiratory infection while in hospital; those in intensive care units receiving ventilation are at particular risk of developing so-called ventilation-associated pneumonia (Fridkin 1997). It is known that intubation/long dwelling catheters are associated with biofilm formation and sepsis from bacteria such as *Staphylococcus aureus* and *Pseudomonas aeruginosa* (Fridkin 1997). Interestingly, such bacteria have been shown to be receptive to the stress hormones that are elevated in acutely ill patients (Freestone et al. 2008a). What is also of significance is that patients in intensive care are frequently prescribed catecholamine inotropes, drugs that have been shown to increase commensal bacteria biofilm production in intravascular catheters (Lyte et al. 2003), as well as resuscitation of antibiotic-damaged pathogenic bacteria (Freestone et al. 2008b).

How might the catecholamines that are released into the respiratory mucosa during acute stress, or applied therapeutically, influence the infectivity of any potentially pathogenic bacteria within the vicinity of the elaborated hormone? The first context is growth. All pathogenic bacteria require iron to grow in vivo. This growth is prevented due to the presence of lactoferrin and transferrin, key innate immune defense proteins the role of which is to bind all free ferric iron, thereby making blood and respiratory secretions too low in iron to support the growth of iron-requiring pathogens. To overcome this usually very effective iron-limitation defense of host fluids, bacteria produce siderophores, very high affinity secreted ferric iron binding molecules, which allow them to acquire the transferrin and lactoferrin-sequestered iron. Work from several groups has shown that catecholamine stress hormones can also function as direct bacterial siderophores (Freestone et al. 2000, 2002) (see also Chap. 3). This stress hormone provision of iron can induce up to a million fold increase in the growth of pathogenic bacteria, including those causing airway infections such as *Klebsiella pneumoniae* and *Pseudomonas aeruginosa* and *Bordetella* species (Freestone et al. 1999; Alverdy et al. 2000; Anderson and Armstrong 2006, 2008). Epinephrine, norepinephrine, dopamine, and their metabolites can all reduce the iron-limiting function of bacteriostatic, key innate defense, proteins lactoferrin, and transferrin (Freestone et al. 2002). Through interaction of these host iron-binding proteins with stress hormones, blood and epithelial secretions are rendered much less bacteriostatic as bacteria are able to access normally inaccessible host sequestered iron sources (Alverdy et al. 2000; Freestone et al. 2000, 2002, 2008a; Anderson and Armstrong 2006, 2008). *Bordetella* species, significant respiratory pathogens of infants (causing whooping cough), also make use of stress hormones to increase transcription of several iron acquisition components (Anderson and Armstrong 2006, 2008). Dopamine, norepinephrine, and epinephrine are all able to induce the transcription of *bfeA*, the gene for the *Bordetella* enterobactin/catechol xenosiderophore receptor. In addition, as has been shown for many enteric and skin bacteria (Freestone et al. 1999), norepinephrine also stimulate growth of *Bordetella bronchiseptica* growth in iron-limiting medium containing serum, via enabling the bacteria to access transferrin bound iron (Anderson and Armstrong 2006, 2008). As already noted, respiratory secretions are markedly iron limited, though the action of host iron binding proteins, and catecholamines may therefore represent a route by which respiratory pathogens, such as *B. bronchiseptica,* could obtain essential iron in the host environment (Anderson and Armstrong 2006, 2008). Exposure of respiratory pathogens to stress-released hormones can also induce the bacteria to synthesize their own growth factors, the mechanism of action of which is non-transferrin dependent (Freestone et al. 1999). Thus, the effects of stress hormone exposure on bacteria can be manifest long after the initial stress event has ended, and catecholamine levels returned to normal.

10.4   Conclusion

Stress can influence susceptibility to respiratory infection through both the modulation of airway immune responses as well as direct effects upon the microbes that cause infections of the respiratory tract. Stress-mediated modulation of upper and lower respiratory tract disease is therefore multifactorial, encompassing host, microorganisms, and environmental influences. Although most of the microbial endocrinology research field has so far focused on bacteria resident in gut and skin tissues, it is now apparent that respiratory pathogens are also able to use stress hormone as direct environmental cues (Lyte et al. 2003; Anderson and Armstrong 2006, 2008; Freestone et al. 2008a). Because of the relative newness of the microbial endocrinology respiratory research field, it is unclear at present how the presence of stress hormones might shape the signaling events between host and bacterial species, and as a consequence the overall impact on mucosal immunity. However, it is clear that microbial endocrinology has the potential to lead to a better understanding of how emotional states can modulate susceptibility to both upper and lower respiratory tract infections.

References

Alverdy, J., Holbrook, C., Rocha, F., Seiden, L., Wu, R., Musch, M., Chang, E., Ohman, D., and Suh, S. 2000. Gut-derived sepsis occurs when the right pathogen with the right virulence genes meets the right host: evidence for in vivo virulence expression in *Pseudomonas aeruginosa*. Ann. Surg. 232: 480–489.

Anderson, M., and Armstrong, S.K. 2006. The *Bordetella* Bfe system: growth and transcriptional response to siderophores, catechols, and neuroendocrine catecholamines. J. Bacteriol. 188: 5731–5740.

Anderson, M., and Armstrong, S.K. 2008. Norepinephrine mediates acquisition of transferrin-iron in *Bordetella bronchiseptica*. J. Bacteriol. 190: 3940–3947.

Bals, R., and Hiemstra, P.S. 2004. Innate immunity in the lung: how epithelial cells fight against respiratory pathogens. Eur. Respir. J. 23: 327–333.

Barnes, P. 2008. The cytokine network in asthma and chronic obstructive pulmonary disease. J. Clin. Invest. 118: 3546–3556.

Black, P.H. 2002. Stress and inflammatory response: a review of neurogenic inflammation. Brain Behav. Immun. 16: 622–653.

Broug-Holub, E., Persoons, J., Schornagel, K., Mastbergen, S., and Kraal, G. 1998. Effects of stress on alveolar macrophages: a role for the sympathetic nervous system. Am. J. Respir. Cell. Mol. Biol. 19: 842–848.

Burastero, S. 2006. Pollen-cross allergenicity mediated by panallergens: a clue to the pathogenesis of multiple sensitisations. Inflamm. Allergy Drug Targets 5: 203–209.

Chen, E., and Miller, G. 2007. Stress and inflammation in exacerbations of asthma. Brain Behav. Immun. 21: 993–999

Cohen, S., Doyle, W., Turner, R., Alper, C., and Skoner, D. 2003. Sociability and susceptibility to the common cold. Psychol. Sci. 14: 389–395.

Constant, S., Brogdon, J., Piggott, D., Herrick, C., Visintin, I., Ruddle, N., and Bottomly, K. 2002. Resident lung antigen-presenting cells have the capacity to promote Th2 T cell differentiation in situ. J. Clin. Invest. 110: 1441–1448.

Davril, M., Degroote, S., Humbert, P., Galabert, C., Dumur, V., Lafitte, J.J., Lamblin, G., and Roussel, P. 1999. The sialylation of bronchial mucins secreted by patients suffering from cystic fibrosis or from chronic bronchitis is related to the severity of airway infection. Glycobiology 9: 311–321.

Drummond, P., and Hewson-Bower, B. 1997. Increased psychosocial stress and decreased mucosal immunity in children with recurrent upper respiratory tract infections. J. Psychosom. Res. 43: 271–278.

Elenkov, I., Wilder, R., Chrousos, G., and Vizi, S. 2000. The sympathetic nerve – an integrative interface between two supersystems: the brain and the immune system. Pharmacol. Rev. 52: 595–638.

Flierl, M., Rittirsch, D., Nadeau, B., Chen, A., Sarma, J., Zetoune, F., McGuire, S., List, R., Day, D., Hoesel, L., Gao, H., Van Rooijen, N., Huber-Lang, M., Neubig, R., and Ward, P. 2007. Phagocyte-derived catecholamines enhance acute inflammatory injury. Nature 449: 721–725.

Freestone, P., Haigh, R., Williams, P., and Lyte, M. 1999. Stimulation of bacterial growth by heat-stable, norepinephrine-induced autoinducers. FEMS Microbiol. Lett. 172: 53–60.

Freestone, P., Lyte, M., Neal, C., Maggs, A., Haigh, R., and Williams, P. 2000. The mammalian neuroendocrine hormone norepinephrine supplies iron for bacterial growth in the presence of transferrin or lactoferrin. J. Bacteriol. 182: 6091–6098.

Freestone, P., Williams, P., Haigh, R., Maggs, A., Neal, C., and Lyte, M. 2002. Growth stimulation of intestinal commensal *Escherichia coli* by catecholamines: a possible contributory factor in trauma-induced sepsis. Shock 18: 465–470.

Freestone P., Sandrini S., Haigh, R., and Lyte, M. 2008a. Microbial endocrinology: how stress influences susceptibility to infection. Trends Microbiol. 16: 55–64.

Freestone, P., Haigh, R., and Lyte, M. 2008b. Catecholamine inotrope resuscitation of antibiotic-damaged staphylococci and its blockade by specific receptor antagonists. J Infect. Dis. 197: 1044–1052.

Fridkin, S. 1997. Magnitude and prevention of nosocomial infections in the intensive care unit. Infect. Dis. Clin. North Am. 11: 479–496.

Gonzales, X., Deshmukh, A., Pulse, M., Johnson, K., and Jones, H. 2008. Stress-induced differences in primary and secondary resistance against bacterial sepsis corresponds with diverse corticotropin releasing hormone receptor expression by pulmonary CD11C+ MHC+ and CD11C- MHC+ APCS. Brain Behav. Immun. 22: 552–564.

Grubor, B., Meyerholz, D.K., and Ackermann, M.R. 2006. Collectins and cationic antimicrobial peptides of the respiratory epithelia. Vet. Pathol. 43: 595–612.

Holt, P., Strickland, D., Wikström, M., and Jahnsen, F. 2008. Regulation of immunological homeostasis in the respiratory tract. Nat. Rev. Immunol. 8: 142–152.

Ichimiya, I., Kawauchi, H., Fujiyoshi, T., Tanaka, T., and Mogi, G. 1991. Distribution of immunocompetent cells in normal nasal mucosa: comparisons among germ-free, specific pathogen-free, and conventional mice. Ann. Otol. Rhinol. Laryngol. 100: 638–642.

James, D., and Nijkamp, F.P. 2000. Neuroendocrine and immune interactions with airway macrophages. Inflamm. Res. 49: 254–265.

Joachim, R.A., Cifuentes, L.B., Sagach, V., Quarcoo, D., Hagen, E., Arck, P.C., Fischer, A., Klapp, B.F., and Dinh, Q.T. 2006. Stress induces substance P in vagal sensory neurons innervating the mouse airways. Clin. Exp. Allergy 36: 1001–1010.

Joachim, R.A., Noga, O., Sagach, V., Hanf, G., Fliege, H., Kocalevent, R., Peters, E., and Klapp, B. 2007. Correlation between immune and neuronal parameters and stress perception in allergic asthmatics. Clin. Exp. Allergy 38: 283–290.

Joos, G. 2001. The role of neuroeffector mechanisms in the pathogenesis of asthma. Curr. Allergy Asthma Rep. 1: 134–143.

Kemeny, M., and Schedlowski, M. 2007. Understanding the interaction between psychosocial stress and immune-related diseases: a stepwise progression. Brain Behav. Immun. 21: 1009–1018.

Lyte, M., Freestone, P., Neal, C., Olson, B., Haigh, R., Bayston, R., and Williams, P. 2003. Stimulation of *Staphylococcus epidermidis* growth and biofilm formation by catecholamine inotropes. Lancet 361: 130–135.

Malarkey, W.B., and Mills, P.J. 2007. Endocrinology: the active partner in PNI research. Brain Behav. Immun. 21: 161–168.

Marsland, A., Cohen, S., Rabin, B., and Manuck, S. 2006. Trait positive affect and antibody response to hepatitis B vaccination. Brain Behav. Immun. 20: 261–269.

Merten, M.D., Tournier, J.M., Meckler, Y., and Figarella, C. 1993. Epinephrine promotes growth and differentiation of human tracheal gland cells in culture. Am J Respir Cell Mol Biol. 9: 172–178.

Nadel, J.A., and Barnes, P.J. 1984. Autonomic regulation of the airways. Annu. Rev. Med. 35: 451–467.

Perez Vidakovics, M., Lamberti, Y., Serra, D., Berbers, G., van der Pol, W.-L., and Rodriguez, M. 2007. Iron stress increases *Bordetella pertussis* mucin-binding capacity and attachment to respiratory epithelial cells. FEMS Immunol. Med. Microbiol. 51: 414–421.

Randell, S., Boucher, R., and University of North Carolina Virtual Lung group. 2006. Effective mucus clearance is essential for respiratory health. Am. J. Respir. Cell. Mol. Biol. 35: 20–28.

Reiche, E., Nunes, S., and Morimoto, H. 2004. Stress, depression, the immune system, and cancer. Lancet Oncol. 5: 617–625.

Schleimer, R.P., Kato, A., Kern, R., Kuperman, D., and Avila, P. 2007. Epithelium: at the interface of innate and adaptive immune responses. J. Allergy Clin. Immunol. 120: 1279–1284.

Van der Vliet, A., Eiserich, J., and Cross C. 2000. Nitric oxide: a pro-inflammatory mediator in lung disease? Respir. Res. 1: 67–72.

Vlisidou, I., Lyte, M., van Diemen, P., Hawes, P., Monaghan, P., Wallis, T., and Stevens, M. 2004. The neuroendocrine stress hormone norepinephrine augments *Escherichia coli* O157:H7-induced enteritis and adherence in a bovine ligated ileal loop model of infection. Infect. Immun. 72: 5446–5451.

Wilson, R., Dowling, R., and Jackson, A. 1996. The biology of bacterial colonization and invasion of the respiratory mucosa. Eur. Respir. J. 9: 1523–1530.

Yamazakki, Y., Danelishvili, L., Wu, M., Hidaka, E., Katsuyama, T., Stang, B., Petrofsky, M., Bildfell, R., and Bermudez, L.E. 2006. The ability to form biofilm influences *Mycobacterium avium* invasion and translocation of bronchial epithelial cells. Cell Microbiol. 8: 806–814.

Zuercher, A.W., Coffin, S.E., Thurnheer, M.C., Fundova, P., and Cebra, J.J. 2002. Nasal-associated lymphoid tissue is a mucosal inductive site for virus-specific humoral and cellular immune responses. J. Immunol. 168: 1796–1803.

Chapter 11
Psychological Stress, Immunity, and the Effects on Indigenous Microflora

Michael T. Bailey

11.1 Introduction

It is well known that bidirectional communication exists between the brain and the peripheral organs such that the central nervous system (CNS) can impact organ functioning, and physiological changes in the body can affect the CNS. However, the extent of this communication, the mechanisms through which they occur, and the impact on health are still only beginning to be defined. Current research within the field of PsychoNeuroImmunology (PNI) has clearly shown that different emotional states, or exposure to psychological stressors, are associated with enhanced susceptibility or increased severity of diseases through nervous system-induced alterations in innate and adaptive immunity. And, it is becoming evident that other more primitive defenses, such as the intestinal microflora, are also affected by exposure to psychological stressors (Freestone et al. 2008). Moreover, stressor-induced bacterial translocation of microflora from mucosal surfaces to secondary lymphoid organs may lead to inflammation and/or altered activation of adaptive immunity. This chapter describes the effects of psychological stressors on the gastrointestinal (GI) tract and presents data showing that the stress response affects the number of bacteria residing as part of the intestinal microflora and their ability to translocate to regional lymph nodes. These findings will be discussed within the context of host defense against infectious diseases.

M.T. Bailey (✉)
Division of Oral Biology, College of Dentistry, The Ohio State University,
305 W. 12th Ave, Columbus, OH 43210, USA
e-mail: bailey.494@osu.edu

M. Lyte and P.P.E. Freestone (eds.), *Microbial Endocrinology*,
Interkingdom Signaling in Infectious Disease and Health,
DOI 10.1007/978-1-4419-5576-0_11, © Springer Science+Business Media, LLC 2010

11.2   Psychological Stress, the Stress Response, and the Impact on Immunity

Stress is an intrinsic part of life, and successfully adapting to stimuli that induce stress is necessary for the survival of an organism in its environment that is constantly changing. Although there is not a commonly used definition of stress, the concept of stress is often broken down into the challenge (called the stressor) and the behavioral and physiological responses to this challenge (called the stress response). A stressor is any stimulus that disrupts internal homeostasis, and can involve psychological, physical, or physiological stimuli. Initiation of the response to physiological and physical stressors is often subconscious and completely biological in nature. But, psychological stressors evoke an additional cognitive processing where the stressors must first be encoded as exceeding the organism's ability to cope with the demand. This cognitive processing sets into motion a coordinated behavioral and physiological response that is similar to the response to physiological and physical stressors. Two neuroendocrine pathways are major contributors to the stress response, namely, the hypothalamic–pituitary–adrenal (HPA) axis and the sympathetic nervous system (SNS). Activation of the HPA results in increased circulatory levels of adrenocorticotrophic hormone (ACTH) produced by the pituitary gland as well as mineralcorticoid and glucocorticoid hormones derived from the adrenal cortex. In contrast, SNS activation results in the release of norepinephrine (NE) from sympathetic nerve termini in SNS innervated tissues, including the GI tract and lymphoid tissues. As such, periods of stress are associated with increases in circulating glucocorticoid hormones (primarily cortisol in humans and corticosterone in rodents) as well as increased circulating and tissue levels of NE. These hormones have a variety of effects throughout the body, such as mobilizing energy for the well known "fight-or-flight" response, that are all aimed at helping the body respond to the demands being placed on it.

Research in the field of PNI has amply demonstrated that stressful periods are associated with exacerbations of a variety of different diseases. For example, it has been demonstrated that individuals reporting higher levels of stress in their daily lives are more likely to develop clinical symptoms during experimental respiratory viral infection (Cohen 2005). To determine if these effects are due to stressor-induced immunosuppression, many researchers have studied the immune response to vaccination during stressful situations, and have found that stressors influence antibody and T-cell responses to vaccines. For example, it was demonstrated in medical students that responsiveness to hepatitis B vaccination was significantly reduced during final exams, an effect found to be associated with stress perception and feelings of loneliness (Glaser et al. 1992; Jabaaij et al. 1996). Likewise, the chronic stress associated with caring for a spouse with Alzheimer's disease (AD) resulted in lower antibody responses to influenza vaccination (Kiecolt-Glaser et al. 1996). Determining the mechanisms through which these stressors affect immune reactivity in humans is difficult, but many animal studies demonstrate that stressor-induced hormones are in fact responsible for the stressor-induced exacerbations of infectious diseases. For example, stressor-induced elevations in corticosterone have been found to suppress lymphocyte trafficking and cytokine production during influenza viral infection (Dobbs et al. 1996; Hermann et al. 1995), as well as antigen processing and presentation by dendritic cells infected with recombinant vaccinia virus (Elftman et al. 2007; Truckenmiller et al. 2005, 2006). The anti-inflammatory effects of glucocorticoid hormones are now well known, and it is evident that glucocorticoid hormones suppress inflammatory cytokine production in part through negative regulation of NF-κB activation and function (Sternberg 2006).

The catecholamines can also have immunomodulatory effects through activation of adrenergic receptors. Animal models have demonstrated that adrenergic signaling is responsible for stressor-induced suppression of cytolytic CD8+ T cell responses during influenza viral infection (Dobbs et al. 1993). Likewise, an acute cold/restraint stressor significantly suppressed the CD4+ T cell response to *Listeria monocytogenes* infection through a β1-adrenergic receptor mediated mechanism (Cao et al. 2003). Ex vivo and in vitro data has revealed that catecholamine stimulation of β-adrenergic receptors at the time of immune challenge, suppresses cytokine production, NK cell activity, and T cell proliferation. In this case, cAMP is thought to be involved in this catecholamine induced immunosuppression (Padgett and Glaser 2003).

Under some circumstances, though, stressors can also enhance certain components of the immune response, particularly the innate immune response. For example, Lyte et al. (1990) demonstrated that exposing mice to a social stressor, called Social Conflict, significantly increased the phagocytic capacity of elicited peritoneal macrophages (Lyte et al. 1990). And, rats exposed to acute shock as a stressor produce higher levels of nitric oxide upon subcutaneous bacterial challenge (Campisi et al. 2002). Because in vitro studies have shown that culturing macrophages with NE increases phagocytosis (Garcia et al. 2003) and the production of nitric oxide (Chi et al. 2003), it is likely that stressor-induced increases in phagocyte activity are NE dependent.

These studies reflect the complex nature of the impact of neuroendocrine hormones on the immune response. The field of microbial endocrinology (Lyte 2004) has added an additional layer of complexity by demonstrating that microbes themselves can be influenced by stressor-induced hormones. Moreover, research by our group and by others have shown that more primitive defense mechanisms, such as microbial barrier defenses at cutaneous and mucosal surfaces, can also be affected by the stress response. These studies are a logical extension of previous findings within the fields of PNI and microbial endocrinology, and will be discussed within the context of stress physiology and infectious disease.

11.3 Overview of the Indigenous Microflora

The human body harbors an enormous microflora that even in the healthy host grossly outnumbers cells of the body by a factor of 10 (i.e., approximately 10^{14} bacterial cells:10^{13} human cells) (Berg 1996, 1999). These bacteria are generally referred to as the microflora and colonize all external surfaces of the body, such as the skin, oral and nasal cavities, upper respiratory tract, urinary tract, and reproductive tract. The GI tract, however, is the main reservoir of bacteria and harbors roughly 90% of the microflora. Molecular analysis of the intestinal microflora using 16s ribosomal RNA have increased previous culture-based estimates of between 400–500 species in the intestines to as high as 1,800 genera and 15,000–36,000 different individual species (Frank et al. 2007). As a result of this high bacterial load and great diversity, the microflora genome is estimated to contain more than 100 times as many genes as the human genome (Gill et al. 2006).

The microflora of the body are not simply opportunistic colonizers or potential pathogens. Rather, the microflora are true symbiotic organisms that have many beneficial effects on the host. Although metabolic activities have been attributed to the intestinal microflora, such as the synthesis of vitamin K and vitamin B complex and the conversion of precarcinogens and carcinogens to noncarcinogens, many studies have focused on the importance of the intestinal microflora for maintenance of mucosal immunity. These effects have been well studied using germ free mice, which are known to have reduced levels of serum immunoglobulins, smaller Peyer's patches, fewer intraepithelial lymphocytes, and a diminished capacity to produce cytokines (reviewed in (Shanahan 2002)). Interestingly, introducing intestinal microflora to these germ free mice restores many (but not all) components of the mucosal immune system. (Stepankova et al. 1998; Gordon et al. 1997; Umesaki et al. 1993, 1995).

In addition to stimulating GI physiology and mucosal immunity, the intestinal microflora can directly prevent diseases by creating a barrier to potential pathogens. Colonization exclusion of new strains of bacteria from the external environment is an essential function of the microflora and disruption of this barrier can facilitate pathogen colonization. Two bacterial types are often associated with colonization exclusion, members of the genus *Bifidobacterium*, and members of the genus *Lactobacillus*. Ely Metchnikoff speculated nearly 100 years ago that lactic acid bacteria (such as *Lactobacillus* spp.) were health-promoters, able to limit pathogen colonization and proliferation (Metchnikoff 1908). The development of reliable in vitro models has helped to define the mechanisms through which the microflora provide protection. And, it is now known that attachment of *Lactobacillus acidophilus*, *Bifidobacterium breve*, and *B. infantis* to intestinal cells creates a physical barrier to enteric pathogens, such as enteropathogenic *Escherichia coli*, *Yersinia pseudotuberculosis*, and *Salmonella typhimurium* (Bernet et al. 1993; Coconnier et al. 1993a, b). Moreover, ingestion of probiotic bacteria, i.e., bacteria ingested for their beneficial effects, such as probiotic lactobacilli, significantly affects the LD_{50} of many enteric pathogens (Coconnier et al. 1998; Bernet-Camard et al. 1997; Hudault et al. 1997) and reduces the severity of experimental infection with *Helicobacter pylori* or *Citrobacter rodentium* (Johnson-Henry et al. 2004, 2005). As such, our studies focused primarily on assessing the impact of psychological stress on the levels of *Lactobacillus* spp. and *Bifidobacterium* spp.

11.4   Stress-Induced Alterations in Intestinal Microflora

The number and types of bacteria that reside as part of the indigenous microflora are thought to be relatively stable, but environmental and physiological challenges have been shown to disrupt this stability. For example, early studies by Schaedler and Dubos (1962) demonstrated that rehousing mice into new cages significantly decreased lactobacilli levels (Schaedler and Dubos 1962). And, chronic sleep deprivation in rats was shown to induce a significant overgrowth of microflora in the ileum and cecum (Everson and Toth 2000), with more recent studies indicating that intrinsic factors such as age and gender can also affect the composition of the microflora of laboratory animals (Ge et al. 2006). Fewer studies have focused on environmental affects on the microflora of humans, but an early study in cosmonauts demonstrated that the intestinal microflora were significantly affected during space flight (Lizko 1987), with others suggesting that some of the effects could be due to the stress of confinement (Holdeman et al. 1976). To further study the potential impact of psychological stress on the stability of the intestinal microflora, we assessed the microflora of young rhesus monkeys that were being separated from their mothers for husbandry purposes (Bailey and Coe 1999).

In captive colonies, rhesus monkeys are routinely separated from their mothers at approximately 6 months of age. At this age, the monkeys are no longer nursing and are eating solid foods. Yet, they still show a strong physiological and emotional reaction to separation from their mothers. This transition from living with the mother to living with other peer monkeys is associated with an increased incidence of diseases, including GI diseases. While much of this can be explained by exposure to new contagion or the actions of the nervous system on the immune system, we hypothesized that the stress response during maternal separation could significantly affect microflora levels in the infants, and thus reduce the barrier effects of the intestinal microflora.

Culture-based enumeration of shed microflora revealed significant alterations in bacterial levels the week following maternal separation compared to levels when the infants were still residing with their mothers. This was evident for Gram-negative and total aerobic and facultatively anaerobic microflora, but only reached statistical significance when a single genus of bacteria was enumerated. The number of aerobically grown lactobacilli was significantly altered after maternal separation (Fig. 11.1). In most cases, the alterations followed a standard profile of increased levels immediately after separation, followed by significantly lower levels 3 days after separation and a return to baseline by the end of the week. Interestingly, the magnitude of the reduction in microflora 3 days after maternal separation could be predicted by the infants' behavior on day 2 post-separation. Three stress-indicative behaviors, cooing, barking, and lip smacking, were associated with microflora levels; in general, those animals that had the highest number of stress-indicative behaviors shed the fewest lactobacilli and total aerobic and facultatively anaerobic bacteria on day 3 post-separation (Fig. 11.2) (Bailey and Coe 1999).

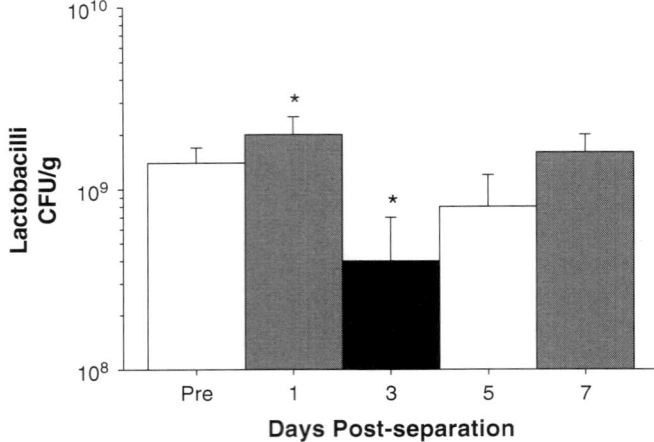

Fig. 11.1 Aerobically grown lactobacilli were enumerated from coprocultures before and for 1 week following maternal separation. Results are mean (S.E.) number of colony forming units (CFU) per gram of fecal matter (wet weight). *$p < 0.05$ versus preseparation values. Reproduced from Developmental Psychobiology, 1999 with permission from Wiley

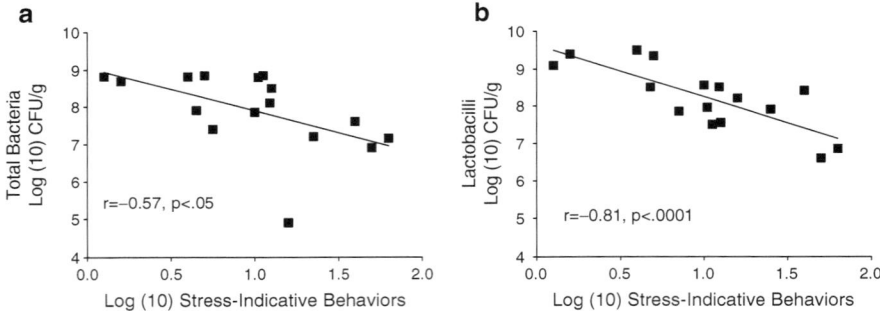

Fig. 11.2 Log transformed stress-indicative behaviors were significantly associated with log(10) CFU/g of intestinal microflora. (**a**) Total aerobic and facultatively anaerobic microflora. (**b**) Aerobically grown lactobacilli. Reproduced from Developmental Psychobiology, 1999 with permission from Wiley

Lactic acid bacteria, such as members of the genus *Lactobacillus*, are thought to be important contributors to microflora-mediated colonization exclusion. Thus, stressor-induced reductions in lactobacilli would be hypothesized to be associated with enhanced susceptibility to enteric infection. In this experiment, none of the monkeys were intentionally infected, but many nonhuman primate colonies have endemic levels of enteric pathogens, notably *Shigella flexneri* and *Campylobacter jejuni*. And, 45% (i.e., 9/20) of the infant monkeys became colonized with either *S. flexneri* or *C. jejuni* during the week following maternal separation. On the first day that pathogen colonization was observed there was a weak, marginally significant ($p = 0.07$) inverse association between the number of lactobacilli and pathogens shed from the intestines (Bailey and Coe 1999). These data are consistent with the idea that lactobacilli are important in colonization resistance against enteric pathogens, but further studies are needed to conclude that stressor-induced alterations in microflora result in increased susceptibility to enteric infection.

Stressor-induced reductions in lactobacilli have also been found in college students during stressful periods (Knowles et al. 2008). In this study, lactobacilli levels were determined during a low stress period (i.e., the first week of the semester) and a high stress period (i.e., final exam week). The exam period was associated with significantly higher levels of perceived daily stress and weekly stress, as well as an increase in GI upset. Moreover, when compared to the low stress period, levels of lactic acid bacteria, primarily lactobacilli, shed in the stool were significantly lower for up to 5 days following examination, with differences in bacterial levels reaching one half log unit in magnitude (e.g., baseline values of 6×10^7 CFU/ml vs. 1×10^7 on day 5 post-examination). It should be noted, however, that significant differences in diet did occur across the two time periods; most notable were significant reductions in vegetable consumption and a significant increase in coffee consumption (Knowles et al. 2008). But, given that stressor exposure alters lactobacilli levels in laboratory animals fed a standardized diet, it is likely that stress-associated changes in human microflora reflect an impact of the stressor as well as potential effects of diet.

Healthy adults are somewhat resistant to the impact of stressors on various physiological systems. For example, stressor-induced alterations in the immune response tend to return to baseline upon termination of the stress response. However, the stress response can have a more prolonged effect on immunity in the very old and the very young (Coe and Lubach 2003). And, stressor exposure in the very young, or even during the prenatal period, is thought to set the infant on a significantly different developmental trajectory, resulting in larger stressor induced effects later in life (Coe and Lubach 2003). One of the most consistent findings in regards to exposure to prenatal stressors is that fetal growth and birth weight are reduced after women experience stressful situations during pregnancy (Field et al. 1985; Lederman et al. 1981; Lederman 1986). Rhesus monkeys have been used extensively to investigate the influence of prenatal stress on infant development. And, it has been shown that prenatal stress affects nueromotor development (Schneider and Coe 1993), emotional reactivity to stressors (Clarke and Schneider 1993), brain monoamine levels (Schneider et al. 1998), cell density in the brain (Coe et al. 2002, 2003), and immune reactivity (Coe et al. 1996, 1999, 2007). Our studies focused on the impact of gestational stress on the intestinal microflora across the four phases of microflora development.

Bacteria colonize the GI tract of newborns in a sequential pattern that is tightly related to developmental milestones in the infant (Cooperstock and Zed 1983). The first phase of colonization begins at birth when bacteria from the mother's reproductive tract colonize the otherwise sterile newborn. These bacteria do not predominate for long and are quickly overcome by maternal aerobic intestinal microflora, which are thought to persist in the intestines for the first few days of life (Tannock et al. 1990). These aerobic species, such as *E. coli* and *Streptococcus* spp. consume molecular oxygen as they grow and begin to reduce the oxidation–reduction potential in the intestines creating a more favorable environment for the growth of anaerobic species (Meynell 1963). As a result, high levels of Enterobacteriaceae are evident 1 day after birth, but anaerobes, such as bifidobacteria, predominate by 6 days of age and throughout the period of exclusive breast feeding (Sakata et al. 1985).

Members of the genus *Bifidobacterium* thrive in breastfed infants and are the predominant bacteria in the intestines due to growth factors found in human milk that bifidobacteria readily use for energy, such as lactose. As bifidobacteria grow, they produce pronounced levels of lactic and acetic acids that can not be buffered by human milk, thus inhibiting the growth of acid sensitive microbes. Breast milk also contains large amounts of immune factors, such as secretory immunoglobulins, lactoferrin, lysozymes, and even leukocytes that can inhibit colonization of certain bacteria (Balmer and Wharton 1991; Wharton et al. 1994a, b). The combination of immune factors and acidic fermentation products gives bifidobacteria a tremendous ecological advantage over other species (Heine et al. 1992; Beerens et al. 1980).

The initiation of weaning from breast milk is associated with a resurgence of aerobic and facultatively anaerobic species, such as *E. coli*, *Streptococci*, and *Clostridia* spp., that are naturally found in newly ingested foods. The concentrations of these newly arrived bacteria fluctuate greatly during this period, but as the diet becomes more consistent, microbial populations in the intestines also stabilize and will remain quite stable throughout the lifespan. This stability is important for maintaining intestinal homeostasis (O'Hara and Shanahan 2006), and if disrupted could contribute to the development of GI infections or cancers (O'Hara and Shanahan 2006).

11.5   Prenatal Stressor-Induced Alterations to Microflora Development

To determine the impact of a prenatal stressor on microflora development, an acoustical startle stressor (i.e., 3 random 110 dB beeps over a 10 min period occurring 5 days per week) was used to evoke a stress response from pregnant rhesus monkeys either early (days 50–92) or late (days 105–147) in the 169 day gestational period. These periods represent crucial time periods in nervous system and GI system development, thus making it likely that disruption of physiological homeostasis at these time points affects fetal development. This stressor resulted in a significant increase in cortisol in the pregnant mothers, but did not appear to significantly affect the number of miscarriages, gestational length, or birth weight (Bailey et al. 2004b). The stressor did, however, significantly affect the development of the intestinal microflora.

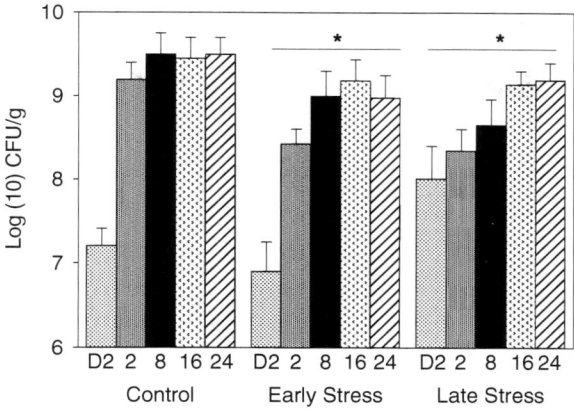

Published in the *Journal of Pediatric Gastroenterology and Nutrition*, ©Lippincott, Williams & Wilkins, 2004

Fig. 11.3 Anaerobically grown *Lactobacillus* spp. during the first 24 weeks of life. Data are the mean (SE) of log(10) transformed number of colony forming units per gram of fecal matter (CFU/g). Concentrations on day 2 of life were not significantly different between pregnancy conditions. * Both Early Stress and Late Stress infants had significantly fewer anaerobic lactobacilli than did control infants across the first 24 weeks of life ($p < 0.05$). In addition, there was a developmental trend for increasing titers across the 24 week period in both control and prenatally stressed infants ($p < 0.05$). Reproduced from Journal of Pediatric Gastroenterology and Nutrition, 2004, with permission from Lipincott, Williams, & Wilkins

During the first 6 months of life, lactobacilli levels in the monkeys born from mothers exposed to the stressor during gestation were significantly lower than levels found in infants from non-stressed control mothers, with the biggest differences in mean levels found at 2 weeks of age (Fig. 11.3). As successful nursing progressed, bifidobacteria began to predominate in the intestines. And, as with the lactobacilli, bifidobacteria levels were significantly lower in the intestines of infant monkeys from mothers that were exposed to the acoustical startle stressor during gestation. This effect, however, was only evident in the offspring from mothers exposed to the stressor late in gestation (Fig. 11.3) (Bailey et al. 2004b). As with the previous study involving maternal separation, none of the monkeys in this study were intentionally infected with enteric pathogens. However, approximately 43% of the infants from mothers stressed early in gestation and 12% of infants from mothers stressed late in gestation became subclinically colonized with *Shigella flexneri*, an endemic pathogen in the monkey colony. Importantly, Shigella were not detected in any of the infants born from the non-stressed control condition (Bailey et al. 2004b), suggesting that prenatal stress, particularly late in gestation, disrupted the development of natural resistance to the enteric pathogen, *S. flexneri*.

11.6   Psychological Stress and the GI Tract: Toward a Mechanism of Stressor-Induced Alterations in Microflora

It is tempting to speculate on the mechanisms through which the intestinal microflora could have been altered by stressful pregnancy conditions. For example, it is known that cortisol can affect many aspects of infant development, and many of the effects of prenatal stress on the immune system can be mimicked by administration of ACTH or the synthetic glucocorticoid, dexamethasone (Coe et al. 1996). And, others have found that giving corticosterone to pregnant rats significantly reduced the concentrations of total and Gram-negative aerobes and facultative anaerobes (Schiffrin et al. 1993). The mechanisms through which glucocorticoids might affect the microflora are not known, but fetal development of the gi tract is thought to be influenced by glucocorticoids. For example, maturation of the intestines occurs concomitantly with the prepartum surge in cortisol in precocial species, such as pigs, sheep, and humans (Trahair and Sangild 1997). Moreover, very high levels of glucocorticoids adversely affect intestinal development, such as the ability to secrete gastric acid and the densitivity of villi and crypts (Sangild et al. 1994), thus changing the microenvironment in the intestines and opening the possibility of shifts in ecological competition.

Altering the microenvironment may also affect established microflora populations in older hosts. The complete set of factors controlling the types of bacteria that can reside as part of the intestinal microflora are not well understood, but it is thought that the host plays a role in "selecting" the microflora. This was elegantly demonstrated by Rawls et al. (2006), who reciprocally transplanted gut bacteria between mice and zebrafish. After transplantation, gut microbial populations shifted to reflect the proportions of bacteria found in the microflora of conventionally reared recipient animals, and no longer reflected the microflora of the original donor animal (Rawls et al. 2006). This may in part be due to the physiology of the host GI tract, since certain aspects of GI physiology are known to influence microbial populations in the GI tract. For example, it is well known that for bacteria to take up residence in the gi tract, they must first survive the low pH of the stomach. Therefore, it is not surprising that reduced production of gastric acid (as occurs with hypochlorhydria) results in overgrowth of bacteria in the gi tract (Drasar et al. 1969). Some species, however, such as members of the genera *Lactobacillus* and *Bifidobacteria* are acid tolerant and are able to grow in the low pH (Drasar et al. 1969). This acid tolerance gives the genera an ecological advantage over other species that compete to colonize the gi tract. Therefore, a logical hypothesis is that any stimulus that disrupts gastric acid production will in turn affect intestinal microflora levels.

The influences of emotional states on the secretion and motility of the GI system were documented as early as 1833, when the surgeon William Beaumont noted that the secretion of gastric juice was decreased or abolished during periods of anger or fear in his patient with a gastric fistula (Beaumont 1838). Experimental data has confirmed this observation, and it is now known that secretion of gastric acid can be suppressed by experimental stressors, such as the cold pressure task and mental arithmetic (Badgley et al. 1969; Holtmann et al. 1990). In animals, different stressors have differential effects on acid secretion, with restraint stress reported to significantly increase or decrease gastric acidity depending upon temperature (Murakami et al. 1985; Lenz et al. 1988). These differences are due to different levels of activation of the sympathetic and parasympathetic nervous systems; activating the SNS suppressed whereas activating the PNS enhanced acid secretion (Yang et al. 2000). Research is needed to determine whether gi acidity plays a role in stressor-induced alterations of microflora.

There are, of course, additional secretory products that can affect microflora levels and are themselves influenced by the stress response such as additional digestive products like bile, and immune products like secretory immunoglobulin A (sIgA) and antimicrobial peptides. The use of secretory immunoglobulin deficient mice has shown the importance of this immunoglobulin in influencing microbial populations; sIgA deficient mice have significantly increased populations of anaerobic microflora in the small intestine (Fagarasan et al. 2002). Moreover, antimicrobial peptides, such as the defensins, have been suggested to modify the types and numbers of bacteria colonizing the GI tract (Salzman et al. 2007). Because these molecules can be affected upon exposure to a stressor (Jarillo-Luna et al. 2007; Korneva et al. 1997), an additional plausible hypothesis is that stress-associated alterations of the microflora are dependent upon stressor-induced alterations in sIgA and/or defensins.

Perhaps the most well-studied effects of stress on the gi tract are the effects on GI motility. Animal models have established that stress reduces gastric emptying (Taché et al. 2001; Nakade et al. 2005) and slows transit in the small intestine (Lenz et al. 1988; Kellow et al. 1992) through stressor-induced elevations of corticotrophin releasing hormone (Taché et al. 2001; Nakade et al. 2005). In contrast to the inhibitory effects in the stomach and small intestine, stress tends to enhance motility in the colon due to increased sacral parasympathetic outflow to the large intestine through a CRH dependent circuit (Lenz et al. 1988; Martinez et al. 1997).

Gastrointestinal motility has long been thought to influence microbial populations in the GI tract. For example, slowing peristalsis, and thus motility, by administering high doses of morphine causes significant bacterial overgrowth in the small intestines of rats (MacFarlane et al. 2000; Scott and Cahall 1982). Moreover, data from humans show an association between surgical trauma, stagnation of intestinal motility, and bacterial overgrowth, thus supporting the notion that delayed intestinal motility can result in bacterial overgrowth (Marshall et al. 1988; Nieuwenhuijzen et al. 1996a, b). Interestingly, increased GI motility can also affect microflora levels, with some studies showing a direct correlation between small intestine microflora levels and the rate of peristalsis.

An equally likely explanation is that the intestinal microflora were directly affected by stressor-induced increases in intestinal hormones, such as NE. The primary focus of this book is the exciting finding that bacteria can change their growth characteristics when exposed to hormones. And, the growth of many types of microflora has been shown to be significantly enhanced upon culture with NE (Freestone et al. 2002). Despite the many studies showing bacterial growth enhancement by NE in vitro, demonstrating that these interactions occur in vivo has been challenging. Neuroendocrine–bacterial interactions, however, undoubtedly occur in vivo when NE levels reach high levels. This was evident with the use of the neurotoxin 6-hydroxydopamine, which lyses the nerve terminals of sympathetic neurons resulting in the release of NE that is stored in the nerve terminals (Lyte and Bailey 1997). Thus, even though 6-OHDA is a useful way to chemically sympathectomize laboratory rodents, its initial effect is the release of a large bolus of NE 24 h after injection (Porlier et al. 1977; De Champlain 1971). Interestingly, bacterial levels in the cecums of mice were found to be significantly increased 24 h after administration of 6-OHDA, with *E. coli* showing the greatest increase (Lyte and Bailey 1997). Since the growth of commensal *E. coli* is strongly affected by exposure to NE (Freestone et al. 2002), the data suggest that overgrowth of *E. coli* in the cecums of chemically sympathectomized mice results from direct enhancement of bacterial growth by NE.

Overgrowth of bacteria in the family Enterobacteriaceae is also evident in the intestines of mice exposed to psychological stressors. Our recent studies indicate that restraining mice for prolonged periods (i.e., 16 h per day for 7 days) result in an overgrowth of Enterobacteriaceae in both the small and large intestines (Bailey et al. manuscript under review) as well as in the cecum (Bailey et al. 2006). This overgrowth may have important health implications, since bacterial overgrowth is a precipitating factor in the translocation of bacteria from the gi tract to the rest of body. In fact, the translocation of some species in the family Enterobacteriaceae was found to be directly related to levels in the small intestine and cecum (Steffen and Berg 1983). The finding that exposing mice to psychological stressors can enhance *E. coli* levels in the intestines prompted the determination of the impact of psychological stressors on bacterial transloction.

11.7   Stressor-Induced Bacterial Translocation

Indigenous microflora are not invasive bacteria, which is one property that allows them to reside with their host. Moreover, the external surfaces of the body, i.e., cutaneous and mucosal surfaces, maintain a barrier to external substances, including microbes. In mucosal tissues, transport of solutes into the body is controlled in part through tight junctions between intestinal epithelial cells that prevent the passive transfer of molecules and microbes. Bacteria from mucosal surfaces, however, are routinely sampled by specialized phagocytic cells, called M cells, which engulf mucosal bacteria and pass them to regional lymph nodes in order to initiate an immune response or to maintain tolerance. Most of these bacteria are killed en route, resulting in low to undetectable levels of culturable bacteria in regional lymph nodes.

The epidermal layer of the skin is also well known to provide a permeability barrier that primarily serves to prevent water loss in a potentially desiccating environment. This barrier, however, is also a potent barrier to the passage of cutaneous microflora into the body. As a result, normal mice rarely have detectable levels of bacteria in lymph nodes that drain cutaneous surfaces. In our studies, less than 15% of non-stressed control mice were found to have bacteria in the inguinal lymph nodes that lie under skin in the lower back (Bailey et al. 2006). This percentage was significantly increased when mice were exposed to prolonged restraint or the social stressor, social disruption (SDR), with 82% of mice in both groups identified as having bacteria in these draining lymph nodes. To try to determine whether this effect could simply be due to mechanical breaches in the skin (such as from biting during SDR, or abrasions from the restraint tube), a separate group of mice received full thickness skin biopsies on the lower back. Interestingly, only 36% of mice in this group were found to have bacteria in the inguinal lymph nodes, which was significantly less than the 82% occurrence in the stressed animals (Bailey et al. 2006). These data indicate that the stress response, rather than mechanical barrier breaches, is responsible for the bacterial translocation of cutaneous microflora.

The percentage of mice with bacteria cultured from mesenteric lymph nodes, which drains the GI tract, was higher than the percentage of mice found to have bacteria in the inguinal lymph nodes. We found that 48% of non-stressed control mice were culture-positive for bacteria in the mesenteric lymph nodes. However, exposure to either restraint or SDR increased the occurrence of bacteria in the mesenteric lymph nodes to over 80% (i.e., 82% of SDR mice were culture positive; 91% of restrained mice were culture positive). Interestingly, depriving the mice of food and water did not significantly affect the translocation of indigenous microflora, indicating that the stress response, rather than other physiological variables significantly enhanced bacterial translocation in the gut (Bailey et al. 2006).

There are now several reports indicating that barrier defenses in both the skin and the GI tract can be disrupted by exposure to psychological stressors. Acute experimental stressors in human participants, such as a public speaking tasks and sleep deprivation, were shown to disrupt the permeability barrier in the skin as determined by measuring transepidermal water loss (TEWL) and by determining the water content of the outermost layer of the skin, i.e., the stratum corneum (Altemus et al. 2001). In mice, TEWL was also found to be affected by exposure to different housing conditions and by immobilization stress (Denda et al. 2000). This effect was later found to be due to the impact of glucocorticoid hormones on the stratum corneum (Choi et al. 2006). In addition to physical barrier properties, the skin also produces many antimicrobial peptides, such as β-defensins and cathelicidins. And, these antimicrobial peptides have been shown to be suppressed during stressor exposure through the actions of stressor-induced glucocorticoids and local production of corticotrophin releasing hormone (Aberg et al. 2007). Thus, stressor-induced alterations in the skin permeability barrier, as well as innate defenses, may explain why bacteria were found in the inguinal lymph nodes of stressed mice.

Exposure to psychological stressors can have similar effects in the GI tract, with stressor-induced changes in gut permeability being well defined (Soderholm and Perdue 2001). These effects have primarily been described in laboratory animals, since studying the impact of psychological stressors on GI permeability in humans has been challenging. Several studies have demonstrated that immobilizing rats in a cold environment significantly increased jejunal permeability to ^{51}Cr-EDTA and mannitol (Saunders et al. 1994) via a cholinergic dependent mechanism (Saunders et al. 1997; Soderholm and Perdue 2001). In the colon, permeability was also increased by immobilization in a cold environment, an effect that could be mimicked by peripheral injection of CRH (Saunders et al. 2002; Soderholm and Perdue 2001).

Enhanced microflora growth and increased permeability of barrier defenses may not be sufficient to result in bacterial translocation. In fact, an additional important component of bacterial translocation is the ability of enteric bacteria to adhere to intestinal tissue. Interestingly, stressor-induced neuroendocrine hormones can also enhance the attachment of enteric bacteria. For example, culturing pathogenic *E. coli* O157:H7 with NE significantly increased the ability of the bacteria to adhere to colonic tissue (Chen et al. 2003; Green et al. 2004). Moreover, internalization of pathogenic (i.e., *Salmonella choleraesuis* and *E. coli O157:H7*), but not necessarily commensal, bacteria was enhanced by treating porcine Peyer's patch mucosa with NE in an Ussing chamber paradigm (Green et al. 2003).

These studies demonstrate that exposure to psychological stressors affects many aspects of host physiology. And, many of these effects have the capacity to alter the commensal microflora. When considered together, a likely scenario emerges from the data in which exposure to a psychological stressor results in a neuroendocrine response that has the potential to directly or indirectly affect commensal microflora populations, the integrity of barrier defenses, and the internalization of microbes. Delineating whether the effects of stress on the microflora are direct effects, i.e., whether stress hormones themselves affect microflora in vivo, or indirect, i.e., through modulation of the microenvironment in which commensals interact with their host, will be a challenge for future studies. However, as animal models to study the interactions between the microflora and their host continue to be developed, insight into the impact of stress on these interactions will undoubtedly follow.

11.8 An Integrative Hypothesis of Stress, Infection, and Immunity

The importance of stressor-induced alterations in commensal microbial populations and translocation to regional lymph nodes is only beginning to be understood. Although the ability of microbial populations to limit pathogen colonization and invasion has been known for many years, it is now thought that these commensal microbes help to regulate the immune system as well. In GI tissue, for example, inflammation is low despite the enormous antigenic potential of the billions of colonized bacteria. These commensal bacteria, though, may actually be active players in maintaining homeostasis through the suppression of innate pattern recognition receptor signaling. For example, signaling through the Toll-like receptors (TLR), which results in cytokine production, is negatively regulated through Toll-interacting protein (Tollip) and single immunoglobulin IL-1R-related molecule (SIGGR) (O'Hara and Shanahan 2006). Importantly, Tollip expression is directly correlated with microflora levels; the highest levels of Tollip are found in healthy colonic tissue that also has the highest microflora levels (O'Hara and Shanahan 2006; Otte et al. 2004). Thus, as long as healthy levels of microflora are maintained within the GI lumen, inflammation may be actively suppressed.

If the stress response facilitates the passage of bacteria from the lumen into the intestinal tissues, however, the TLRs found on the basolateral surface of enterocytes would then be able to respond to translocating bacteria to initiate an inflammatory response (Abreu et al. 2001; Otte et al. 2004). Moreover, there is an immense network of phagocytes and antigen presenting cells residing below the enterocytes within the GI tissue that can respond to translocating microbes and cause a local or systemic inflammatory response. Thus, it is possible that innate receptors, such as the TLRs, are silenced in the presence of normal luminal levels of microflora, but activated when microflora levels are altered or when they translocate into the tissue.

The ability of stressors to induce and/or enhance the inflammatory response is now well recognized. For example, stressor-induced elevations in circulating inflammatory markers have been found in uninfected humans (Brydon et al. 2005, 2006; Steptoe et al. 2007; Coussons-Read et al. 2007; Bierhaus et al. 2003) as well as in uninfected rodents (Avitsur et al. 2001, 2002; Bailey et al. 2007; Engler et al. 2008; Stark et al. 2001). In rodents, it has been shown that exposure to certain stressors causes leukocytes, especially monocytes/macrophages, to become resistant to the suppressive effects of corticosterone (Avitsur et al. 2001; Bailey et al. 2004a; Engler et al. 2005; Stark et al. 2001). Moreover, these cells have a primed phenotype and show exaggerated inflammatory responses to ex vivo stimulation with LPS or even intact bacteria (Avitsur et al. 2003; Bailey et al. 2007), effects that were associated with the inability of glucocorticoids to suppress the activation of the transcription factor NF-κB (Quan et al. 2003). Human studies employing gene chip technology corroborate findings from murine studies, and have found an underrepresentation of genes containing a glucocorticoid response element and an overrepresentation of genes controlled by the transcription factor NF-κB (which controls the transcriptional expression of inflammatory cytokines) in uninfected individuals reporting high levels of psychological stress (Miller et al. 2008; Cole et al. 2007). These data suggest that periods of psychological stress in humans are associated with an inflammatory profile that is not able to be controlled by endogenous glucocorticoids.

The question remains, however, why cells of the innate immune system become activated in the absence of an active infection. While it has been shown that treating cells in culture with NE can result in the production of cytokines (Tan et al. 2007), most immunologists would argue that activation of cells with inflammatory stimuli, such as microbes or microbe-associated molecules, is necessary for the production of appreciable amounts of cytokines. Given the impact of stressors on microbial populations and translocation of microbes or microbe-associated molecules, like

LPS, into the body, a reasonable hypothesis is that stressor-induced alterations and translocation of the indigenous microflora activates and/or primes the immune system and are partly responsible for stressor-induced elevations in circulating inflammatory cytokines. One remaining challenge for the field of microbial endocrinology is to test this hypothesis and determine whether stressor-induced alterations of commensal microflora can shift the balance from health to disease.

References

Aberg, K. M., Radek, K. A., Choi, E. H., Kim, D. K., Demerjian, M., Hupe, M., Kerbleski, J., Gallo, R. L., Ganz, T., Mauro, T., Feingold, K. R., and Elias, P. M. 2007. Psychological stress downregulates epidermal antimicrobial peptide expression and increases severity of cutaneous infections in mice. J. Clin. Invest. 117:3339–3349.

Abreu, M. T., Vora, P., Faure, E., Thomas, L. S., Arnold, E. T., and Arditi, M. 2001. Decreased expression of Toll-like receptor-4 and MD-2 correlates with intestinal epithelial cell protection against dysregulated proinflammatory gene expression in response to bacterial lipopolysaccharide. J. Immunol. 167:1609–1616.

Altemus, M., Rao, B., Dhabhar, F. S., Ding, W., and Granstein, R. D. 2001. Stress-induced changes in skin barrier function in healthy women. J. Invest. Dermatol. 117:309–317.

Avitsur, R., Stark, J. L., and Sheridan, J. F. 2001. Social stress induces glucocorticoid resistance in subordinate animals. Horm. Behav. 39:247–257.

Avitsur, R., Stark, J. L., Dhabhar, F. S., and Sheridan, J. F. 2002. Social stress alters splenocyte phenotype and function. J. Neuroimmunol. 132:66–71.

Avitsur, R., Padgett, D. A., Dhabhar, F. S., Stark, J. L., Kramer, K. A., Engler, H., and Sheridan, J. F. 2003. Expression of glucocorticoid resistance following social stress requires a second signal. J. Leukoc. Biol. 74:507–513.

Badgley, L. E., Spiro, H. M., and Senay, E. C. 1969. Effect of mental arithmetic on gastric secretion. Psychophysiology 5:633–637.

Bailey, M. T., and Coe, C. L. 1999. Maternal separation disrupts the integrity of the intestinal microflora in infant rhesus monkeys. Dev. Psychobiol. 35:146–155.

Bailey, M. T., Avitsur, R., Engler, H., Padgett, D. A., and Sheridan, J. F. 2004a. Physical defeat reduces the sensitivity of murine splenocytes to the suppressive effects of corticosterone. Brain Behav. Immun. 18:416–424.

Bailey, M. T., Lubach, G. R., and Coe, C. L. 2004b. Prenatal stress alters bacterial colonization of the gut in infant monkeys. J. Pediatr. Gastroenterol. Nutr. 38:414–421.

Bailey, M. T., Engler, H., and Sheridan, J. F. 2006. Stress induces the translocation of cutaneous and gastrointestinal microflora to secondary lymphoid organs of C57BL/6 mice. J. Neuroimmunol. 171:29–37.

Bailey, M. T., Engler, H., Powell, N. D., Padgett, D. A., and Sheridan, J. F. 2007. Repeated social defeat increases the bactericidal activity of splenic macrophages through a Toll-like receptor-dependent pathway. Am. J. Physiol. Regul. Integr. Comp. Physiol. 293:R1180–R1190.

Balmer, S. E., and Wharton, B. A. 1991. Diet and faecal flora in the newborn: iron. Arch. Dis. Child. 66:1390–1394.

Beaumont, W. 1838. Experiments and observations on the gastric juice and the physiology of digestion. London: Edinburgh.

Beerens, H., Romond, C., and Neut, C. 1980. Influence of breast-feeding on the bifid flora of the newborn intestine. Am. J. Clin. Nutr. 33:2434–2439.

Berg, R. D. 1996. The indigenous gastrointestinal microflora. Trends Microbiol. 4:430–435.

Berg, R. D. 1999. Bacterial translocation from the gastrointestinal tract. Adv. Exp. Med. Biol. 473:11–30.

Bernet, M. F., Brassart, D., Neeser, J. R., and Servin, A. L. 1993. Adhesion of human bifidobacterial strains to cultured human intestinal epithelial cells and inhibition of enteropathogen–cell interactions. Appl. Environ. Microbiol. 59:4121–4128.

Bernet-Camard, M. F., Lievin, V., Brassart, D., Neeser, J. R., Servin, A. L., and Hudault, S. 1997. The human *Lactobacillus acidophilus* strain LA1 secretes a nonbacteriocin antibacterial substance(s) active in vitro and in vivo. Appl. Environ. Microbiol. 63:2747–2753.

Bierhaus, A., Wolf, J., Andrassy, M., Rohleder, N., Humpert, P. M., Petrov, D., Ferstl, R., von Eynatten, M., Wendt, T., Rudofsky, G., Joswig, M., Morcos, M., Schwaninger, M., McEwen, B., Kirschbaum, C., and Nawroth, P. P. 2003. A mechanism converting psychosocial stress into mononuclear cell activation. Proc. Natl. Acad. Sci. USA 100:1920–1925.

Brydon, L., Edwards, S., Jia, H., Mohamed-Ali, V., Zachary, I., Martin, J. F., and Steptoe, A. 2005. Psychological stress activates interleukin-1beta gene expression in human mononuclear cells. Brain Behav. Immun. 19:540–546.

Brydon, L., Magid, K., and Steptoe, A. 2006. Platelets, coronary heart disease, and stress. Brain Behav. Immun. 20:113–119.

Campisi, J., Leem, T. H., and Fleshner, M. 2002. Acute stress decreases inflammation at the site of infection. A role for nitric oxide. Physiol. Behav. 77:291–299.

Cao, L., Hudson, C. A., and Lawrence, D. A. 2003. Acute cold/restraint stress inhibits host resistance to *Listeria monocytogenes* via beta1-adrenergic receptors. Brain Behav. Immun. 17:121–133.

Chen, C., Brown, D. R., Xie, Y., Green, B. T., and Lyte, M. 2003. Catecholamines modulate *Escherichia coli* O157:H7 adherence to murine cecal mucosa. Shock 20:183–188.

Chi, D. S., Qui, M., Krishnaswamy, G., Li, C., and Stone, W. 2003. Regulation of nitric oxide production from macrophages by lipopolysaccharide and catecholamines. Nitric Oxide 8:127–132.

Choi, E. H., Demerjian, M., Crumrine, D., Brown, B. E., Mauro, T., Elias, P. M., and Feingold, K. R. 2006. Glucocorticoid blockade reverses psychological stress-induced abnormalities in epidermal structure and function. Am. J. Physiol. Regul. Integr. Comp. Physiol. 291:R1657–R1662.

Clarke, A. S., and Schneider, M. L. 1993. Prenatal stress has long-term effects on behavioral responses to stress in juvenile rhesus monkeys. Dev. Psychobiol. 26:293–304.

Coconnier, M. H., Bernet, M. F., Chauviere, G., and Servin, A. L. 1993a. Adhering heat-killed human *Lactobacillus acidophilus*, strain LB, inhibits the process of pathogenicity of diarrhoeagenic bacteria in cultured human intestinal cells. J. Diarrhoeal Dis. Res. 11:235–242.

Coconnier, M. H., Bernet, M. F., Kerneis, S., Chauviere, G., Fourniat, J., and Servin, A. L. 1993b. Inhibition of adhesion of enteroinvasive pathogens to human intestinal Caco-2 cells by *Lactobacillus acidophilus* strain LB decreases bacterial invasion. FEMS Microbiol. Lett. 110:299–305.

Coconnier, M. H., Lievin, V., Hemery, E., and Servin, A. L. 1998. Antagonistic activity against *Helicobacter* infection in vitro and in vivo by the human *Lactobacillus acidophilus* strain LB. Appl. Environ. Microbiol. 64:4573–4580.

Coe, C. L., and Lubach, G. R. 2003. Critical periods of special health relevance for psychoneuroimmunology. Brain Behav. Immun. 17:3–12.

Coe, C. L., Lubach, G. R., Karaszewski, J. W., and Ershler, W. B. 1996. Prenatal endocrine activation alters postnatal cellular immunity in infant monkeys. Brain Behav. Immun. 10:221–234.

Coe, C. L., Lubach, G. R., and Karaszewski, J. W. 1999. Prenatal stress and immune recognition of self and nonself in the primate neonate. Biol. Neonate 76:301–310.

Coe, C. L., Lulbach, G. R., and Schneider, M. L. 2002. Prenatal disturbance alters the size of the corpus callosum in young monkeys. Dev. Psychobiol. 41:178–185.

Coe, C. L., Kramer, M., Czeh, B., Gould, E., Reeves, A. J., Kirschbaum, C., and Fuchs, E. 2003. Prenatal stress diminishes neurogenesis in the dentate gyrus of juvenile rhesus monkeys. Biol. Psychiatry 54:1025–1034.

Coe, C. L., Lubach, G. R., and Shirtcliff, E. A. 2007. Maternal stress during pregnancy predisposes for iron deficiency in infant monkeys impacting innate immunity. Pediatr. Res. 61:520–524.

Cohen, S. 2005. Keynote Presentation at the Eight International Congress of Behavioral Medicine: the Pittsburgh common cold studies: psychosocial predictors of susceptibility to respiratory infectious illness. Int. J. Behav. Med. 12:123–131.

Cole, S. W., Hawkley, L. C., Arevalo, J. M., Sung, C. Y., Rose, R. M., and Cacioppo, J. T. 2007. Social regulation of gene expression in human leukocytes. Genome Biol. 8:R189.

Cooperstock, M. S., and Zed, A. J. 1983. Intestinal microflora of infants. In D. Hentges (Ed.), Human Intestinal Microflora in Health and Disease (pp. 79–99). New York: Academic.

Coussons-Read, M. E., Okun, M. L., and Nettles, C. D. 2007. Psychosocial stress increases inflammatory markers and alters cytokine production across pregnancy. Brain Behav. Immun. 21:343–350.

De Champlain, J. 1971. Degeneration and regrowth of adrenergic nerve fibers in the rat peripheral tissues after 6-hydroxydopamine. Can. J. Physiol. Pharmacol. 49:345–355.

Denda, M., Tsuchiya, T., Elias, P. M., and Feingold, K. R. 2000. Stress alters cutaneous permeability barrier homeostasis. Am. J. Physiol. Regul. Integr. Comp. Physiol. 278:R367–R372.

Dobbs, C. M., Vasquez, M., Glaser, R., and Sheridan, J. F. 1993. Mechanisms of stress-induced modulation of viral pathogenesis and immunity. J. Neuroimmunol. 48:151–160.

Dobbs, C. M., Feng, N., Beck, F. M., and Sheridan, J. F. 1996. Neuroendocrine regulation of cytokine production during experimental influenza viral infection: effects of restraint stress-induced elevation in endogenous corticosterone. J. Immunol. 157:1870–1877.

Drasar, B. S., Shiner, M., and McLeod, G. M. 1969. Studies on the intestinal flora. I. The bacterial flora of the gastrointestinal tract in healthy and achlorhydric persons. Gastroenterology 56:71–79.

Elftman, M. D., Norbury, C. C., Bonneau, R. H., and Truckenmiller, M. E. 2007. Corticosterone impairs dendritic cell maturation and function. Immunology 122:279–290.

Engler, H., Engler, A., Bailey, M. T., and Sheridan, J. F. 2005. Tissue-specific alterations in the glucocorticoid sensitivity of immune cells following repeated social defeat in mice. J. Neuroimmunol. 163:110–119.

Engler, H., Bailey, M. T., Engler, A., Stiner-Jones, L. M., Quan, N., and Sheridan, J. F. 2008. Interleukin-1 receptor type 1-deficient mice fail to develop social stress-associated glucocorticoid resistance in the spleen. Psychoneuroendocrinology 33:108–117.

Everson, C. A., and Toth, L. A. 2000. Systemic bacterial invasion induced by sleep deprivation. Am. J. Physiol. Regul. Integr. Comp. Physiol. 278:R905–R916.

Fagarasan, S., Muramatsu, M., Suzuki, K., Nagaoka, H., Hiai, H., and Honjo, T. 2002. Critical roles of activation-induced cytidine deaminase in the homeostasis of gut flora. Science 298:1424–1427.

Field, T., Sandberg, D., Quetel, T. A., Garcia, R., and Rosario, M. 1985. Effects of ultrasound feedback on pregnancy anxiety, fetal activity, and neonatal outcome. Obstet. Gynecol. 66:525–528.

Frank, D. N., St Amand, A. L., Feldman, R. A., Boedeker, E. C., Harpaz, N., and Pace, N. R. 2007. Molecular-phylogenetic characterization of microbial community imbalances in human inflammatory bowel diseases. Proc. Natl. Acad. Sci. USA 104:13780–13785.

Freestone, P. P., Williams, P. H., Haigh, R. D., Maggs, A. F., Neal, C. P., and Lyte, M. 2002. Growth stimulation of intestinal commensal Escherichia coli by catecholamines: a possible contributory factor in trauma-induced sepsis. Shock 18:465–470.

Freestone, P. P., Sandrini, S. M., Haigh, R. D., and Lyte, M. 2008. Microbial endocrinology: how stress influences susceptibility to infection. Trends Microbiol. 16:55–64.

Garcia, J. J., del Carmen, S. M., De la, F. M., and Ortega, E. 2003. Regulation of phagocytic process of macrophages by noradrenaline and its end metabolite 4-hydroxy-3-metoxyphenyl-glycol. Role of alpha- and beta-adrenoreceptors. Mol. Cell. Biochem. 254:299–304.

Ge, Z., Feng, Y., Taylor, N. S., Ohtani, M., Polz, M. F., Schauer, D. B., and Fox, J. G. 2006. Colonization dynamics of altered Schaedler flora is influenced by gender, aging, and Helicobacter hepaticus infection in the intestines of Swiss Webster mice. Appl. Environ. Microbiol. 72:5100–5103.

Gill, S. R., Pop, M., Deboy, R. T., Eckburg, P. B., Turnbaugh, P. J., Samuel, B. S., Gordon, J. I., Relman, D. A., Fraser-Liggett, C. M., and Nelson, K. E. 2006. Metagenomic analysis of the human distal gut microbiome. Science 312:1355–1359.

Glaser, R., Kiecolt-Glaser, J. K., Bonneau, R. H., Malarkey, W., Kennedy, S., and Hughes, J. 1992. Stress-induced modulation of the immune response to recombinant hepatitis B vaccine. Psychosom. Med. 54:22–29.

Gordon, J. I., Hooper, L. V., McNevin, M. S., Wong, M., and Bry, L. 1997. Epithelial cell growth and differentiation. III. Promoting diversity in the intestine: conversations between the microflora, epithelium, and diffuse GALT. Am. J. Physiol. 273:G565–G570.

Green, B. T., Lyte, M., Kulkarni-Narla, A., and Brown, D. R. 2003. Neuromodulation of enteropathogen internalization in Peyer's patches from porcine jejunum. J. Neuroimmunol. 141:74–82.

Green, B. T., Lyte, M., Chen, C., Xie, Y., Casey, M. A., Kulkarni-Narla, A., Vulchanova, L., and Brown, D. R. 2004. Adrenergic modulation of *Escherichia coli* O157:H7 adherence to the colonic mucosa. Am. J. Physiol Gastrointest. Liver Physiol. 287:G1238–G1246.

Heine, W., Mohr, C., and Wutzke, K. D. 1992. Host–microflora correlations in infant nutrition. Prog. Food Nutr. Sci. 16:181–197.

Hermann, G., Beck, F. M., and Sheridan, J. F. 1995. Stress-induced glucocorticoid response modulates mononuclear cell trafficking during an experimental influenza viral infection. J. Neuroimmunol. 56:179–186.

Holdeman, L. V., Good, I. J., and Moore, W. E. 1976. Human fecal flora: variation in bacterial composition within individuals and a possible effect of emotional stress. Appl. Environ. Microbiol. 31:359–375.

Holtmann, G., Kriebel, R., and Singer, M. V. 1990. Mental stress and gastric acid secretion. Do personality traits influence the response? Dig. Dis. Sci. 35:998–1007.

Hudault, S., Lievin, V., Bernet-Camard, M. F., and Servin, A. L. 1997. Antagonistic activity exerted in vitro and in vivo by *Lactobacillus casei* (strain GG) against *Salmonella typhimurium* C5 infection. Appl. Environ. Microbiol. 63:513–518.

Jabaaij, L., van Hattum, J., Vingerhoets, J. J., Oostveen, F. G., Duivenvoorden, H. J., and Ballieux, R. E. 1996. Modulation of immune response to rDNA hepatitis B vaccination by psychological stress. J. Psychosom. Res. 41:129–137.

Jarillo-Luna, A., Rivera-Aguilar, V., Garfias, H. R., Lara-Padilla, E., Kormanovsky, A., and Campos-Rodriguez, R. 2007. Effect of repeated restraint stress on the levels of intestinal IgA in mice. Psychoneuroendocrinology 32:681–692.

Johnson-Henry, K. C., Mitchell, D. J., Avitzur, Y., Galindo-Mata, E., Jones, N. L., and Sherman, P. M. 2004. Probiotics reduce bacterial colonization and gastric inflammation in *H. pylori*-infected mice. Dig. Dis. Sci. 49:1095–1102.

Johnson-Henry, K. C., Nadjafi, M., Avitzur, Y., Mitchell, D. J., Ngan, B. Y., Galindo-Mata, E., Jones, N. L., and Sherman, P. M. 2005. Amelioration of the effects of *Citrobacter rodentium* infection in mice by pretreatment with probiotics. J. Infect. Dis. 191:2106–2117.

Kellow, J. E., Langeluddecke, P. M., Eckersley, G. M., Jones, M. P., and Tennant, C. C. 1992. Effects of acute psychologic stress on small-intestinal motility in health and the irritable bowel syndrome. Scand. J. Gastroenterol. 27:53–58.

Kiecolt-Glaser, J. K., Glaser, R., Gravenstein, S., Malarkey, W. B., and Sheridan, J. 1996. Chronic stress alters the immune response to influenza virus vaccine in older adults. Proc. Natl. Acad. Sci. USA 93:3043–3047.

Knowles, S. R., Nelson, E. A., and Palombo, E. A. 2008. Investigating the role of perceived stress on bacterial flora activity and salivary cortisol secretion: a possible mechanism underlying susceptibility to illness. Biol. Psychol. 77:132–137.

Korneva, E. A., Rybakina, E. G., Orlov, D. S., Shamova, O. V., Shanin, S. N., and Kokryakov, V. N. 1997. Interleukin-1 and defensins in thermoregulation, stress, and immunity. Ann. NY Acad. Sci. 813:465–473.

Lederman, R. P. 1986. Maternal anxiety in pregnancy: relationship to fetal and newborn health status. Annu. Rev. Nurs. Res. 4:3–19.

Lederman, E., Lederman, R. P., Work, B. A., Jr., and McCann, D. S. 1981. Maternal psychological and physiologic correlates of fetal-newborn health status. Am. J. Obstet. Gynecol. 139:956–958.

Lenz, H. J., Raedler, A., Greten, H., Vale, W. W., and Rivier, J. E. 1988. Stress-induced gastrointestinal secretory and motor responses in rats are mediated by endogenous corticotropin-releasing factor. Gastroenterology 95:1510–1517.

Lizko, N. N. 1987. Stress and intestinal microflora. Nahrung 31:443–447.

Lyte, M. 2004. Microbial endocrinology and infectious disease in the 21st century. Trends Microbiol. 12:14–20.

Lyte, M., and Bailey, M. T. 1997. Neuroendocrine–bacterial interactions in a neurotoxin-induced model of trauma. J. Surg. Res. 70:195–201.

Lyte, M., Nelson, S. G., and Thompson, M. L. 1990. Innate and adaptive immune responses in a social conflict paradigm. Clin. Immunol. Immunopathol. 57:137–147.

MacFarlane, A. S., Peng, X., Meissler, J. J., Jr., Rogers, T. J., Geller, E. B., Adler, M. W., and Eisenstein, T. K. 2000. Morphine increases susceptibility to oral *Salmonella typhimurium* infection. J. Infect. Dis. 181:1350–1358.

Marshall, J. C., Christou, N. V., Horn, R., and Meakins, J. L. 1988. The microbiology of multiple organ failure. The proximal gastrointestinal tract as an occult reservoir of pathogens. Arch. Surg. 123:309–315.

Martinez, V., Rivier, J., Wang, L., and Tache, Y. 1997. Central injection of a new corticotropin-releasing factor (CRF) antagonist, astressin, blocks CRF- and stress-related alterations of gastric and colonic motor function. J. Pharmacol. Exp. Ther. 280:754–760.

Metchnikoff, E. 1908. The prolongation of life. New York: GP Putnam's Sons.

Meynell, G. G. 1963. Antibacterial mechanisms of the mouse gut. II. The role of Eh and volatile fatty acids in the normal gut. Br. J. Exp. Pathol. 44:209–219.

Miller, G. E., Chen, E., Sze, J., Marin, T., Arevalo, J. M., Doll, R., Ma, R., and Cole, S. W. 2008. A functional genomic fingerprint of chronic stress in humans: blunted glucocorticoid and increased NF-kappaB signaling. Biol. Psychiatry 64(4):266–272.

Murakami, M., Lam, S. K., Inada, M., and Miyake, T. 1985. Pathophysiology and pathogenesis of acute gastric mucosal lesions after hypothermic restraint stress in rats. Gastroenterology 88:660–665.

Nakade, Y., Tsuchida, D., Fukuda, H., Iwa, M., Pappas, T. N., and Takahashi, T. 2005. Restraint stress delays solid gastric emptying via a central CRF and peripheral sympathetic neuron in rats. Am. J. Physiol. Regul. Integr. Comp. Physiol. 288:R427–R432.

Nieuwenhuijzen, G. A., Deitch, E. A., and Goris, R. J. 1996a. Infection, the gut and the development of the multiple organ dysfunction syndrome. Eur. J. Surg. 162:259–273.

Nieuwenhuijzen, G. A., Deitch, E. A., and Goris, R. J. 1996b. The relationship between gut-derived bacteria and the development of the multiple organ dysfunction syndrome. J. Anat. 189 (Pt 3):537–548.

O'Hara, A. M., and Shanahan, F. 2006. The gut flora as a forgotten organ. EMBO Rep. 7:688–693.

Otte, J. M., Cario, E., and Podolsky, D. K. 2004. Mechanisms of cross hyporesponsiveness to Toll-like receptor bacterial ligands in intestinal epithelial cells. Gastroenterology 126: 1054–1070.

Padgett, D. A., and Glaser, R. 2003. How stress influences the immune response. Trends Immunol. 24:444–448.

Porlier, G. A., Nadeau, R. A., De, C. J., and Bichet, D. G. 1977. Increased circulating plasma catecholamines and plasma renin activity in dogs after chemical sympathectomy with 6-hydroxydopamine. Can. J. Physiol. Pharmacol. 55:724–733.

Quan, N., Avitsur, R., Stark, J. L., He, L., Lai, W., Dhabhar, F., and Sheridan, J. F. 2003. Molecular mechanisms of glucocorticoid resistance in splenocytes of socially stressed male mice. J. Neuroimmunol. 137:51–58.

Rawls, J. F., Mahowald, M. A., Ley, R. E., and Gordon, J. I. 2006. Reciprocal gut microbiota transplants from zebrafish and mice to germ-free recipients reveal host habitat selection. Cell 127:423–433.

Sakata, H., Yoshioka, H., and Fujita, K. 1985. Development of the intestinal flora in very low birth weight infants compared to normal full-term newborns. Eur. J. Pediatr. 144:186–190.

Salzman, N. H., Underwood, M. A., and Bevins, C. L. 2007. Paneth cells, defensins, and the commensal microbiota: a hypothesis on intimate interplay at the intestinal mucosa. Semin. Immunol. 19:70–83.

Sangild, P. T., Hilsted, L., Nexo, E., Fowden, A. L., and Silver, M. 1994. Secretion of acid, gastrin, and cobalamin-binding proteins by the fetal pig stomach: developmental regulation by cortisol. Exp. Physiol. 79:135–146.

Saunders, P. R., Kosecka, U., McKay, D. M., and Perdue, M. H. 1994. Acute stressors stimulate ion secretion and increase epithelial permeability in rat intestine. Am. J. Physiol. 267:G794–G799.

Saunders, P. R., Hanssen, N. P., and Perdue, M. H. 1997. Cholinergic nerves mediate stress-induced intestinal transport abnormalities in Wistar-Kyoto rats. Am. J. Physiol. 273:G486–G490.

Saunders, P. R., Santos, J., Hanssen, N. P., Yates, D., Groot, J. A., and Perdue, M. H. 2002. Physical and psychological stress in rats enhances colonic epithelial permeability via peripheral CRH. Dig. Dis. Sci. 47:208–215.

Schaedler, R. W., and Dubos, R. J. 1962. The fecal flora of various strains of mice. Its bearing on their susceptibility to endotoxin. J. Exp. Med. 115:1149–1160.

Schiffrin, E. J., Carter, E. A., Walker, W. A., Frieberg, E., Benjamin, J., and Israel, E. J. 1993. Influence of prenatal corticosteroids on bacterial colonization in the newborn rat. J. Pediatr. Gastroenterol. Nutr. 17:271–275.

Schneider, M. L., and Coe, C. L. 1993. Repeated social stress during pregnancy impairs neuromotor development of the primate infant. J. Dev. Behav. Pediatr. 14:81–87.

Schneider, M. L., Clarke, A. S., Kraemer, G. W., Roughton, E. C., Lubach, G. R., Rimm-Kaufman, S., Schmidt, D., and Ebert, M. 1998. Prenatal stress alters brain biogenic amine levels in primates. Dev. Psychopathol. 10:427–440.

Scott, L. D., and Cahall, D. L. 1982. Influence of the interdigestive myoelectric complex on enteric flora in the rat. Gastroenterology 82:737–745.

Shanahan, F. 2002. The host-microbe interface within the gut. Best Pract. Res. Clin. Gastroenterol. 16:915–931.

Soderholm, J. D., and Perdue, M. H. 2001. Stress and gastrointestinal tract. II. Stress and intestinal barrier function. Am. J. Physiol. Gastrointest. Liver Physiol. 280:G7–G13.

Stark, J. L., Avitsur, R., Padgett, D. A., Campbell, K. A., Beck, F. M., and Sheridan, J. F. 2001. Social stress induces glucocorticoid resistance in macrophages. Am. J. Physiol. Regul. Integr. Comp. Physiol. 280:R1799–R1805.

Steffen, E. K., and Berg, R. D. 1983. Relationship between cecal population levels of indigenous bacteria and translocation to the mesenteric lymph nodes. Infect. Immun. 39:1252–1259.

Stepankova, R., Sinkora, J., Hudcovic, T., Kozakova, H., and Tlaskalova-Hogenova, H. 1998. Differences in development of lymphocyte subpopulations from gut-associated lymphatic tissue (GALT) of germfree and conventional rats: effect of aging. Folia Microbiol. (Praha) 43:531–534.

Steptoe, A., Hamer, M., and Chida, Y. 2007. The effects of acute psychological stress on circulating inflammatory factors in humans: a review and meta-analysis. Brain Behav. Immun. 21:901–912.

Sternberg, E. M. 2006. Neural regulation of innate immunity: a coordinated nonspecific host response to pathogens. Nat. Rev. Immunol. 6:318–328.

Tache, Y., Martinez, V., Million, M., and Wang, L. 2001. Stress and the gastrointestinal tract III. Stress-related alterations of gut motor function: role of brain corticotropin-releasing factor receptors. Am. J. Physiol. Gastrointest. Liver Physiol. 280:G173–G177.

Tan, K. S., Nackley, A. G., Satterfield, K., Maixner, W., Diatchenko, L., and Flood, P. M. 2007. Beta2 adrenergic receptor activation stimulates pro-inflammatory cytokine production in macrophages via PKA- and NF-kappaB-independent mechanisms. Cell Signal. 19:251–260.

Tannock, G. W., Fuller, R., Smith, S. L., and Hall, M. A. 1990. Plasmid profiling of members of the family Enterobacteriaceae, lactobacilli, and bifidobacteria to study the transmission of bacteria from mother to infant. J. Clin. Microbiol. 28:1225–1228.

Trahair, J. F., and Sangild, P. T. 1997. Systemic and luminal influences on the perinatal development of the gut. Equine Vet. J. Suppl. (24):40–50.

Truckenmiller, M. E., Princiotta, M. F., Norbury, C. C., and Bonneau, R. H. 2005. Corticosterone impairs MHC class I antigen presentation by dendritic cells via reduction of peptide generation. J. Neuroimmunol. 160:48–60.

Truckenmiller, M. E., Bonneau, R. H., and Norbury, C. C. 2006. Stress presents a problem for dendritic cells: corticosterone and the fate of MHC class I antigen processing and presentation. Brain Behav. Immun. 20:210–218.

Umesaki, Y., Setoyama, H., Matsumoto, S., and Okada, Y. 1993. Expansion of alpha beta T-cell receptor-bearing intestinal intraepithelial lymphocytes after microbial colonization in germ-free mice and its independence from thymus. Immunology 79:32–37.

Umesaki, Y., Okada, Y., Matsumoto, S., Imaoka, A., and Setoyama, H. 1995. Segmented filamentous bacteria are indigenous intestinal bacteria that activate intraepithelial lymphocytes and induce MHC class II molecules and fucosyl asialo GM1 glycolipids on the small intestinal epithelial cells in the ex-germ-free mouse. Microbiol. Immunol. 39:555–562.

Wharton, B. A., Balmer, S. E., and Scott, P. H. 1994a. Faecal flora in the newborn. Effect of lactoferrin and related nutrients. Adv. Exp. Med. Biol. 357:91–98.

Wharton, B. A., Balmer, S. E., and Scott, P. H. 1994b. Sorrento studies of diet and fecal flora in the newborn. Acta Paediatr. Jpn. 36:579–584.

Yang, H., Yuan, P. Q., Wang, L., and Tache, Y. 2000. Activation of the parapyramidal region in the ventral medulla stimulates gastric acid secretion through vagal pathways in rats. Neuroscience 95:773–779.

Chapter 12
The Epinephrine/Norepinephrine/Autoinducer-3 Interkingdom Signaling System in *Escherichia coli* O157:H7

Cristiano G. Moreira and Vanessa Sperandio

12.1   *Escherichia coli* O157:H7

E. coli is one of the most well-studied bacterium in the microbiology field due to its frequent incidence in different environments and hosts, as well as its use as a tool in molecular biology. Currently, there are several categories of *E. coli* known to cause disease, mainly diarrhea in humans, also named as diarrheiogenic *E. coli*. Among those, enterohemorrhagic *E. coli* O157:H7 (EHEC) is one of the most important pathogenic *E. coli*. EHEC has been associated with several recent food-borne outbreaks of bloody diarrhea and hemolytic uremic syndrome (HUS) throughout the world. EHEC has an unusually low infectious dose when compared with other enteric bacterial pathogens such as *Vibrio cholera* and *Salmonella enterica* Typhimurium. EHEC colonizes the large intestine and produces a potent toxin, Shiga toxin (Stx), responsible for the hemorrhagic colitis and HUS, which can culminate in kidney failure and leads to the mortality associated with EHEC outbreaks (Kaper et al. 2004).

EHEC causes a histopathological lesion on intestinal epithelial cells called attaching and effacing (AE). The AE lesion is characterized by the destruction of the microvilli and the rearrangement of the cytoskeleton to form a unique pedestal structure that cups the bacterium individually (Fig. 12.1a).

Chemical signaling through the AI-3/Epinephrine/Norepineprhine signals activates expression of virulence genes in EHEC. Most of these virulence genes are involved in the formation of the AE lesion and are contained within a pathogenicity island named the locus of enterocyte effacement (LEE) (McDaniel et al. 1995) (Fig. 12.2).

The EHEC LEE region contains 41 genes, most of which are organized into five major operons: *LEE1, LEE2, LEE3, LEE5,* and *LEE4* (Elliott et al. 1998; Elliott et al. 1999; Mellies et al. 1999). The LEE encodes a type III secretion system

C.G. Moreira (✉) and V. Sperandio
Molecular Microbiology Department, University of Texas Southwestern Medical Center,
6000 Harry Hines Bvld, Dallas 75390, TX, USA
e-mail: cristiano.moreira@utsouthwestern.edu

M. Lyte and P.P.E. Freestone (eds.), *Microbial Endocrinology,*
Interkingdom Signaling in Infectious Disease and Health,
DOI 10.1007/978-1-4419-5576-0_12, © Springer Science+Business Media, LLC 2010

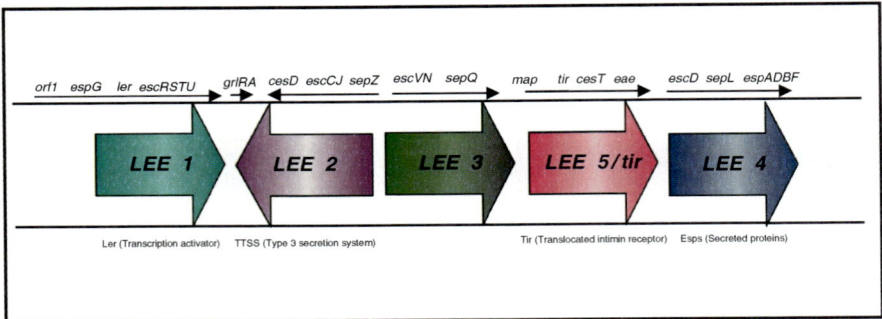

Fig. 12.1 (**a**) The locus of enterocyte effacement (LEE) pathogenicity Island found in EHEC, which encodes factors responsible for type III secretion and pedestal formation. *LEE1* encodes for *ler*, the LEE-encoded regulator. *LEE1*, *LEE2*, and *LEE3* encode for factors involved in type III secretion. *LEE4* encodes for EspA, EspB, and EspD. The *LEE5/tir* operon encodes for intimin and Tir (McDaniel et al. 1995; Kenny et al. 1997a; Mellies et al. 1999)

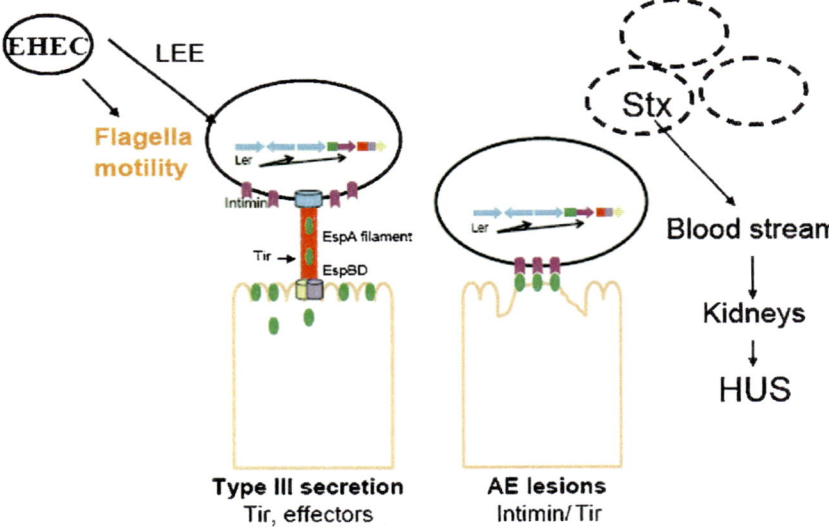

Fig. 12.2 General model for EHEC pathogenesis. LEE region of EHEC encodes most of T3SS effectors, which are essential for EHEC pathogenesis, as well as flagella motility. Ler is master regulator of LEE, intimin is an outer membrane protein, through T3SS EHEC translocates Tir, Intimin-Tir binding culminates in histopathologic lesion called attaching and effacing lesion (or lesion AE). Shiga toxin in EHEC plays late role during infection that can cause hemolictic uremic sindrome (Kaper et al. 2004)

(TTSS) (Jarvis et al. 1995), an adhesin (intimin) (Jerse et al. 1990), and this adhesin's receptor, the translocated intimin receptor (Tir) (Kenny et al. 1997b), which is translocated into the epithelial cell through the bacterial TTSS (Elliott et al. 1998; Elliott et al. 1999; Mellies et al. 1999) (Fig. 12.1a).

The TTSS is an apparatus that spans the inner and outer bacterial membranes forming a microscopic "needle." Several proteins, including EscD, EscR, EscU, EscV, EscS, and EscT span the inner membrane and associate with a cytoplasmic ATPase, EscN, which is required for secretion of proteins (Roe et al. 2003). EscC is predicted to form the main protein ring in the outer membrane to which the EscF "needle" is connected (Wilson et al. 2001). EscF comprises the syringe connected to the filament of the translocon. The translocon consists of EspA, which creates a sheath around the EscF needle. EspB and EspD are located at the distal end of the TTSS and form 3–5 nm pores in the host cell membrane (Ide et al. 2001) through which translocated proteins are secreted.

The *eae* gene (*E. coli* attaching and effacing) encodes for intimin, an outer membrane protein that acts as an intestinal adherence factor (Jerse et al. 1990). Mutants of the *eae* gene are defective in intimate adherence to intestinal epithelial cells, which prevents the concentration of polymerized actin necessary for the development of AE lesions. The translocated intimin receptor (Tir), which is also encoded in the LEE, is translocated from the bacterium through the TTSS into the host cell to serve as a receptor for intimin (DeVinney et al. 1999; Kenny and Finlay 1995; Rosenshine et al. 1996). In the host cell membrane, Tir adopts a hairpin loop conformation and serves as a receptor for the bacterial surface adhesin, intimin (Deibel et al. 1998). Binding of intimin to Tir promotes the clustering of N- and C-terminal cytoplasmic regions and leads to the initiation of localized actin assembly beneath the plasma membrane (Campellone et al. 2004). The EHEC Tir recruits the host protein N-WASP (Goosney et al. 2001) through an interaction with EspFu, another bacterial protein encoded within a prophage, which is also translocated through the TTSS into the host cell (Campellone et al. 2004).

The TTSS encoded by the LEE translocates LEE-encoded and non-LEE encoded effectors. The mitochondrial associated protein, *map*, affects the integrity of the host mitochondrial membrane (Kenny and Jepson 2000) and is encoded directly upstream of *tir*. Another effector, EspF, is responsible for the disruption of the intestinal barrier function and induces cell death by an unknown mechanism (McNamara and Donnenberg 1998; McNamara et al. 2001). EspG is responsible for the disruption of microtubule formation and plays a role in virulence in the rabbit enteropathogenic *E. coli* (REPEC) model (Tomson et al. 2005), while EspH, which is encoded in *LEE3*, is responsible for the modulation of the host cell cytoskeleton through the inhibition of cell cycle signals (Tu et al. 2003). Although encoded outside the LEE pathogenicity island, several effector proteins have been recently shown to be secreted through the EHEC TTSS (Tobe et al. 2006). These include Cif, which induces host cell cycle arrest and reorganization of host actin cytoskeleton (Charpentier and Oswald 2004), and NleA, which has been shown to localize to the Golgi and play a key role in virulence in an animal model (Gruenheid et al. 2004).

EHEC also produces a powerful Shiga toxin (Stx) that is responsible for the major symptoms of hemorrhagic colitis and HUS. The Stx family contains two subgroups, Stx1 and Stx2. Stx1 shows little sequence variation between strains (Zhang et al. 2002), whereas antigenic divergence has been observed among the Stx2s, including Stx2, Stx2c, Stx2d, and Stx2e (Perera et al. 1988; Schmitt et al. 1991;

Zhang et al. 2002). Stx2 has been more associated epidemiologically with severe human disease than Stx1 (Boerlin et al. 1999), with Stx2 and Stx2c being most frequently found in patients with HUS (Ritter et al. 1997).

The genes encoding Stx1 and Stx2 are located within the late genes of a λ-like bacteriophage and are transcribed when the phage enters its lytic cycle (Neely and Friedman 1998). Once the phage replicates, Shiga toxin is produced, and the phage lyse the bacteria, thereby releasing the toxin into the host. The bacteriophage enters its lytic cycle during an SOS response triggered by disturbances in the bacterial membrane, DNA replication, or protein synthesis (Kimmitt et al. 1999; Kimmitt et al. 2000). These triggers are all common targets of conventional antibiotics and may contribute to the controversy surrounding the use of antibiotics to treat EHEC-mediated disease. Shiga toxins consist of a 1A:5B noncovalently associated subunit structure (Donohue-Rolfe et al. 1984). The B subunit of Stx is known to form a pentamer that binds to the eukaryotic glycolipid receptor, globotriaosylceramide (Jacewicz et al. 1986; Lindberg et al. 1987; Waddell et al. 1988). The A subunit is then cleaved by trypsin and reduced, resulting in a polypeptide that causes depurination of a residue in the 28S rRNA of 60S ribosomes (Endo et al. 1988). This leads to the inhibition of protein synthesis, injury of renal glomerular endothelial cells, and the initiation of a pathophysiological cascade that leads to HUS (Fig. 12.2).

12.2 Transcriptional Regulation of the LEE Region

Chemical signaling plays an important role in LEE and flagella expression (Sperandio et al. 1999; Sperandio et al. 2001). The bacterial signal, autoinducer-3 (AI-3), produced by the human intestinal microbial flora, the epinephrine and norepinephrine host stress hormones, as well as the flagella regulon activate expression of the LEE genes (Sperandio et al. 2003). These signals act agonistically to increase LEE gene expression (Walters and Sperandio 2006) Fig. 12.3.

In addition to the regulation via stress hormones, other transcription factors are involved on LEE regulation. Iyoda and Watanabe (2004) have observed that EHEC encodes the genes *pchA, pchB,* and *pchC* (PerC homologs) that positively activate the expression of the LEE genes (Iyoda and Watanabe 2004). *LEE1* encodes for Ler, the LEE-encoded regulator, which was shown to be required for the expression of all genes within the LEE (Bustamante et al. 2001; Elliott et al. 2000; Kaper et al. 2004; Mellies et al. 1999). Another important factor in the regulation of the LEE is the integration-host factor, or IHF. IHF has been shown to be required for the expression of the entire LEE through the direct activation of *ler* transcription (Friedberg et al. 1999). Additionally, EtrA and EivF are two negative regulators of the LEE region, acting possibly through *ler* transcriptional repression. The *etrA* and *eivF* genes are found within a second pathogenicity island that encodes a cryptic type III secretion system (ETT2) (Zhang et al. 2004). The histone-like nucleoid-structuring protein, H-NS, is responsible for the repression of *LEE2, LEE3,* and *LEE5* transcription in the absence of Ler (Bustamante et al. 2001; Haack et al. 2003). RpoS, a sta-

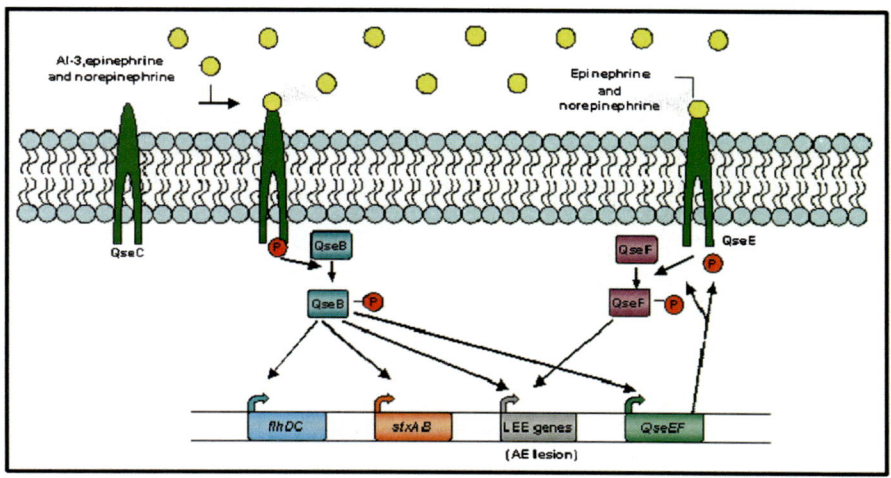

Fig. 12.3 Model for regulation of autoinducer (AI)-3, epinephrine, and norepinephrine (NE) bind the bacterial membrane receptor QseC, which results in its autophosphorylation. QseC then phosphorylates its response regulator QseB and initiates a complex phosphorelay signaling cascade that activates the expression of a second two-component system (QseEF), the locus of enterocyte effacement (LEE) genes, which encode various proteins, including the components of a type III section system that are involved in attaching and effacing (AE) lesion formation, the motility genes (flhDC), and Shiga toxin (stxAB). The QseEF two-component system is also involved in the expression of the LEE genes, and although its activators have not yet been elucidated, it is possible that it senses epinephrine and/or NE (Hughes and Sperandio 2008)

tionary phase sigma factor, activates the transcription of the *LEE3* operon (Sperandio et al. 1999). Finally, Hha has been reported to repress the transcription of the *LEE4* operon (Sharma and Zuerner 2004).

Two previously uncharacterized genes in the LEE region, *orf10* and *orf11*, were recently renamed GrlR, global regulator of LEE repressor, and GrlA, global regulator of LEE activator (Barba et al. 2005; Deng et al. 2004). This study suggested that GrlA is responsible for the transcriptional activation of *ler*, while GrlR represses *ler*. Additionally, it is known that Ler activates the transcription of *grlRA* (Barba et al. 2005; Elliott et al. 2000) and that GrlRA activates the expression of *LEE2* and *LEE4*, independently of Ler (Russell et al. 2007).

12.3   Quorum Sensing in EHEC

Surette and Bassler reported in 1998 quorum sensing signaling in *E coli* K12 and *S. enterica* Typhimurium (Surette and Bassler 1998) through the production of autoinducer-2 (AI-2) by these bacteria (Surette et al. 1999). A common gene on these bacterial species was cloned and identified as responsible for AI-2 production, and it was named *luxS* (Surette et al. 1999). Later on, LuxS was characterized as the enzyme involved in the metabolism of S-adenosylmethionine and shown to convert ribose–homocysteine into homocysteine and 4,5-dihydroxy-2,3-pentanedione, which is the precursor of AI-2 (Schauder et al. 2001). Currently, the autoinducer referred to as AI-2 is a furanosylborate-diester in *Vibrio harveyi* (Chen et al. 2002), and a 2R, 4S-2-methyl-2,3,3,4-tetrahydrofuran (R-THMF) for *Salmonella* sp.(Miller et al. 2004).

However, at that time, no phenotypes have been shown to be regulated by quorum sensing in *E. coli* and *Salmonella* sp. In 1999, Sperandio et al. (1999) described that quorum sensing signaling activated transcription of the LEE genes and type three secretion in pathogenic *E. coli*, both in EHEC and in enteropathogenic *E. coli* (EPEC). It was then proposed that signaling through AI-2 might be involved in virulence gene regulation in *E. coli*. In 2001, quorum sensing was shown to be a global regulatory mechanism in EHEC, involved in AE lesion formation, motility, metabolism, growth, and Shiga toxin expression (Sperandio et al. 2001). The EHEC's quorum sensing regulatory cascade started to be unraveled with the description of the first three quorum sensing regulators (Sperandio et al. 2002a; Sperandio et al. 2002b), which were named Quorum sensing *E. coli*, Qse regulators. QseA was reported as a transcription factor from the widespread LysR family, directly involved in activation of *LEE1* (*ler*) transcription and EHEC's pathogenesis (Russell et al. 2007; Sharp and Sperandio 2007; Sperandio et al. 2002a). The QseBC regulatory system was described as a novel two-component system involved in quorum sensing regulation of flagella expression and motility (Clarke and Sperandio 2005b; Sperandio et al. 2002b).

A year later, it was shown that AI-2 was not the autoinducer involved in EHEC virulence regulation. The signal identified was a yet undescribed autoinducer, which was named autoinducer 3 (AI-3). In addition, it was reported that the eukaryotic hormones epinephrine and/or norepinephrine could substitute for the bacterial AI-3 signal to activate expression of the EHEC virulence genes. Among them, *ler*, the master regulator of LEE, was shown to be activated in presence of AI-3. Moreover, flagella motility of EHEC regulated via QseBC system was demonstrated to be affected by AI-3. Given that eukaryotic cell-to-cell signaling occurs through hormones, like epinephrine and norepinephrine, this cross talk between bacteria and host seem to happen through cross-signaling between quorum sensing signals and host hormones (Sperandio et al. 2003).

12.4   Infectious Disease and Hormones

Since the 1920s, hormones have been used for treatment of a variety of illnesses (Yamashima 2003). Following studies have shown the role of stress hormones directly on bacterial functions.

Catecholamines, such as epinephrine and norepinephrine, are chemical compounds derived from the amino acid tyrosine containing catechol and amine groups. They are stress hormones involved in the fight or flight response.

They are synthesized from L-dopa into dopamine, then norepinephrine, and epinephrine. Both norepinephrine and dopamine containing sympathetic nerve terminals are distributed throughout the body, including the intestinal tract, where they are part of the enteric nervous system. Epinephrine synthesis is restricted to the central nervous system and adrenal glands. However, epinephrine is released into the bloodstream, especially during stress, acting systemically in the whole body (Furness 2000).

Both epinephrine and norepinephrine are recognized by adrenergic receptors in mammalian cells. Adrenergic receptors are a subset of the G-protein coupled receptors (GPCRs) family, which are transmembrane receptors coupled to heterotrimeric guanine-binding proteins (G proteins). Adrenergic receptors are divided in two main subtypes: α and β. Both epinephrine and norepinephrine are recognized by a very similar ligand-binding pocket in all adrenergic receptors (Cherezov et al. 2007; Freddolino et al. 2004).

12.5 QseC: A Bacterial Functional Analog of an Adrenergic Receptor

Unlike mammalian cells, EHEC does not encode a GPCR receptor in its genome; hence, sensing of adrenergic hormones by this bacterium was occurring through another type of receptor. The main signaling transduction systems in bacterial cells are the two-component systems. In these systems, the sensor for environmental cues is a histidine kinase (usually membrane bound), which upon autophosphorylation transfers its phosphate to an aspartate residue in the response regulator. This response regulator is usually a transcription factor that is activated by phosphorylation. QseC is a histidine sensor kinase that augments its phosphorylation specifically in response to the AI-3 bacterial quorum sensing signal and the host hormones epinephrine and/or norepinephrine (Clarke et al. 2006). Importantly, tritiated norepinephrine was shown to specifically bind to QseC, and this binding could be antagonized only by a λ-adrenergic antagonist (phentolamine), while a β-adrenergic antagonist (propranolol) did not have any effect on QseC activity (Clarke et al. 2006). QseC then transfers its phosphate to its cognate response regulator QseB, which only upon phosphorylation binds to its target promoters to regulate gene expression (Clarke et al. 2006; Clarke and Sperandio 2005a; b).

12.6 QseBC Two-Component System

The QseBC system was initially described as a two-component system regulated by quorum sensing, which shares homology with *Salmonella enterica* serovar Typhimurium PmrAB (Sperandio et al. 2002b). The same study showed that QseBC was involved in regulation of flagella and motility in EHEC.

In a detailed study (Clarke and Sperandio 2005a), it was shown that QseBC constituted a two-component system, and that the *qseBC* genes were cotranscribed forming an operon. Moreover, it was reported that QseB autoactivated transcription of the *qseBC* operon (Clarke and Sperandio 2005a). Using primer extension, the start site of the *qseBC* transcript was mapped, and through nested deletion analysis it was determined the minimal region necessary for QseB transcriptional auto-activation. Also, electrophoretic mobility shift assays, competition experiments, and DNAse I footprints showed that QseB directly binds to 2 sites in its own promoter.

Additionally, Clarke and Sperandio (2005b) described that QseBC regulates the flagella expression and motility through *flhDC*, the master regulator of flagella. Using electrophoretic mobility shift assays, competition experiments, and DNAse I footprints, it was shown that QseB directly binds to the *flhDC* promoter both in low and high affinity binding sites. In this study, it was also reported that the promoter of *flhDC* responsive to QseBC had a σ^{28} consensus. In summary, these studies suggested that transcription of the *flhDC* promoter by QseBC is a complex system and is dependent on the presence of FliA (σ^{28}).

QseC's homologs are found at least in 25 species of bacteria, including animals and plant pathogens (Table 12.1), suggesting that this signaling system is not restricted to *E. coli* (Rasko et al. 2008). A few recent reports have implicated these non-EHEC *qseC* homologues in virulence gene activation presumably through a combination of AI-3, norepinephrine, and epinephrine activation.

Table 12.1 Some of bacterial genera and species where QseBC (Rasko et al. 2008) and QseEF two-components systems are conserved, other than EHEC

QseBC	QseEF
Enteropathogenic *E. coli*	Uropathogenic *E. coli*
Enteroaggregative *E. coli*	*E. coli* k12
E. coli k12	*S. enterica* Typhimurium
Salmonella	*Salmonella enterica* Typhi
Vibrio	*Salmonella enterica* Paratyphi
Shigella	*Shigella flexneri*
Enterococcus	*Shigella boydii*
Campylobacter	*Shigella dysenteriae*
Yersinia	*Shigella sonnei*
Psychrobacter	*Bacillus subtilis*
Fransciella	*Bacillus amyloliquetaciens*
Erwinia carotovora	
Hemophilus influenzae	
Pasteurella multocida	
Actinobacillus pleuropneumoniae	
Chromobacterium violaceium	
Rubrivivax gelatinosus	
Thiobacillus denitrificans	
Ralstonia eutropa	
Ralstonia metallidurans	

The chemical structure of AI-3 was not completely elucidated yet. However, the role of epinephrine and norepinephrine has been extensively reported in different strains of *E. coli, Salmonella, Staphylococcus*, among others. A *qseC* homologue was shown to contribute to virulence in a mouse infection model of the class A bioterrorism threat agent *Francisella tularensis* (Weiss et al. 2007). A *qseC* homologue was also demonstrated to be involved in norepinephrine dependent enhancement of motility and colonization of juvenile pigs by *S. enterica* serovar *Typhimurium* (Bearson and Bearson, 2008) and virulence gene expression in this bacterium (Merighi et al. 2006; Merighi et al. 2009).

12.7   The QseEF Two-Component System

Recently, a second two-component system in the AI-3/epinephrine/norepinephrine signaling cascade was described by Reading et al. (2007) and named QseEF. The *qseE* and *qseF* genes are part of a polycistronic operon that also contain in this cluster the *yfhG* gene, which encodes for uncharacterized protein, and *glnB*, which encodes the PII protein involved in nitrogen regulation (Reading et al. 2007). Transcription of *qseEF* is activated by epinephrine through QseC, suggesting that QseEF comes second to QseBC in the hierarchy of gene expression in this signaling cascade.

In this system, QseE is the sensor kinase, and QseF the response regulator. QseF activates transcription of the recently described gene encoding EspFu (Reading et al. 2007), an effector protein of EHEC, which is encoded outside the LEE region. EspFu is translocated into the host cell by EHEC where it mimics an eukaryotic SH2/SH3 adapter protein to engender actin polymerization during pedestal formation (Campellone et al. 2004). In silico analysis indicates that QseF contains a σ^{54} activator domain. However, the *espFu* gene contains a conserved extended σ^{70} consensus promoter region, suggesting that QseF indirectly activates transcription of *espFu*. In agreement to this hypothesis, eletrophoretic mobility shift experiments demonstrated that QseF does not bind to the *espFu* promoter region. These data showed that QseF-dependent activation of *espFu* transcription is indirect and involves an intermediary factor transcribed in a σ^{54}-dependent fashion.

Reading et al. (2007) then showed that the QseEF two-component system, involved in *espFu* transcriptional activation, is essential for AE lesion formation. This system is also found in other bacterial species. However, it is restricted to enteric bacteria not being as widespread as QseBC (Table 12.1).

12.8   Qse A Regulator

Sperandio et al. (1999) reported that the LEE genes from EHEC were activated by quorum sensing through Ler, which is encoded by the first gene of the *LEE1* operon (Sperandio et al. 1999). QseA belongs to the LysR family of transcription factors and shares homology with the AphB and PtxR regulators of *V. cholerae* and *Pseudomonas aeruginosa*, respectively (Sperandio et al. 2002a). This study reported that QseA activates transcription of *ler* and consequently all the LEE genes. The *LEE1* operon has two promoters, a distal promoter P1 (163 base pairs upstream of the translational start site), and a proximal promoter P2 (32 base pairs upstream of the translational start site). QseA acts on the distal (P1) promoter of *LEE1*. An EHEC *qseA* mutant also presented a remarkable reduction on the type three secretion system. Hence, QseA is part of the AI-3/epinephrine/norepinephrine regulatory cascade that regulates the EHEC LEE region via transcriptional activation of *ler*. Altogether, this study showed that QseA activates the transcription of *LEE1* by directly binding upstream of its P1 promoter region.

In 2007, Russell et al. (2007) reported yet another level of LEE regulation through QseA. They reported that *grlRA* transcription is activated by QseA in both a Ler-dependent as well as a Ler-independent fashion, adding another layer of complexity to the regulation of the LEE genes (Russell et al. 2007).

12.9 Future Implications of the AI-3/Epinephrine/ Norepinephrine InterKingdom Signaling in EHEC Pathogenesis and Development of Therapeutics

The growing worldwide challenge of antimicrobial resistance and the paucity of novel antibiotics underscore the urgent need for innovative therapeutics. The increasing understanding of bacterial pathogenesis and intercellular communication, when combined with contemporary drug discovery tools and technologies, provides a powerful platform for translating such basic science into therapeutic applications to combat bacterial infections. Interference with bacterial cell-to-cell signaling via the quorum-sensing pathway constitutes an especially compelling and novel strategy since it also obviates the development of bacterial resistance. Quorum sensing allows bacteria to respond to hormone-like molecules called autoinducers and is responsible for controlling a plethora of virulence genes in several bacterial pathogens. Because quorum sensing is not directly involved in essential processes such as growth of the bacteria, inhibition of quorum sensing should not yield a selective pressure for development of resistance. Quorum sensing antagonists confuse or obfuscate signaling between bacteria and bacteria and host and, unlike antibiotics, do not kill or hinder bacterial growth. Hence, quorum sensing antagonists should be viewed as blockers of pathogenicity rather than as antimicrobials.

QseC is a receptor for the AI-3/epinephrine/norepinephrine signals and is central for the pathogenesis of EHEC. In addition, QseC homologs are present in many bacterial pathogens of animals and plants (Table 12.1). Finally, QseC is involved in quorum sensing and adrenergic signaling, and this signaling is not directly involved in processes essential for bacterial growth. Thus, in theory, inhibitors of QseC would not induce selective pressures promoting evolution of bacterial resistance.

Indeed a recent report demonstrated proof of principal that small molecule inhibitors of QseC-mediated signaling markedly inhibit the virulence of several pathogens in vitro and in vivo in animal models. This study utilized a high throughput screen to identify a potent small molecule, LED209, which inhibits binding of signals to QseC, preventing QseC's autophosphorylation, and consequently inhibiting QseC-mediated activation of virulence gene expression in enterohemorrhagic *E. coli* (EHEC), *Salmonella Typhimurium*, and *Francisella tularensis*. LED209 also prevented formation of lesions on epithelial cells by EHEC, and *F. tularensis* survival within macrophages. Moreover, LED209 treatment protected mice from lethal *S. Typhimurium* and *F. tularensis* infection. LED209 is not toxic and does not inhibit pathogen growth. Extensively, pharmacologic studies have been performed to first of all show that LED209 had not been cytotoxic in the animal models used in the study, such as mice and rabbits. This study demonstrated that inhibition of interkingdom intercellular signaling constitutes a novel and effective strategy for the development of a new generation of broad spectrum antimicrobial agents (Rasko et al. 2008).

References

Barba, J., V. H. Bustamante, M. A. Flores-Valdez, W. Deng, B. B. Finlay, and J. L. Puente. A Positive Regulatory Loop Controls Expression of the Locus of Enterocyte Effacement-Encoded Regulators Ler and Grla. *J Bacteriol* 187, no. 23 (2005): 7918–30.

Bearson, B. L., and S. M. Bearson. The Role of the Qsec Quorum-Sensing Sensor Kinase in Colonization and Norepinephrine-Enhanced Motility of *Salmonella enterica* Serovar Typhimurium. *Microb Pathog* 44, no. 4 (2008): 271–8.

Boerlin, P., S. A. McEwen, F. Boerlin-Petzold, J. B. Wilson, R. P. Johnson, and C. L. Gyles. Associations between Virulence Factors of Shiga Toxin-Producing *Escherichia coli* and Disease in Humans. *J Clin Microbiol* 37, no. 3 (1999): 497–503.

Bustamante, V. H., F. J. Santana, E. Calva, and J. L. Puente. Transcriptional Regulation of Type Iii Secretion Genes in Enteropathogenic *Escherichia coli*: Ler Antagonizes H-Ns-Dependent Repression. *Mol Microbiol* 39, no. 3 (2001): 664–78.

Campellone, K. G., D. Robbins, and J. M. Leong. Espfu Is a Translocated Ehec Effector That Interacts with Tir and N-Wasp and Promotes Nck-Independent Actin Assembly. *Dev Cell* 7, no. 2 (2004): 217–28.

Charpentier, X., and E. Oswald. Identification of the Secretion and Translocation Domain of the Enteropathogenic and Enterohemorrhagic *Escherichia coli* Effector Cif, Using Tem-1 Beta-Lactamase as a New Fluorescence-Based Reporter. *J Bacteriol* 186, no. 16 (2004): 5486–95.

Chen, X., S. Schauder, N. Potier, A. Van Dorssealaer, I. Pelczer, B.L. Bassler, and F.M. Hughson. Structural Identification of a Bacterial Quorum-Sensing Signal Containing Boron. *Nature* 415 (2002): 545–49.

Cherezov, V., D. M. Rosenbaum, M. A. Hanson, S. G. Rasmussen, F. S. Thian, T. S. Kobilka, H. J. Choi, P. Kuhn, W. I. Weis, B. K. Kobilka, and R. C. Stevens. High-Resolution Crystal Structure of an Engineered Human Beta2-Adrenergic G Protein-Coupled Receptor. *Science* 318, no. 5854 (2007): 1258–65.

Clarke, M. B., D. T. Hughes, C. Zhu, E. C. Boedeker, and V. Sperandio. The Qsec Sensor Kinase: A Bacterial Adrenergic Receptor. *Proc Natl Acad Sci U S A* 103, no. 27 (2006): 10420–5.

Clarke, M. B., and V. Sperandio. Transcriptional Autoregulation by Quorum Sensing *Escherichia coli* Regulators B and C (Qsebc) in Enterohaemorrhagic *E. coli* (Ehec). *Mol Microbiol* 58, no. 2 (2005): 441–55.

Clarke, M. B., and V. Sperandio. Transcriptional Regulation of Flhdc by Qsebc and Sigma (Flia) in Enterohaemorrhagic *Escherichia coli*. *Mol Microbiol* 57, no. 6 (2005): 1734–49.

Deibel, C., S. Kramer, T. Chakraborty, and F. Ebel. Espe, a Novel Secreted Protein of Attaching and Effacing Bacteria, Is Directly Translocated into Infected Host Cells, Where It Appears as a Tyrosine-Phosphorylated 90 Kda Protein. *Mol Microbiol* 28, no. 3 (1998): 463–74.

Deng, W., J. L. Puente, S. Gruenheid, Y. Li, B. A. Vallance, A. Vazquez, J. Barba, J. A. Ibarra, P. O'Donnell, P. Metalnikov, K. Ashman, S. Lee, D. Goode, T. Pawson, and B. B. Finlay. Dissecting Virulence: Systematic and Functional Analyses of a Pathogenicity Island. *Proc Natl Acad Sci U S A* 101, no. 10 (2004): 3597–602.

DeVinney, R., A. Gauthier, A. Abe, and B. B. Finlay. Enteropathogenic *Escherichia coli*: A Pathogen That Inserts Its Own Receptor into Host Cells. *Cell Mol Life Sci* 55, no. 6–7 (1999): 961–76.

Donohue-Rolfe, A., G. T. Keusch, C. Edson, D. Thorley-Lawson, and M. Jacewicz. Pathogenesis of *Shigella* Diarrhea. Ix. Simplified High Yield Purification of *Shigella* Toxin and Characterization of Subunit Composition and Function by the Use of Subunit-Specific Monoclonal and Polyclonal Antibodies. *J Exp Med* 160, no. 6 (1984): 1767–81.

Elliott, S. J., V. Sperandio, J. A. Giron, S. Shin, J. L. Mellies, L. Wainwright, S. W. Hutcheson, T. K. McDaniel, and J. B. Kaper. The Locus of Enterocyte Effacement (Lee)-Encoded Regulator Controls Expression of Both Lee- and Non-Lee-Encoded Virulence Factors in Enteropathogenic and Enterohemorrhagic *Escherichia coli*. *Infect Immun* 68, no. 11 (2000): 6115–26.

Elliott, S. J., L. A. Wainwright, T. K. McDaniel, K. G. Jarvis, Y. K. Deng, L. C. Lai, B. P. McNamara, M. S. Donnenberg, and J. B. Kaper. The Complete Sequence of the Locus of Enterocyte Effacement (Lee) from Enteropathogenic *Escherichia coli* E2348/69. *Mol Microbiol* 28, no. 1 (1998): 1–4.

Elliott, S. J., J. Yu, and J. B. Kaper. The Cloned Locus of Enterocyte Effacement from Enterohemorrhagic *Escherichia coli* O157:H7 Is Unable to Confer the Attaching and Effacing Phenotype Upon *E. coli* K-12. *Infect Immun* 67, no. 8 (1999): 4260–3.

Endo, Y., K. Tsurugi, T. Yutsudo, Y. Takeda, T. Ogasawara, and K. Igarashi. Site of Action of a Vero Toxin (Vt2) from *Escherichia coli* O157:H7 and of Shiga Toxin on Eukaryotic Ribosomes. Rna N-Glycosidase Activity of the Toxins. *Eur J Biochem* 171, no. 1–2 (1988): 45–50.

Freddolino, P. L., M. Y. Kalani, N. Vaidehi, W. B. Floriano, S. E. Hall, R. J. Trabanino, V. W. Kam, and W. A. Goddard, 3rd. Predicted 3d Structure for the Human Beta 2 Adrenergic Receptor and Its Binding Site for Agonists and Antagonists. *Proc Natl Acad Sci U S A* 101, no. 9 (2004): 2736–41.

Friedberg, D., T. Umanski, Y. Fang, and I. Rosenshine. Hierarchy in the Expression of the Locus of Enterocyte Effacement Genes of Enteropathogenic *Escherichia coli*. *Mol Microbiol* 34, no. 5 (1999): 941–52.

Furness, J. B. Types of Neurons in the Enteric Nervous System. *J Auton Nerv Syst* 81, no. 1–3 (2000): 87–96.

Goosney, D. L., R. DeVinney, and B. B. Finlay. Recruitment of Cytoskeletal and Signaling Proteins to Enteropathogenic and Enterohemorrhagic *Escherichia coli* Pedestals. *Infect Immun* 69, no. 5 (2001): 3315–22.

Gruenheid, S., I. Sekirov, N. A. Thomas, W. Deng, P. O'Donnell, D. Goode, Y. Li, E. A. Frey, N. F. Brown, P. Metalnikov, T. Pawson, K. Ashman, and B. B. Finlay. Identification and Characterization of Nlea, a Non-Lee-Encoded Type Iii Translocated Virulence Factor of Enterohaemorrhagic *Escherichia coli* O157:H7. *Mol Microbiol* 51, no. 5 (2004): 1233–49.

Haack, K. R., C. L. Robinson, K. J. Miller, J. W. Fowlkes, and J. L. Mellies. Interaction of Ler at the Lee5 (Tir) Operon of Enteropathogenic *Escherichia coli*. *Infection & Immunity* 71, no. 1 (2003): 384–92.

Hughes, D. T., and V. Sperandio. Inter-Kingdom Signalling: Communication between Bacteria and Their Hosts. *Nat Rev Microbiol* 6, no. 2 (2008): 111–20.

Ide, T., S. Laarmann, L. Greune, H. Schillers, H. Oberleithner, and M. A. Schmidt. Characterization of Translocation Pores Inserted into Plasma Membranes by Type Iii-Secreted Esp Proteins of Enteropathogenic *Escherichia coli*. *Cell Microbiol* 3, no. 10 (2001): 669–79.

Iyoda, S., and H. Watanabe. Positive Effects of Multiple Pch Genes on Expression of the Locus of Enterocyte Effacement Genes and Adherence of Enterohaemorrhagic *Escherichia coli* O157 : H7 to Hep-2 Cells. *Microbiology* 150, no. Pt 7 (2004): 2357–571.

Jacewicz, M., H. Clausen, E. Nudelman, A. Donohue-Rolfe, and G. T. Keusch. Pathogenesis of *Shigella* Diarrhea. Xi. Isolation of a *Shigella* Toxin-Binding Glycolipid from Rabbit Jejunum and Hela Cells and Its Identification as Globotriaosylceramide. *J Exp Med* 163, no. 6 (1986): 1391–404.

Jarvis, K. G., J. A. Giron, A. E. Jerse, T. K. McDaniel, M. S. Donnenberg, and J. B. Kaper. Enteropathogenic *Escherichia coli* Contains a Putative Type Iii Secretion System Necessary for the Export of Proteins Involved in Attaching and Effacing Lesion Formation. *Proc Natl Acad Sci U S A* 92, no. 17 (1995): 7996–8000.

Jerse, A. E., J. Yu, B. D. Tall, and J. B. Kaper. A Genetic Locus of Enteropathogenic *Escherichia coli* Necessary for the Production of Attaching and Effacing Lesions on Tissue Culture Cells. *Proc Natl Acad Sci U S A* 87, no. 20 (1990): 7839–43.

Kaper, J. B., J. P. Nataro, and H. L. Mobley. Pathogenic *Escherichia coli*. *Nat Rev Microbiol* 2, no. 2 (2004): 123–40.

Kenny, B., A. Abe, M. Stein, and B. B. Finlay. Enteropathogenic *Escherichia coli* Protein Secretion Is Induced in Response to Conditions Similar to Those in the Gastrointestinal Tract. *Infect Immun* 65, no. 7 (1997): 2606–12.

Kenny, B., R. DeVinney, M. Stein, D. J. Reinscheid, E. A. Frey, and B. B. Finlay. Enteropathogenic *E. coli* (Epec) Transfers Its Receptor for Intimate Adherence into Mammalian Cells. *Cell* 91, no. 4 (1997): 511–20.

Kenny, B., and B. B. Finlay. Protein Secretion by Enteropathogenic *Escherichia coli* Is Essential for Transducing Signals to Epithelial Cells. *Proc Natl Acad Sci U S A* 92, no. 17 (1995): 7991–5.

Kenny, B., and M. Jepson. Targeting of an Enteropathogenic *Escherichia coli* (Epec) Effector Protein to Host Mitochondria. *Cell Microbiol* 2, no. 6 (2000): 579–90.

Kimmitt, P. T., C. R. Harwood, and M. R. Barer. Induction of Type 2 Shiga Toxin Synthesis in *Escherichia coli* O157 by 4-Quinolones. *Lancet* 353, no. 9164 (1999): 1588–9.

Kimmitt, P. T., C. R. Harwood, and M. R. Barer. Toxin Gene Expression by Shiga Toxin-Producing *Escherichia coli*: The Role of Antibiotics and the Bacterial Sos Response. *Emerg Infect Dis* 6, no. 5 (2000): 458–65.

Lindberg, A. A., J. E. Brown, N. Stromberg, M. Westling-Ryd, J. E. Schultz, and K. A. Karlsson. Identification of the Carbohydrate Receptor for Shiga Toxin Produced by *Shigella dysenteriae* Type 1. *J Biol Chem* 262, no. 4 (1987): 1779–85.

McDaniel, T. K., K. G. Jarvis, M. S. Donnenberg, and J. B. Kaper. A Genetic Locus of Enterocyte Effacement Conserved among Diverse Enterobacterial Pathogens. *Proc Natl Acad Sci U S A* 92, no. 5 (1995): 1664–8.

McNamara, B. P., and M. S. Donnenberg. A Novel Proline-Rich Protein, Espf, Is Secreted from Enteropathogenic *Escherichia coli* Via the Type Iii Export Pathway. *FEMS Microbiol Lett* 166, no. 1 (1998): 71–8.

McNamara, B. P., A. Koutsouris, C. B. O'Connell, J. P. Nougayrede, M. S. Donnenberg, and G. Hecht. Translocated Espf Protein from Enteropathogenic *Escherichia coli* Disrupts Host Intestinal Barrier Function. *J Clin Invest* 107, no. 5 (2001): 621–9.

Mellies, J. L., S. J. Elliott, V. Sperandio, M. S. Donnenberg, and J. B. Kaper. The Per Regulon of Enteropathogenic *Escherichia coli*: Identification of a Regulatory Cascade and a Novel Transcriptional Activator, the Locus of Enterocyte Effacement (Lee)-Encoded Regulator (Ler). *Mol Microbiol* 33, no. 2 (1999): 296–306.

Merighi, M., A. Carroll-Portillo, A. N. Septer, A. Bhatiya, and J. S. Gunn. Role of *Salmonella Enterica* Serovar Typhimurium Two-Component System Prea/Preb in Modulating Pmra-Regulated Gene Transcription. *J Bacteriol* 188, no. 1 (2006): 141–9.

Merighi, M., A. N. Septer, A. Carroll-Portillo, A. Bhatiya, S. Porwollik, M. McClelland, and J. S. Gunn. Genome-Wide Analysis of the Prea/Preb (Qseb/Qsec) Regulon of *Salmonella enterica* Serovar Typhimurium. *BMC Microbiol* 9, no. 1 (2009): 42.

Miller, S. T., K. B. Xavier, S. R. Campagna, M. E. Taga, M. F. Semmelhack, B. L. Bassler, and F. M. Hughson. *Salmonella typhimurium* Recognizes a Chemically Distinct Form of the Bacterial Quorum-Sensing Signal Ai-2. *Mol Cell* 15, no. 5 (2004): 677–87.

Neely, M. N., and D. I. Friedman. Functional and Genetic Analysis of Regulatory Regions of Coliphage H-19b: Location of Shiga-Like Toxin and Lysis Genes Suggest a Role for Phage Functions in Toxin Release. *Mol Microbiol* 28, no. 6 (1998): 1255–67.

Perera, L. P., L. R. Marques, and A. D. O'Brien. Isolation and Characterization of Monoclonal Antibodies to Shiga-Like Toxin Ii of Enterohemorrhagic *Escherichia coli* and Use of the Monoclonal Antibodies in a Colony Enzyme-Linked Immunosorbent Assay. *J Clin Microbiol* 26, no. 10 (1988): 2127–31.

Rasko, D. A., C. G. Moreira, R. Li de, N. C. Reading, J. M. Ritchie, M. K. Waldor, N. Williams, R. Taussig, S. Wei, M. Roth, D. T. Hughes, J. F. Huntley, M. W. Fina, J. R. Falck, and V. Sperandio. Targeting Qsec Signaling and Virulence for Antibiotic Development. *Science* 321, no. 5892 (2008): 1078–80.

Reading, N. C., A. G. Torres, M. M. Kendall, D. T. Hughes, K. Yamamoto, and V. Sperandio. A Novel Two-Component Signaling System That Activates Transcription of an Enterohemorrhagic *Escherichia coli* Effector Involved in Remodeling of Host Actin. *J Bacteriol* 189, no. 6 (2007): 2468–76.

Ritter, A., D. L. Gally, P. B. Olsen, U. Dobrindt, A. Friedrich, P. Klemm, and J. Hacker. The Pai-Associated Leux Specific Trna5(Leu) Affects Type 1 Fimbriation in Pathogenic *Escherichia coli* by Control of Fimb Recombinase Expression. *Mol Microbiol* 25, no. 5 (1997): 871–82.

Roe, A. J., D. E. Hoey, and D. L. Gally. Regulation, Secretion and Activity of Type Iii-Secreted Proteins of Enterohaemorrhagic *Escherichia coli* O157. *Biochem Soc Trans* 31, no. Pt 1 (2003): 98–103.

Rosenshine, I., S. Ruschkowski, and B. B. Finlay. Expression of Attaching/Effacing Activity by Enteropathogenic *Escherichia coli* Depends on Growth Phase, Temperature, and Protein Synthesis Upon Contact with Epithelial Cells. *Infect Immun* 64, no. 3 (1996): 966–73.

Russell, R. M., F. C. Sharp, D. A. Rasko, and V. Sperandio. Qsea and Grlr/Grla Regulation of the Locus of Enterocyte Effacement Genes in Enterohemorrhagic *Escherichia coli*. *J Bacteriol* 189, no. 14 (2007): 5387–92.

Schauder, S., K. Shokat, M. G. Surette, and B. L. Bassler. The Luxs Family of Bacterial Autoinducers: Biosynthesis of a Novel Quorum-Sensing Signal Molecule. *Mol Microbiol* 41, no. 2 (2001): 463–76.

Schmitt, C. K., M. L. McKee, and A. D. O'Brien. Two Copies of Shiga-Like Toxin Ii-Related Genes Common in Enterohemorrhagic *Escherichia coli* Strains Are Responsible for the Antigenic Heterogeneity of the O157:H- Strain E32511. *Infect Immun* 59, no. 3 (1991): 1065–73.

Sharma, V. K., and R. L. Zuerner. Role of Hha and Ler in Transcriptional Regulation of the Esp Operon of Enterohemorrhagic *Escherichia coli* O157:H7. *J Bacteriol* 186, no. 21 (2004): 7290–301.

Sharp, F. C., and V. Sperandio. Qsea Directly Activates Transcription of Lee1 in Enterohemorrhagic *Escherichia coli*. *Infect Immun* 75, no. 5 (2007): 2432–40.

Sperandio, V., C. C. Li, and J. B. Kaper. Quorum-Sensing *Escherichia coli* Regulator A: A Regulator of the Lysr Family Involved in the Regulation of the Locus of Enterocyte Effacement Pathogenicity Island in Enterohemorrhagic *E. coli*. *Infect Immun* 70, no. 6 (2002): 3085–93.

Sperandio, V., J. L. Mellies, W. Nguyen, S. Shin, and J. B. Kaper. Quorum Sensing Controls Expression of the Type Iii Secretion Gene Transcription and Protein Secretion in Enterohemorrhagic and Enteropathogenic *Escherichia coli*. *Proc Natl Acad Sci U S A* 96, no. 26 (1999): 15196–201.

Sperandio, V., A. G. Torres, J. A. Giron, and J. B. Kaper. Quorum Sensing Is a Global Regulatory Mechanism in Enterohemorrhagic *Escherichia coli* O157:H7. *J Bacteriol* 183, no. 17 (2001): 5187–97.

Sperandio, V., A. G. Torres, B. Jarvis, J. P. Nataro, and J. B. Kaper. Bacteria-Host Communication: The Language of Hormones. *Proc Natl Acad Sci U S A* 100, no. 15 (2003): 8951–6.

Sperandio, V., A. G. Torres, and J. B. Kaper. Quorum Sensing *Escherichia coli* Regulators B and C (Qsebc): A Novel Two-Component Regulatory System Involved in the Regulation of Flagella and Motility by Quorum Sensing in *E. coli*. *Mol Microbiol* 43, no. 3 (2002): 809–21.

Surette, M. G., and B. L. Bassler. Quorum Sensing in *Escherichia coli* and *Salmonella typhimurium*. *Proc Natl Acad Sci U S A* 95, no. 12 (1998): 7046–50.

Surette, M. G., M. B. Miller, and B. L. Bassler. Quorum Sensing in *Escherichia coli*, *Salmonella typhimurium*, and *Vibrio harveyi*: A New Family of Genes Responsible for Autoinducer Production. *Proc Natl Acad Sci U S A* 96, no. 4 (1999): 1639–44.

Tobe, T., S. A. Beatson, H. Taniguchi, H. Abe, C. M. Bailey, A. Fivian, R. Younis, S. Matthews, O. Marches, G. Frankel, T. Hayashi, and M. J. Pallen. An Extensive Repertoire of Type Iii Secretion Effectors in *Escherichia coli* O157 and the Role of Lambdoid Phages in Their Dissemination. *Proc Natl Acad Sci U S A* 103, no. 40 (2006): 14941–6.

Tomson, F. L., V. K. Viswanathan, K. J. Kanack, R. P. Kanteti, K. V. Straub, M. Menet, J. B. Kaper, and G. Hecht. Enteropathogenic *Escherichia coli* Espg Disrupts Microtubules and in Conjunction with Orf3 Enhances Perturbation of the Tight Junction Barrier. *Mol Microbiol* 56, no. 2 (2005): 447–64.

Tu, X., I. Nisan, C. Yona, E. Hanski, and I. Rosenshine. Esph, a New Cytoskeleton-Modulating Effector of Enterohaemorrhagic and Enteropathogenic *Escherichia coli*. *Mol Microbiol* 47, no. 3 (2003): 595–606.

Waddell, T., S. Head, M. Petric, A. Cohen, and C. Lingwood. Globotriosyl Ceramide Is Specifically Recognized by the *Escherichia coli* Verocytotoxin 2. *Biochem Biophys Res Commun* 152, no. 2 (1988): 674–9.

Walters, M., and V. Sperandio. Autoinducer 3 and Epinephrine Signaling in the Kinetics of Locus of Enterocyte Effacement Gene Expression in Enterohemorrhagic *Escherichia coli*. *Infect Immun* 74, no. 10 (2006): 5445–55.

Weiss, D. S., A. Brotcke, T. Henry, J. J. Margolis, K. Chan, and D. M. Monack. In Vivo Negative Selection Screen Identifies Genes Required for *Francisella* Virulence. *Proc Natl Acad Sci U S A* 104, no. 14 (2007): 6037–42.

Wilson, R. K., R. K. Shaw, S. Daniell, S. Knutton, and G. Frankel. Role of Escf, a Putative Needle Complex Protein, in the Type Iii Protein Translocation System of Enteropathogenic *Escherichia coli*. *Cellular Microbiology* 3, no. 11 (2001): 753–62.

Yamashima, T. Jokichi Takamine (1854-1922), the Samurai Chemist, and His Work on Adrenalin. *J Med Biogr* 11, no. 2 (2003): 95–102.

Zhang, L., R. R. Chaudhuri, C. Constantinidou, J. L. Hobman, M. D. Patel, A. C. Jones, D. Sarti, A. J. Roe, I. Vlisidou, R. K. Shaw, F. Falciani, M. P. Stevens, D. L. Gally, S. Knutton, G. Frankel, C. W. Penn, and M. J. Pallen. "Regulators Encoded in the *Escherichia coli* Type Iii Secretion System 2 Gene Cluster Influence Expression of Genes within the Locus for Enterocyte Effacement in Enterohemorrhagic *E. coli* O157:H7." *Infect Immun* 72, no. 12 (2004): 7282-93.

Zhang, W., M. Bielaszewska, T. Kuczius, and H. Karch. Identification, Characterization, and Distribution of a Shiga Toxin 1 Gene Variant (Stx(1c)) in *Escherichia coli* Strains Isolated from Humans. *J Clin Microbiol* 40, no. 4 (2002): 1441-6.

Chapter 13
Molecular Profiling: Catecholamine Modulation of Gene Expression in Enteropathogenic Bacteria

Bradley L. Bearson and Scot E. Dowd

13.1   Introduction

Bacteria have been shown to respond to stress-related neuroendocrine hormones, of which the catecholamines are the principal class, by increasing bacterial growth rates (Lyte and Ernst 1992), production of virulence-related properties, such as toxin production (Rahman et al. 2000), and epithelial cell adherence (Vlisidou et al. 2004; Chen et al. 2006). Although the first description of catecholamine-induced potentiation of bacterial infection was described as early as 1930 (Renaud and Miget 1930), it was not until 1992 (Lyte and Ernst 1992) that the first reports of direct action of catecholamines on bacterial growth in vivo and in vitro (in mammalian-like model systems) were published and the growing discipline of microbial endocrinology eventually formulated (Lyte 2004). Although it has been demonstrated that both Gram-positive and Gram-negative bacteria respond to catecholamines (Chap. 3), this chapter will focus on current knowledge of transcriptional profiling in enteropathogenic bacteria including *Escherichia coli* O157:H7 (O157:H7), *Salmonella enterica* serovar Typhimurium (*S.* Typhimurium), and *Vibrio parahaemolyticus* (*V. parahaemolyticus*).

13.2   *E. coli* O157:H7

Norepinephrine (NE) has been shown to be a beneficial growth adjuvant to O157:H7 in serum-based media (Lyte and Ernst 1993; Kinney et al. 2000) with a primary effect of enhancing iron bioavailability (Bowdre et al. 1976; Freestone

B.L. Bearson (✉)
Agroecosystems Management Research Unit
USDA, ARS, National Laboratory for Agriculture and the Environment, Ames, IA 50011, USA
e-mail: brad.bearson@ars.usda.gov

S.E. Dowd
Medical Biofilm Research Institute, Lubbock, TX 79407, USA

M. Lyte and P.P.E. Freestone (eds.), *Microbial Endocrinology*,
Interkingdom Signaling in Infectious Disease and Health,
DOI 10.1007/978-1-4419-5576-0_13, © Springer Science+Business Media, LLC 2010

et al. 2000, 2003). A connection between the virulence systems of O157:H7 with an NE-influenced gastrointestinal (GI) tract indicates that NE enhances growth and iron acquisition, effects motility, increases toxin expression, and promotes attachment of O157:H7 in vivo (Lyte et al. 1996; Freestone et al. 1999, 2000; Chen et al. 2003; Green et al. 2004; Vlisidou et al. 2004). Thus, a link between the noradrenergic stress response in the host and the pathogenesis of O157:H7 has been described.

Recently, microarray analyses have evaluated the effects of NE and epinephrine (EPI) on O157:H7 gene expression to elucidate the mechanisms by which these phenotypic changes are modulated (Bansal et al. 2007; Dowd 2007; Kendall et al. 2007). This section will discuss each of these studies as well as present unpublished results from the Dowd lab, which not only expands upon these findings but further defines those changes in gene expression that occur specific to NE, apart from the effects of enhanced iron bioavailability.

13.2.1   Conditions of O157:H7 Experiments

Dowd (2007) evaluated the expression of planktonic phenotype O157:H7 (EDL933) after 5 h exposure to 50 μM NE in Serum-SAPI medium (Lyte and Ernst 1992) at 37°C, 0.05% CO_2, and 95% humidity. Bansal et al. (Bansal et al. 2007) evaluated gene expression in biofilm phenotype O157:H7 (EDL933) cells after 7 h in rich medium (Luria Bertani + glucose) at 37°C exposed to 50 μM NE. Kendall et al. (2007) evaluated gene expression after exposure of a *luxS* enterohemorrhagic *E. coli* (EHEC) mutant isolate (vs94) to 50 μM of EPI grown in Dulbecco's modified Eagle's medium (DMEM) until late exponential phase.

13.2.2   Shiga Toxin

One of the most important findings associated with growth of O157:H7 in the presence of 50 μM NE is the increased production of Shiga toxin (Lyte et al. 1996). Using the same medium conditions, Dowd (2007) confirmed these results using microarrays and demonstrated that *stx1* and *stx2* gene expression was comparatively induced during exposure to 50 μM NE. This study also hypothesized that NE exposure induces a positive adaptive state (Dowd et al. 2007) in O157:H7 since error-prone DNA pathways were concurrently enriched. Such pathways are associated with *recA* induction and are known to induce phage production (Plunkett et al. 1999). Dowd demonstrated significant induction of *umuD*, *recB*, and lambdoid phage-encoded transcripts, which may also indicate an initiated SOS response due to NE exposure (Keller et al. 2001; Bjedov et al. 2003; Hare et al. 2006). The *umuD* gene product is a known regulator of the RecA-dependent DNA repair SOS response (Hare et al. 2006) and has also been noted as an inducer of the SOS *recA* response (Keller et al. 2001). Since the *stx2* genes are located on a lambdoid phage, induction of the phage by a

RecA-dependent SOS response (Plunkett et al. 1999) would account for NE induction of this toxin. A similar profile was observed in vivo when O157:H7 was exposed to NE (Dowd and Lyte 2007).

Recent work by Dowd has investigated whether NE is an inducer of a novel positive adaptive state (Dowd 2007) in *E. coli* O157:H7, which could have dramatic implications in pathogenesis (LeClerc and Cebula 2000). Results of mutagenesis assays have shown that NE exposure does have the potential to enhance mutagenesis. As part of a modified Ames assay (Ames et al. 1975), NE at concentrations greater than 50 μM (e.g., 100 μM) induced a significant increase in mutagenesis (Dowd, unpublished results). Similarly, in an SOS induction assay, an SOS response is induced in *S. enterica* after only 2 h exposure to 100 μM NE (unpublished results).

The Bansal et al. study did not identify regulation of the *stx1* or *stx2* genes in the biofilm phenotype (rich media growth conditions), and the Kendall et al. study of the *luxS* mutant strain in DMEM showed a repression of *stx2a* at late growth phase (Bansal et al. 2007; Kendall et al. 2007). The differences in the experimental conditions or isolates used in these three studies make it difficult to draw comparative conclusions about gene expression associated with *stx1* or *stx2* expression. It was hypothesized that the oxidation of NE over time induces DNA damage in bacteria, which subsequently induces DNA damage repair systems. One of the systems that has been shown to be induced is the error-prone DNA repair system. The results of the Dowd studies go far toward showing that this is indeed one of the potential mechanisms by which *stx2* expression is enhanced. Recent studies in the Dowd lab have shown that at 30 min of exposure using conditions similar to those described in the original Dowd study (above) there is no induction of the error prone response while after 90 min this system is induced. Thus, induction of *stx2* expression associated with NE could be due to oxidation of the NE molecule over time and subsequent oxidative damage to the DNA of the bacterial cell.

13.2.3   Intimate Attachment

The locus of enterocyte effacement (LEE) region of EHEC O157:H7 strain EDL933 (Perna et al. 2001) is made up of 41 open reading frames (ORFs) organized into several operons (Elliott et al. 1998). The product of the *ler* gene is noted as the essential activator of the LEE-related operons (Elliott et al. 2000), and the *hha* gene has been associated with repression of *ler* transcription and subsequent LEE repression (Sharma and Zuerner 2004). Dowd found that exposure to NE increased the expression of the *eae* gene encoding the intimate attachment protein, but no notable alteration in *hha* gene regulation occurred (Dowd 2007). This would suggest an induction of LEE through an alternate pathway or posttranscriptional regulation of *hha*. Walters and Sperandio (2006) have also identified several factors and pathways that control LEE expression including stringent response activation of LEE due to NE (Tobe et al. 2005; Nakanishi et al. 2006). The expression results in the Dowd study also indicate a stringent response activation of LEE due to NE

(Nakanishi et al. 2006) with repression of *spoT* but an induction of *relA* and *rcsC* (Tobe et al. 2005). Thus, Dowd found both agreement and disagreement with several of the known mechanisms of LEE induction when O157:H7 was exposed to NE. Because NE is associated with increased iron bioavailability and all of the mechanisms noted are also intimately linked with the iron regulon (Gaille et al. 2003; Kim et al. 2005; Vinella et al. 2005; Laaberki et al. 2006; Muller et al. 2006; Diggle et al. 2007), it was deduced that iron bioavailability could be the primary governing factor of LEE regulation in the presence of NE.

Kendall et al. (2007) showed that EPI induced the expression of all LEE operons (LEE1-5) in their *luxS* mutant. Their study also found induction of genes encoding H-NS, HU, FIS, and Hha nucleoid proteins. Each of these transcriptional factors have also been linked to LEE expression. Similar to results seen with *stx2* expression, Bansal et al. (2007) did not document any regulation of LEE- or *eae*-related genes in the rich medium grown biofilm phenotype for O157:H7. The Dowd and Kendall et al. studies both demonstrated induction of intimate attachment-related genes in planktonic cells, suggesting there is some catecholamine-associated regulation of these operons.

13.2.4   Motility, Curli, LPS, and Fimbriae

Genes associated with motility, curli, LPS, and fimbriae are of great importance in the pathogenesis of O157:H7 (Albert et al. 1996; Elliott and Kaper 1997; Kaper et al. 1997; Kaper 1998; Giron et al. 2002; Jordan et al. 2004; Torres et al. 2004). Norepinephrine is thought to enhance motility and act as a chemoattractant for O157:H7. Bansal et al. found a comprehensive array of motility, curli, and fimbrial genes regulated by both NE and EPI (Bansal et al. 2007). Those genes associated with chemotaxis and flagella were repressed in the biofilm phenotype; yet, Bansal et al. noted in laboratory studies an increased motility associated with both NE and EPI. Curli genes were also repressed while fimbrial-related genes were both repressed and induced in the Bansal et al. study. Kendall et al. noted induction of flagellar genes in their *luxS* mutant strain (Kendall et al. 2007). Dowd did not notice any regulation of flagellar genes but did see both repression and induction of several fimbrial genes and repression of curli-related genes in agreement with the Bansal et al. study (Dowd 2007). Dowd also noted a comprehensive induction of genes associated with LPS modification not noted by either the Kendall et al. or Bansal et al. studies.

13.2.5   Iron Acquisition

The effects of NE enhancing iron bioavailability in serum-based media (Freestone et al. 2000) were highly evident in the Dowd study (Dowd 2007). In the presence of NE, there was a repression of genes associated with iron scavenging, including genes associated with enterobactin. Conversely, there was induction of genes in the ferrous iron ion transport and ABC iron transport related systems. The Bansal et al. study also noted induction of ABC and ferrous iron transport systems (Bansal et al. 2007). These results support the role of NE in increasing iron bioavailability to O157:H7 (Freestone et al. 2003). In medium where NE is present, iron availability is greater, which would comparatively reduce the need to synthesize iron scavenging proteins. Dowd suggested that much of the gene expression seen in O157:H7 in the presence of NE could be related to the differential iron bioavailability and growth phase (Dowd 2007). However, specific genes may respond to NE independent of iron-related gene expression, and future studies of gene expression can be directed to normalize out the effects of iron bioavailability.

13.2.6 Cold Shock Proteins

One of the most striking similarities in the Dowd and Bansal et al. studies was the induction of cold shock proteins, especially *cspHG* (Bansal et al. 2007; Dowd 2007). In the Bansal et al. study, *cspH* was the most highly induced gene, while in the Dowd study, the *cspHG* genes were in the top 10 most highly induced genes suggesting a potentially critical role for these genes in response to NE. The *cspH* and other cold and heat shock genes are induced in a variety of situations and are usually associated with stabilization of mRNA and DNA. The *cspH* gene product, for instance, is noted for its role in stabilizing mRNA at lower temperatures and has also been noted to be induced when there are nutrient upshifts in growth medium (Kim et al. 2004). Dowd also observed induction of the primary heat shock proteins encoded by *ibpAB*. These genes later are also induced in the presence of oxidants, which ties in well with the discussion related to *stx2* expression (see above) in which oxidation of NE over time results in DNA damage and subsequent induction of bacterial stress responses.

13.2.7 Positive Adaptive State

Positive adaptive state is a condition in bacterial physiology in which there is active replication of the bacterial chromosome combined with the influence of environmental conditions that induce mutations at a moderate to high rate. The results obtained in recent studies by Dowd and colleagues (Dowd 2007; Dowd et al. 2007; Dowd and Lyte 2007) represent the first reports suggestive of the ability of NE to induce a positive adaptive state (Dowd et al. 2007) in O157:H7, and therefore represents a newly described mechanism by which neuroendocrine hormones, either derived from the host or provided in the dietary intake (Lyte 2004), may contribute to the ability of a pathogen to rapidly adapt to a new niche and establish a productive infection. These results suggest that microbes with highly plastic genomes and growth phase-dependent virulence factors may benefit during host stress states.

From a microbial endocrinology perspective, these findings have several implications and advance our understanding of the role that neuroendocrine hormones may play in the infection process. First and foremost, the ability of NE to affect bacterial growth is not solely limited to its ability to aid in the sequestration of iron from iron-restrictive tissues, such as those found in tissue, blood, and gastrointestinal environments, to facilitate bacterial growth (Freestone et al. 2003). Indeed, exposure of O157:H7 to NE resulted in a dynamic transcription profile that can be described as inducing a unique and population-wide positive adaptive state (Dowd 2007). The results described herein combine the two tenants of infectious pathogens, high replication rate and high adaptation rate (LeClerc et al. 1996; LeClerc and Cebula 2000), and demonstrates that neuroendocrine hormones, particularly those such as NE that are involved in the host stress response, can contribute to the virulence and adaptability of pathogens and ultimately disease causation.

13.3 *S. enterica* serovar Typhimurium

Most research investigations on the exposure of *S. enterica* serovar Typhimurium to NE has focused on enhanced bacterial growth in Serum-SAPI minimal medium (Freestone et al. 1999; Williams et al. 2006; Bearson et al. 2008). However, recent research on *S.* Typhimurium has also investigated enhanced bacterial motility in the presence of NE (Bearson and Bearson 2008).

13.3.1 *Norepinephrine-Enhanced Motility*

Norepinephrine-enhanced motility of *S.* Typhimurium has been demonstrated on DMEM medium containing 0.3% agar (Bearson and Bearson 2008). Furthermore, transcriptional analysis of *S.* Typhimurium in the presence of 2 mM NE (Bearson and Bearson 2008) in Serum-SAPI minimal medium using DNA microarrays and real-time RT-PCR analyses indicated that a number of flagellar and chemotaxis genes were upregulated during NE exposure. Flagellar assembly is a complex process and requires the coordination of a cascade of early, middle, and late genes for the production of gene products that ultimately result in flagellar assembly for motility (Chilcott and Hughes 2000). This hierarchy results in the amplification of the signal such that the relative level of expression is "late">"middle">"early" genes. For example, in the presence of 2 mM NE, the *S.* Typhimurium genes *fljB* (late), *fliY* (middle), *fliA* (middle), and *flhC* (early) are induced 15.4-, 4.2-, 3.6-, and 1.4-fold compared to the absence of NE, respectively (Bearson and Bearson 2008). Norepinephrine-enhanced motility is decreased in a *S.* Typhimurium *qseC* mutant, indicating that, similar to *E. coli*, the QseC sensor kinase of *S.* Typhimurium is involved in NE-enhanced motility (Sperandio et al. 2003; Bearson and Bearson 2008). Although *qseC* is involved in NE-enhanced motility, its transcription is not significantly increased in the presence of NE (Bearson and Bearson 2008). The α-adrenergic antagonist phentolamine interferes with the motility-enhancing property that has been demonstrated for 50 μM NE in *S*. Typhimurium (Bearson and Bearson 2008). However, transcriptional analysis comparing *S*. Typhimurium exposed to NE and NE/phentolamine has not been performed.

13.3.2 Norepinephrine-Enhanced Growth

Norepinephrine-enhanced growth of *S*. Typhimurium in Serum-SAPI minimal medium is due to the siderophore-like activity of NE in the presence of transferrin (Freestone et al. 1999). The ability of transferrin to sequester iron from the bacterial cell creates an iron-deplete environment resulting in no or slow growth depending on the concentration of bacterial cells. Transcriptional analysis using microarrays to monitor gene expression of *S*. Typhimurium grown in Serum-SAPI medium containing 2 mM NE compared to the absence of NE confirms the iron-deplete environment of Serum-SAPI medium, since transcription of genes encoding iron uptake and utilization pathways are decreased in the presence of NE Fig. 13.2, Transport and binding proteins; (Bearson et al. 2008). This indicates that NE scavenges iron from the environment in a siderophore-like manner and increases iron availability to the bacterial cell, resulting in NE-enhanced growth of *S*. Typhimurium and a decreased need for the production of iron uptake and utilization proteins. Interestingly, although the expression of genes encoding iron acquisition proteins is downregulated in the presence of NE, a subset of these iron uptake proteins are required for NE-enhanced growth of *S*. Typhimurium (Williams et al. 2006; Bearson et al. 2008). This suggests that the relative expression of iron acquisition genes is lower in the presence of NE, but these genes are not in a transcriptional "off" state. In addition, the microarray experiments were performed during the exponential phase (O.D.$_{600}$=0.4) of bacterial growth when enterochelin, salmochelin, and their breakdown products are accumulating in the growth medium. Since enterochelin/salmochelin production is necessary for NE-enhanced growth of *S*. Typhimurium (Bearson et al. 2008), the accumulation of these siderophores in the presence of NE increases iron availability, which concomitantly reduces the expression of genes encoding iron acquisition proteins via the iron regulator Fur.

Norepinephrine-enhanced growth in Serum-SAPI minimal medium is a model of in vivo growth during systemic infection of the mammalian host. As previously described, transferrin present in serum sequesters iron from the bacterial cell and prevents bacterial growth. However, NE can assist the bacterial cell by providing iron in the presence of enterochelin/salmochelin (Freestone et al. 2000, 2003; Burton et al. 2002). A countermeasure is deployed by the host immune system using lipocalin 2 to sequester enterochelin (Goetz et al. 2002). *S*. Typhimurium's defense is the synthesis of salmochelin via glucosylation of enterochelin, which prevents binding to lipocalin 2 (Hantke et al. 2003; Fischbach et al. 2006; Smith 2007). In addition to the binding by lipocalin 2, enterochelin has a high membrane affinity, which results in membrane sequestration (Luo et al. 2006). Glucosylation of enterochelin to salmochelin by IroB and hydrolysis of salmochelin by the periplasmic hydrolase IroE decreases membrane affinity and increases the iron acquisition rate for *S.* Typhimurium. Therefore, the ability of *S.* Typhimurium to produce, transport, and breakdown salmochelin via products of the *iroA* gene cluster is a virulence hallmark that assists *S.* Typhimurium in causing systemic disease (Fischbach et al. 2005; Luo et al. 2006; Smith 2007). This is in contrast to most *E. coli* strains except a subset that also contains the *iroA* gene cluster including uropathogenic *E. coli* (Hantke et al. 2003).

The ability of NE to enhance the growth of *S.* Typhimurium in Serum-SAPI minimal medium (Freestone et al. 1999; Williams et al. 2006; Bearson et al. 2008) suggests that additional biosynthetic pathways would be modulated besides the iron utilization and transport genes. Transcriptional analysis using microarrays on *S.* Typhimurium grown in Serum-SAPI medium containing 2 mM NE indicates that NE exposure increases transcription of genes involved in amino acid biosynthesis, cofactor biosynthesis, central intermediary metabolism, energy metabolism, and synthesis of transport and binding proteins (Bearson et al. 2008), suggesting that in order to increase the growth rate of the bacterial cell due to the increased availability of iron provided by the NE present in Serum-SAPI minimal medium, *S.* Typhimurium modulates the biosynthesis of multiple cellular pathways (Figs. 13.1).

13.3.3   *Salmonella* Pathogenicity Island 2

Genes encoded by *Salmonella* pathogenicity island 2 (SPI2) have been shown to be iron-responsive with transcription upregulated during iron chelation (Zaharik et al. 2002). The *S.* Typhimurium SPI2-associated genes are required for intramacrophage survival and systemic infection, which are iron-deplete environments. Microarray analysis indicates that the transcription of 13 SPI2-associated genes (Fig. 13.2,

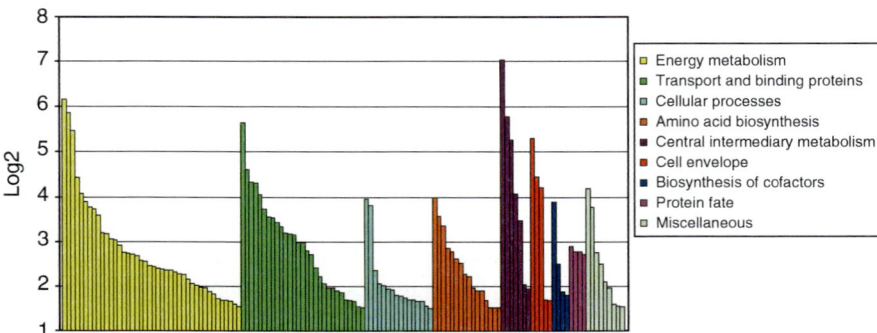

Fig. 13.1   Increased S. Typhimurium gene expression in the presence of 2 mM Norepinephrine

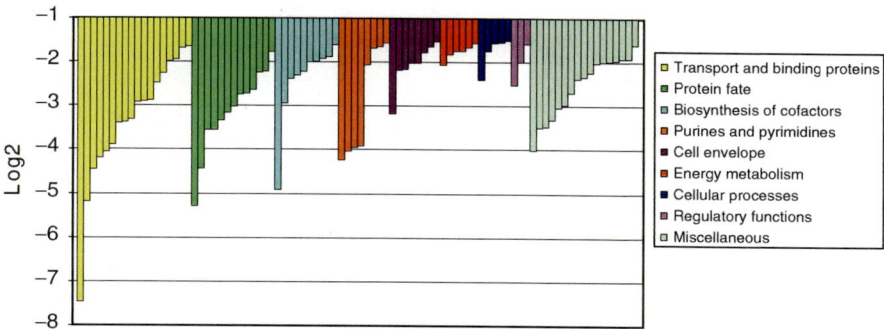

Fig. 13.2 Decreased S. Typhimurium gene expression in the presence of 2 mM Norepinephrine

Protein fate; Bearson and Bearson, unpublished results) is repressed in the presence of 2 mM NE in Serum-SAPI medium compared to the absence of NE, providing additional evidence that Serum-SAPI medium is iron-deplete (Bearson et al. 2008).

13.4 *V. parahaemolyticus*

Nakano et al. investigated NE stimulation of the cytotoxic activity of *V. parahaemolyticus* against cultured cell lines and enterotoxicity in the rat ileal loop model (Nakano et al. 2007). The presence of 50 µM NE enhances the cytotoxicity of *V. parahaemolyticus* against Caco-2 cells and this NE-enhanced cytotoxicity can be prevented by 100 µM phentolamine (Nakano et al. 2007). The cytotoxic activity of *V. parahaemolyticus* on Caco-2 cells requires type III secretion system (TTSS) 1 of the bacterial pathogen. A TTSS is a complex bacterial structure that allows the pathogen to inject effector molecules into a host cell. Transcription of the *V. parahaemolyticus vscQ* and *vscU* genes associated with TTSS1 was increased during infection of Caco-2 cells in the presence of NE. However, an increase in transcription was not observed for genes associated with TTSS2 even though, in the presence of 100 µM NE, the enteropathogenicity of *V. parahaemolyticus* (requiring TTSS2) was enhanced in the rat ileal loop model by increasing fluid accumulation. The increased fluid accumulation due to NE-enhanced enteropathogenicity of *V. parahaemolyticus* is prevented by both phentolamine (100 µM, nonselective α-adrenergic antagonist) and prazosin (10 µM, α_1-adrenergic antagonist).

13.5 Summary

The transcriptional profiles of enteropathogens that have been investigated to date suggest an enhancement of bacterial pathogenesis upon exposure to mammalian catecholamines. Bacterial phenotypes that correspond to the alteration in gene expression profiles include enhanced motility, increased iron acquisition in serum resulting in enhanced growth, epithelial cell attachment, and cytotoxicity. Norepinephrine-enhanced pathogen virulence may occur when the mammalian host is immunologically vulnerable due to the production of catecholamines during periods of stress (e.g., transportation and marketing).

References

Albert, M. J., Faruque, S. M., Faruque, A. S., Bettelheim, K. A., Neogi, P. K., Bhuiyan, N. A. and Kaper, J. B. 1996. Controlled study of cytolethal distending toxin-producing *Escherichia coli* infections in Bangladeshi children. J. Clin. Microbiol. 34:717-719.

Ames, B. N., McCann, J. and Yamasaki, E. 1975. Proceedings: carcinogens are mutagens: a simple test system. Mutat. Res. 33:27-28.

Bansal, T., Englert, D., Lee, J., Hegde, M., Wood, T. K. and Jayaraman, A. 2007. Differential effects of epinephrine, norepinephrine, and indole on *Escherichia coli* O157:H7 chemotaxis, colonization, and gene expression. Infect. Immun. 75:4597-4607.

Bearson, B. L. and Bearson, S. M. 2008. The role of the QseC quorum-sensing sensor kinase in colonization and norepinephrine-enhanced motility of *Salmonella enterica* serovar Typhimurium. Microb. Pathog. 44:271-278.

Bearson, B. L., Bearson, S. M. D., Uthe, J. J., Dowd, S. E., Houghton, J. O., Lee, I. S., Toscano, M. J. and Lay, D. C. 2008. Iron regulated genes of *Salmonella enterica* serovar Typhimurium in response to norepinephrine and the requirement of *fepDGC* for norepinephrine-enhanced growth. Microbes Infect. 10:807-816.

Bjedov, I., Tenaillon, O., Gerard, B., Souza, V., Denamur, E., Radman, M., Taddei, F. and Matic, I. 2003. Stress-induced mutagenesis in bacteria. Science 300:1404-1409.

Bowdre, J. H., Krieg, N. R., Hoffman, P. S. and Smibert, R. M. 1976. Stimulatory effect of dihy-droxyphenyl compounds on the aerotolerance of *Spirillum volutans* and *Campylobacter fetus* subspecies *jejuni*. Appl. Environ. Microbiol. 31:127-133.

Burton, C. L., Chhabra, S. R., Swift, S., Baldwin, T. J., Withers, H., Hill, S. J. and Williams, P. 2002. The growth response of *Escherichia coli* to neurotransmitters and related catecholamine drugs requires a functional enterobactin biosynthesis and uptake system. Infect. Immun. 70:5913-5923.

Chen, C., Brown, D. R., Xie, Y., Green, B. T. and Lyte, M. 2003. Catecholamines modulate *Escherichia coli* O157:H7 adherence to murine cecal mucosa. Shock 20:183-188.

Chen, C., Lyte, M., Stevens, M. P., Vulchanova, L. and Brown, D. R. 2006. Mucosally-directed adrenergic nerves and sympathomimetic drugs enhance non-intimate adherence of *Escherichia coli* O157:H7 to porcine cecum and colon. Eur. J. Pharmacol. 539:116-124.

Chilcott, G. S. and Hughes, K. T. 2000. Coupling of flagellar gene expression to flagellar assembly in *Salmonella enterica* serovar Typhimurium and *Escherichia coli*. Microbiol. Mol. Biol. Rev. 64:694-708.

Diggle, S. P., Matthijs, S., Wright, V. J., Fletcher, M. P., Chhabra, S. R., Lamont, I. L., Kong, X., Hider, R. C., Cornelis, P., Camara, M. and Williams, P. 2007. The *Pseudomonas aeruginosa* 4-quinolone signal molecules HHQ and PQS play multifunctional roles in quorum sensing and iron entrapment. Chem. Biol. 14:87-96.

Dowd, S. E. 2007. *Escherichia coli* O157:H7 gene expression in the presence of catecholamine norepinephrine. FEMS Microbiol. Lett. 273:214-223.

Dowd, S. E., Killinger-Mann, K., Blanton, J., San Francisco, M. and Brashears, M. 2007. Positive adaptive state: microarray evaluation of gene expression in *Salmonella enterica* Typhimurium exposed to nalidixic acid. Foodborne Pathog. Dis. 4:187-200.

Dowd, S. E. and Lyte, M. 2007. Microarray Analysis of Norepinephrine-*Escherichia coli* O157:H7 Gene Expression in a Porcine Ligated Ileal Loop Model and Comparison to In Vitro data: In Vivo Veritas? 107th General Meeting of the American Society for Microbiology B-256.

Elliott, S. J. and Kaper, J. B. 1997. Role of type 1 fimbriae in EPEC infections. Microb. Pathog. 23:113-118.

Elliott, S. J., Sperandio, V., Giron, J. A., Shin, S., Mellies, J. L., Wainwright, L., Hutcheson, S. W., McDaniel, T. K. and Kaper, J. B. 2000. The locus of enterocyte effacement (LEE)-encoded regulator controls expression of both LEE- and non-LEE-encoded virulence factors in enteropathogenic and enterohemorrhagic *Escherichia coli*. Infect. Immun. 68:6115-6126.

Elliott, S. J., Wainwright, L. A., McDaniel, T. K., Jarvis, K. G., Deng, Y. K., Lai, L. C., McNamara, B. P., Donnenberg, M. S. and Kaper, J. B. 1998. The complete sequence of the locus of enterocyte effacement (LEE) from enteropathogenic *Escherichia coli* E2348/69. Mol. Microbiol. 28:1-4.

Fischbach, M. A., Lin, H., Liu, D. R. and Walsh, C. T. 2005. In vitro characterization of IroB, a pathogen-associated C-glycosyltransferase. Proc. Natl. Acad. Sci. U. S. A. 102:571-576.

Fischbach, M. A., Lin, H., Zhou, L., Yu, Y., Abergel, R. J., Liu, D. R., Raymond, K. N., Wanner, B. L., Strong, R. K., Walsh, C. T., Aderem, A. and Smith, K. D. 2006. The pathogen-associated *iroA* gene cluster mediates bacterial evasion of lipocalin 2. Proc. Natl. Acad. Sci. U. S. A. 103:16502-16507.

Freestone, P. P., Haigh, R. D., Williams, P. H. and Lyte, M. 1999. Stimulation of bacterial growth by heat-stable, norepinephrine-induced autoinducers. FEMS Microbiol. Lett. 172:53-60.

Freestone, P. P., Haigh, R. D., Williams, P. H. and Lyte, M. 2003. Involvement of enterobactin in norepinephrine-mediated iron supply from transferrin to enterohaemorrhagic *Escherichia coli*. FEMS Microbiol. Lett. 222:39-43.

Freestone, P. P., Lyte, M., Neal, C. P., Maggs, A. F., Haigh, R. D. and Williams, P. H. 2000. The mammalian neuroendocrine hormone norepinephrine supplies iron for bacterial growth in the presence of transferrin or lactoferrin. J. Bacteriol. 182:6091-6098.

Gaille, C., Reimmann, C. and Haas, D. 2003. Isochorismate synthase (PchA), the first and rate-limiting enzyme in salicylate biosynthesis of *Pseudomonas aeruginosa*. J. Biol. Chem. 278:16893-16898.

Giron, J. A., Torres, A. G., Freer, E. and Kaper, J. B. 2002. The flagella of enteropathogenic *Escherichia coli* mediate adherence to epithelial cells. Mol. Microbiol. 44:361-379.

Goetz, D. H., Holmes, M. A., Borregaard, N., Bluhm, M. E., Raymond, K. N. and Strong, R. K. 2002. The neutrophil lipocalin NGAL is a bacteriostatic agent that interferes with siderophore-mediated iron acquisition. Mol. Cell 10:1033-1043.

Green, B. T., Lyte, M., Chen, C., Xie, Y., Casey, M. A., Kulkarni-Narla, A., Vulchanova, L. and Brown, D. R. 2004. Adrenergic modulation of *Escherichia coli* O157:H7 adherence to the colonic mucosa. Am. J. Physiol. Gastrointest. Liver Physiol. 287:G1238-G1246.

Hantke, K., Nicholson, G., Rabsch, W. and Winkelmann, G. 2003. Salmochelins, siderophores of *Salmonella enterica* and uropathogenic *Escherichia coli* strains, are recognized by the outer membrane receptor IroN. Proc. Natl. Acad. Sci. U. S. A. 100:3677-3682.

Hare, J. M., Perkins, S. N. and Gregg-Jolly, L. A. 2006. A constitutively expressed, truncated *umuDC* operon regulates the *recA*-dependent DNA damage induction of a gene in *Acinetobacter baylyi* strain ADP1. Appl. Environ. Microbiol. 72:4036-4043.

Jordan, D. M., Cornick, N., Torres, A. G., Dean-Nystrom, E. A., Kaper, J. B. and Moon, H. W. 2004. Long polar fimbriae contribute to colonization by *Escherichia coli* O157:H7 in vivo. Infect. Immun. 72:6168-6171.

Kaper, J. B. 1998. Enterohemorrhagic *Escherichia coli*. Curr. Opin. Microbiol. 1:103-108.

Kaper, J. B., McDaniel, T. K., Jarvis, K. G. and Gomez-Duarte, O. 1997. Genetics of virulence of enteropathogenic *E. coli*. Adv. Exp. Med. Biol. 412:279-287.

Keller, K. L., Overbeck-Carrick, T. L. and Beck, D. J. 2001. Survival and induction of SOS in *Escherichia coli* treated with cisplatin, UV-irradiation, or mitomycin C are dependent on the function of the RecBC and RecFOR pathways of homologous recombination. Mutat. Res. 486:21-29.

Kendall, M. M., Rasko, D. A. and Sperandio, V. 2007. Global effects of the cell-to-cell signaling molecules autoinducer-2, autoinducer-3, and epinephrine in a *luxS* mutant of enterohemorrhagic *Escherichia coli*. Infect. Immun. 75:4875-4884.

Kim, B. H., Kim, H. G., Bae, G. I., Bang, I. S., Bang, S. H., Choi, J. H. and Park, Y. K. 2004. Expression of *cspH* upon nutrient up-shift in *Salmonella enterica* serovar Typhimurium. Arch. Microbiol. 182:37-43.

Kim, E. J., Wang, W., Deckwer, W. D. and Zeng, A. P. 2005. Expression of the quorum-sensing regulatory protein LasR is strongly affected by iron and oxygen concentrations in cultures of *Pseudomonas aeruginosa* irrespective of cell density. Microbiology 151:1127-1138.

Kinney, K. S., Austin, C. E., Morton, D. S. and Sonnenfeld, G. 2000. Norepinephrine as a growth stimulating factor in bacteria – mechanistic studies. Life Sci. 67:3075-3085.

Laaberki, M. H., Janabi, N., Oswald, E. and Repoila, F. 2006. Concert of regulators to switch on LEE expression in enterohemorrhagic *Escherichia coli* O157:H7: interplay between Ler, GrlA, HNS and RpoS. Int. J. Med. Microbiol. 296:197-210.

LeClerc, J. E. and Cebula, T. A. 2000. *Pseudomonas* survival strategies in cystic fibrosis. Science 289:391-392.

LeClerc, J. E., Li, B., Payne, W. L. and Cebula, T. A. 1996. High mutation frequencies among *Escherichia coli* and *Salmonella* pathogens. Science 274:1208-1211.

Luo, M., Lin, H., Fischbach, M. A., Liu, D. R., Walsh, C. T. and Groves, J. T. 2006. Enzymatic tailoring of enterobactin alters membrane partitioning and iron acquisition. ACS Chem. Biol. 1:29-32.

Lyte, M. 2004. Microbial endocrinology and infectious disease in the 21st century. Trends Microbiol. 12:14-20.

Lyte, M., Arulanandam, B. P. and Frank, C. D. 1996. Production of Shiga-like toxins by *Escherichia coli* O157:H7 can be influenced by the neuroendocrine hormone norepinephrine. J. Lab. Clin. Med. 128:392-398.

Lyte, M. and Ernst, S. 1992. Catecholamine induced growth of gram negative bacteria. Life Sci. 50:203-212.

Lyte, M. and Ernst, S. 1993. Alpha and beta adrenergic receptor involvement in catecholamine-induced growth of gram-negative bacteria. Biochem. Biophys. Res. Commun. 190:447-452.

Muller, C. M., Dobrindt, U., Nagy, G., Emody, L., Uhlin, B. E. and Hacker, J. 2006. Role of histone-like proteins H-NS and StpA in expression of virulence determinants of uropathogenic *Escherichia coli*. J. Bacteriol. 188:5428-5438.

Nakanishi, N., Abe, H., Ogura, Y., Hayashi, T., Tashiro, K., Kuhara, S., Sugimoto, N. and Tobe, T. 2006. ppGpp with DksA controls gene expression in the locus of enterocyte effacement (LEE) pathogenicity island of enterohaemorrhagic *Escherichia coli* through activation of two virulence regulatory genes. Mol. Microbiol. 61:194-205.

Nakano, M., Takahashi, A., Sakai, Y. and Nakaya, Y. 2007. Modulation of pathogenicity with norepinephrine related to the type III secretion system of *Vibrio parahaemolyticus*. J. Infect. Dis. 195:1353-1360.

Perna, N. T., Plunkett, G., 3rd, Burland, V., Mau, B., Glasner, J. D., Rose, D. J., Mayhew, G. F., Evans, P. S., Gregor, J., Kirkpatrick, H. A., Posfai, G., Hackett, J., Klink, S., Boutin, A., Shao, Y., Miller, L., Grotbeck, E. J., Davis, N. W., Lim, A., Dimalanta, E. T., Potamousis, K. D., Apodaca, J., Anantharaman, T. S., Lin, J., Yen, G., Schwartz, D. C., Welch, R. A. and Blattner, F. R. 2001. Genome sequence of enterohaemorrhagic *Escherichia coli* O157:H7. Nature 409:529-533.

Plunkett, G., 3rd, Rose, D. J., Durfee, T. J. and Blattner, F. R. 1999. Sequence of Shiga toxin 2 phage 933W from *Escherichia coli* O157:H7: Shiga toxin as a phage late-gene product. J. Bacteriol. 181:1767-1778.

Rahman, H., Reissbrodt, R. and Tschape, H. 2000. Effect of norepinephrine on growth of *Salmonella* and its enterotoxin production. Indian J. Exp. Biol. 38:285-286.

Renaud, M. and Miget, A. 1930. Role favorisant des perturbations locales causees par l' adrenaline sur le developpement des infections microbiennes. C. R. Seances Soc. Biol. Fil. 103:1052-1054.

Sharma, V. K. and Zuerner, R. L. 2004. Role of *hha* and *ler* in transcriptional regulation of the *esp* operon of enterohemorrhagic *Escherichia coli* O157:H7. J. Bacteriol. 186:7290-7301.

Smith, K. D. 2007. Iron metabolism at the host pathogen interface: Lipocalin 2 and the pathogen-associated *iroA* gene cluster. Int. J. Biochem. Cell Biol. 39:1776-1780.

Sperandio, V., Torres, A. G., Jarvis, B., Nataro, J. P. and Kaper, J. B. 2003. Bacteria-host communication: the language of hormones. Proc. Natl. Acad. Sci. U. S. A. 100:8951-8956.

Tobe, T., Ando, H., Ishikawa, H., Abe, H., Tashiro, K., Hayashi, T., Kuhara, S. and Sugimoto, N. 2005. Dual regulatory pathways integrating the RcsC-RcsD-RcsB signalling system control enterohaemorrhagic *Escherichia coli* pathogenicity. Mol. Microbiol. 58:320-333.

Torres, A. G., Kanack, K. J., Tutt, C. B., Popov, V. and Kaper, J. B. 2004. Characterization of the second long polar (LP) fimbriae of *Escherichia coli* O157:H7 and distribution of LP fimbriae in other pathogenic *E. coli* strains. FEMS Microbiol. Lett. 238:333-344.

Vinella, D., Albrecht, C., Cashel, M. and D'Ari, R. 2005. Iron limitation induces SpoT-dependent accumulation of ppGpp in *Escherichia coli*. Mol. Microbiol. 56:958-970.

Vlisidou, I., Lyte, M., van Diemen, P. M., Hawes, P., Monaghan, P., Wallis, T. S. and Stevens, M. P. 2004. The neuroendocrine stress hormone norepinephrine augments *Escherichia coli* O157:H7-induced enteritis and adherence in a bovine ligated ileal loop model of infection. Infect. Immun. 72:5446-5451.

Walters, M. and Sperandio, V. 2006. Autoinducer 3 and epinephrine signaling in the kinetics of locus of enterocyte effacement gene expression in enterohemorrhagic *Escherichia coli*. Infect. Immun. 74:5445-5455.

Williams, P. H., Rabsch, W., Methner, U., Voigt, W., Tschape, H. and Reissbrodt, R. 2006. Catecholate receptor proteins in *Salmonella enterica*: role in virulence and implications for vaccine development. Vaccine 24:3840-3844.

Zaharik, M. L., Vallance, B. A., Puente, J. L., Gros, P. and Finlay, B. B. 2002. Host-pathogen interactions: host resistance factor *Nramp1* up-regulates the expression of *Salmonella* pathogenicity island-2 virulence genes. Proc. Natl. Acad. Sci. U. S. A. 99:15705-15710.

Chapter 14
Microbial Signaling Compounds as Endocrine Effectors

Aruna Jahoor, Simon Williams, and Kendra Rumbaugh

14.1 Contribution of Interkingdom Signaling to the Human Microbiome

Microbiologists like to quote the statistic that there are ten times as many microbial cells as human cells in the human body as a way of illustrating the importance of microbes to our overall physiology. However, any effects that our microbial inhabitants elicit on human biology are very poorly understood. One approach to increasing our understanding of host–microbe interactions is the Human Microbiome Project (HMP), which was recently initiated by the National Institutes of Health to elucidate the influence of microbes on human development, physiology, immunity, and nutrition. The mission of the HMP is to generate resources that enable the "comprehensive characterization of the human microbiota and analysis of its role in human health and disease" (http://nihroadmap.nih.gov/hmp/). Among the specific goals of the HMP are to determine whether individuals share a core human microbiome and, if so, to understand whether changes in the human microbiome can be correlated with changes in human health. Apart from raising awareness of the contributions of microbes to human development and health, these studies should introduce novel technologies for studying these interactions. For example, systems biology

A. Jahoor
Department of Cell Biology and Biochemistry, Texas Tech University Health Sciences Center, 3601 4th Street, MS 6540, Lubbock, TX 79430, USA
e-mail: aruna.jahoor@ttuhsc.edu

S. Williams
Associate Professor, Department of Cell Biology and Biochemistry, Texas Tech University Health Sciences Center, 3601 4th Street, MS 6540, Lubbock, TX 79430, USA
e-mail: simon.williams@ttuhsc.edu

K. Rumbaugh (✉)
Department of Surgery, Texas Tech University Health Sciences Center, 3601 4th Street, Lubbock, Texas, 79430, USA
e-mail: kendra.rumbaugh@ttuhsc.edu

M. Lyte and P.P.E. Freestone (eds.), *Microbial Endocrinology*, Interkingdom Signaling in Infectious Disease and Health, DOI 10.1007/978-1-4419-5576-0_14, © Springer Science+Business Media, LLC 2010

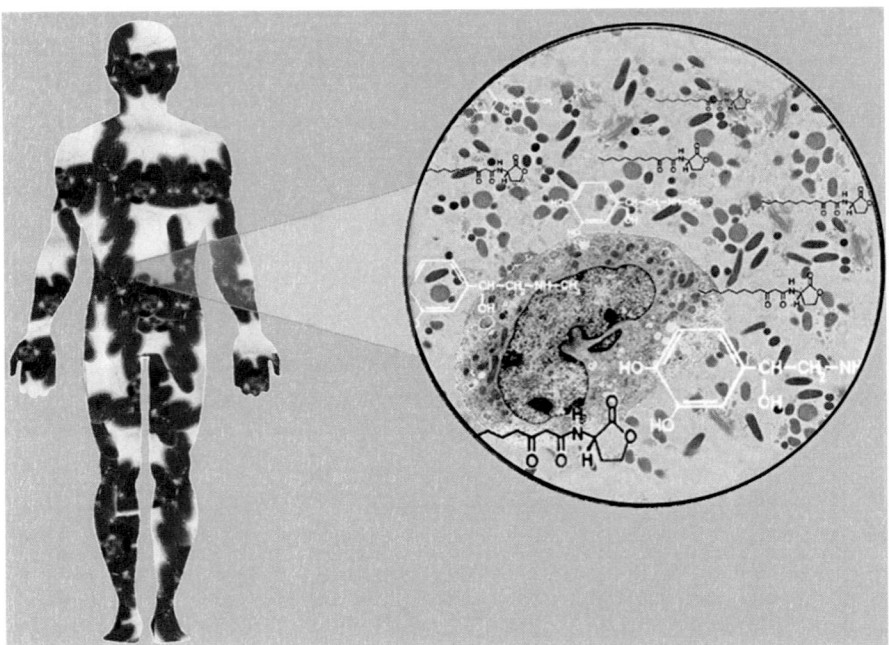

Fig. 14.1 The microbiome is composed of the microflora and transient microbes dwelling in the human host. Interkingdom signaling, or small chemical cell-to-cell signal exchanges, between microbes and their hosts may contribute to the function or dysfunction or of the human microbiome

approaches to understanding the microbiome, such as meta-genomic analyses, should reveal new and potentially instructive microbial/host interactions.

We propose that interkingdom signaling between microbes and their hosts is a process essential to a functional and/or dysfunctional or diseased microbiome (Fig. 14.1). The term interkingdom signaling refers to the concept that signaling molecules produced by organisms from every kingdom of nature can participate in cross-talk, communication, or signal interference across kingdom boundaries. This term was coined independently by two different groups to describe the effects of quorum sensing autoinducers from Gram-negative bacteria on mammalian host cells (Williams et al. 2004; Shiner et al. 2005a, b; Rumbaugh 2007), and to describe the effects of mammalian hormones on the QS systems of bacteria (Sperandio et al. 2003). Here, we use the term interkingdom signaling or IKS to describe the ability of small microbial chemicals to affect signaling cascades and elicit specific cellular responses in host cells.

There are several well-explored examples of microbial products that modulate cell signaling in host cells, the majority of which are mediated via host toll-like receptors (TLRs), which recognize the microbial products as pathogen-associated molecular patterns (PAMPs) (Iwasaki and Medzhitov 2004). Common PAMPs include microbial effectors, such as lipopolysaccharides, peptidoglycan, and flagellin, which are generally potent regulators of host immune responses. The IKS concept suggests that other, more subtle signals mediated by small soluble and/or volatile chemicals produced by microbes may also utilize host receptors to mediate interactions that may be beneficial to one or both participants in the interaction.

14.2 Properties of an Interkingdom Signal

The ability of cells to perceive and correctly respond to signals in their environment is essential for normal cellular function in both prokaryotes and eukaryotes. For example, bacterial cells communicate information about cell number through the process of quorum sensing while classic endocrine signaling in humans involves the transmission of information via secreted hormones transported in the circulation. Errors or deficiencies in signal production, interpretation, or processing can result in congenital or acquired diseases in humans such as cancer and autoimmunity. Although intracellular signaling is normally considered in the context of a single organism, whether it is a single-celled or multicellular organism, cells from different origins may inhabit common niches in which they are exposed to signals from disparate sources. One widely described example of such an interaction is the commensal relationship between bacteria and mammalian cells in the gut epithelium (Hooper and Gordon 2001). Coevolution of organisms from different kingdoms may have resulted in other, as yet poorly characterized, examples of interkingdom signaling, which could be beneficial or harmful to one or both participants involved in signal exchange. The timing and exchange of signals are likely to be crucial factors in symbiotic, commensal, and pathogenic interactions; thus, it is important to define the signals and the messages they convey.

Hormones are the prototypical examples of endogenous mammalian first messengers that normally transmit signals between cells via the bloodstream. Hormones come in a variety of different structures, usually being either peptide or lipid-based, or a combination of the two. Hormones participate in signal transduction through interactions with their cognate receptors, which may be located at the cell membrane, in the cytosol or even in the nucleus. Mammalian hormones can also affect microbial virulence, metabolism, and physiology, and thus can influence the course of infections. The question we address here is whether microbe-derived chemicals can have hormone-like properties and affect mammalian cell responses. We might expect that microbe-derived chemicals would be unlikely to achieve high enough concentrations in the host circulation to function as endocrine signals; thus, paracrine signaling to cells in the vicinity of microbial populations, such as biofilms, might be a more likely scenario. In the following sections, we describe different classes of microbial compounds that have the potential to behave as interkingdom signals.

14.3 Candidate Interkingdom Signals

The recognition that bacteria and other single-celled organisms possess efficient and varied mechanisms for communication has led to the identification of a variety of compounds that function as intercellular signals. These signals are usually associated with the process referred to variously as quorum sensing (Miller and Bassler 2001), diffusion sensing (Redfield 2002) or efficiency sensing (Hense et al. 2007). Regardless of the terminology, the secreted compounds that mediate these communication networks represent excellent candidates for interkingdom signals, particularly in multicellular microbial colonies, such as biofilms, where QS compounds may achieve high concentrations.

We will begin the next sections by examining compounds that have been traditionally classified as mammalian hormones but which have also been shown to elicit responses in microbes. We will follow this discussion by highlighting some compounds produced by microorganisms, such as quorum sensing molecules, that have established effects on host cells.

14.3.1 GABA: A Universal Signal?

Gamma-aminobutyric acid (GABA) is a four carbon, nonprotein amino acid conserved from bacteria to vertebrates. It was first discovered in plants (Thompson et al. 1953), but interest in GABA shifted to animals when it was discovered that GABA occurred in high levels in the mammalian brain (Bouche et al. 2003). Organisms synthesize GABA from glutamate in a reaction catalyzed by the cytosolic enzyme L-glutamic acid decarboxylase (GAD). Subsequently, the mitochondrial enzymes GABA transaminase and succinic semialdehyde dehydrogenase catalyze the conversion of the GABA carbon into succinic semialdehyde and then to succinate (Bown and Shelp 1997).

The identification of biosynthetic and metabolic pathways for GABA showed that its production, release, reuptake, and metabolism occurred within the nervous system (Roberts and Difiglia 1988). The effects of GABA can be mediated by activating ionotropic or metabotropic receptors, which are localized either pre- or postsynaptically. GABA plays an important role in processing information and regulating neuronal excitability, and is the main inhibitory neurotransmitter in the mammalian nervous system (Krnjevic 1984). GABA also attenuates renal vasoconstriction during activation of the sympathetic nervous system in the mammalian kidney (Fujimura et al. 1999), indicating that GABA can act as a signaling molecule outside the mammalian brain to alter peripheral neurotransmission. Since bacteria may utilize GABA for energy production and resistance, the availability of GABA in these sites outside the CNS could potentially increase bacterial growth in these areas of increased GABA production.

14.3.1.1 Bacterial Production of GABA

Pseudomonas fluorescens (ATCC 13430) can utilize the amino acid breakdown products pyrrolidine or putrescine as a sole source of carbon, nitrogen, and energy for growth (Jakoby and Fredericks 1959). In 1959, researchers discovered that bacterial metabolism of these precursors led to the production of GABA (Jakoby and Fredericks 1959). Researchers also found that some bacteria possessed GAD activity, allowing them to convert glutamic acid to GABA (Smith et al. 1992; Ueno et al. 1997). Among these bacteria were *Escherichia coli* and members of the *Lactobacillus* genus, specifically *L. brevis*. In a Japanese alcohol distillery, *L. brevis* was shown to convert almost all the free glutamic acid to GABA (Yokoyama et al. 2002). Since then, several GABA-producing lactic acid bacterial strains have been reported including *L. lactis* from cheese starters (Nomura et al. 1998) and from fermented fish (Komatsuzaki et al. 2008). Researchers have suggested that GABA may confer resistance to bacterial cells under acidic conditions (Cotter and Hill 2003), and the GAD decarboxylation process has been coupled with energy synthesis in a strain of *Lactobacillus* (Higuchi et al. 1997).

14.3.1.2 GABA as an Interkingdom Signal

As discussed in (Shelp et al. 2006), several studies indicate that plant-derived GABA may mediate communication between organisms belonging to different kingdoms. As early as 1979, researchers determined that applied GABA triggered developmental processes in plantonic larvae of *Haliotis rufescens* Swainson, a large red abalone. The larvae are induced to settle and undergo metamorphosis by recognizing GABA mimetic molecules associated with the red algae surface (Morse et al. 1979). GABA receptors were later identified in *H. rufescens* as the receptors responsible for metamorphosis. Thus, GABA facilitates a symbiotic relationship in which abalone gain food and the algae are kept free of epibionts. GABA produced by plants can also influence the growth and development of invertebrates. Reduced growth and survival rates were observed in insect larvae raised on diets containing elevated GABA levels (Bown et al. 2006). Plant-attached tobacco (*Nicotiana tabacum*) and soybean leaves exhibited up to 12-fold increases in GABA concentration within 10 min of physical interaction with insect predator larvae. This phenomenon is associated with wounding of the leaf by hook-like grips attached to larval feet. Studies utilizing transgenic tobacco plants overexpressing full length GAD showed a reduction in tobacco budworm larval feeding and impaired nematode development (MacGregor et al. 2003). Researchers propose that excess GABA produced by plants may disrupt invertebrate development and physiological processes by activating chloride channels at neuromuscular junctions.

GABA may also facilitate interactions between plant hosts and pathogenic fungi. Hyphae of the fungus *Cladosporium fulvum* are restricted to the apoplast of the plant host cell; therefore, the fungus is dependent on nutrients within the apoplast. During the infection process, the GABA concentration within the apoplast increases due to stimulation of GAD activity by pH and Ca^{2+} changes associated with fungal infection. (Solomon and Oliver 2001). In addition, enzymes that degrade GABA, such as GABA transaminase and succinic semialdehyde dehydrogenase, are induced in the fungus, indicating the use of GABA as source of nutrients. The fungal infection leads to enhanced GABA production by the plant host, which plays a role in the induction of fungal enzymes responsible for the breakdown of GABA (Solomon and Oliver 2001). GABA can also mediate the communication between plants and bacteria. The addition of GABA or wounded tomato stems to cultures of *Agrobacterium tumefaciens* stimulates the production of lactonase AttM by attKLM operon induction. Lactonase AttM inactivates $3OC_8$-HSL autoinducer, which is responsible for tumoral severity due to conjugation of the Ti plasmid (Chevrot et al. 2006). These findings indicate that increases in GABA concentration after plant wounding results in decreased *A. tumefaciens* virulence

14.3.2 Insulin

While insulin has long been considered a vertebrate hormone, comparative physiology studies from as far back as 80 years revealed the existence of related proteins in multiple invertebrates including some unicellular organisms. In the 1980s, Le Roith et al. (1985) recovered an immunoactive material from *E. coli* K-12 extracts, whose peak overlapped with that of porcine insulin, and was recognized by an antibody to insulin. Since then, insulin-like antigens from cyanobacterium extracts have been detected using both Elisa and Western Blotting. Further studies by this group and others illustrated that biologically and immunologically active insulin could be found in extracts from the fungi *Aspergillus fumigates*, *Neurospora crassa*, and *Saccharomyces cerevisiae*, the protists *Amoeba proteus*, and *Tetrahymena pyriformis*, and other microbial organisms (de Souza and Lopez 2004). In 1980, extracts from the protozoan *Tetrahymena pyriformis*, grown in defined media contained a material similar to mammalian insulin. This eluate stimulated glucose uptake into lipids in rats, and this activity was mostly neutralized by an antibody against insulin (LeRoith et al. 1985). Later studies provided evidence that this insulin was internalized by *Tetrahymena* cells and localized to the nucleus (Fulop and Csaba 1991), and that pretreating these cells with insulin leads to greater levels of internalized, labeled insulin than when compared to untreated cells, the retention of which was decreased after treatment with unlabeled insulin. Several studies have since attempted to identify bacterial receptors for insulin and insulin-like receptors, the mechanism of action of insulin in bacterial cells, the evolutionary origins of eukaryotic insulin, and potential signaling pathways in lower eukaryotes (de Souza and Lopez 2004Souza and Lopez).

14.3.3 Androgens and Glucocorticoids

Bacteria are exposed to endogenous androgens and glucocorticoids in the host environment, as well as exogenous steroids due to their presence in the environment and clinical uses. Although the effects of these hormones on mammalian cells are well studied, the influence of these steroids on prokaryotes has largely been overlooked. However, these types of interactions between host hormones and bacteria may have treatment implications for people with certain bacterial infections. Both androgens and glucocorticoids have been shown to affect both the growth rates and antibiotic susceptibility of both gram-positive and gram-negative bacteria, including clinically relevant bacterial strains such as *Pseudomonas aeruginosa* and *Enterococcus faecalis* (Plotkin et al. 2003). The authors observed that if the rate of bacterial growth was slowed in the response to the hormones, the minimum inhibitory concentrations for antibiotics would have been increased. These types of studies may have various clinical applications such as recognizing when antibiotics may be more or less effective depending on the hormonal environment.

14.3.4 Ovarian and Urogenital Hormones

Female sex hormones have also been shown to exhibit effects on microbial cells, which may play a role during infection (reviewed by Sonnex 1998). Clinical anec-dotes first led to the recognition that certain sexually transmitted diseases, such as Gonorrhea and Chlamydia, were more likely to occur in females at the time of menstruation (Sweet et al. 1986). While part of the reason may be related to hor-monal influences on the host immune response, there is also evidence to support effects of female sex hormones on the bacteria themselves. Of note, *Neisseria gon-orrhoeae* shows adherence to vaginal epithelial cells, which varies depending upon the stage of the menstrual cycle (Sobel et al. 1981). In addition, a protein that con-tains estrogen-binding sites has been identified in *Trichomonas vaginalis*, along with evidence that estrogen may affect the organisms' motility and adherence to cells (Ford et al. 1987; Sugarman and Mummaw 1988; 1990). Estrogen also has direct effects on the growth of the yeast *Candidiasis*, which is discussed in more detail later in this chapter. Estradiol treatment causes an increase in the endogenous levels of vaginal bacteria in mice, and some *Bacteroides* strains may utilize proges-terone or estradiol as growth factors (Sonnex 1998). During urinary tract infections, fluctuations in occurrence have also been linked to whether a woman is pre- or postmenopausal and her stage in the menstrual cycle (Sonnex 1998). *E. coli* adher-ence has been shown to vary with the greatest adherence occurring in the early phase of the cycle versus less adherence postovulation (Schaeffer et al. 1979). Studies with a rat model have shown that estradiol treatment reduces the ability of *E. coli* to bind to endometrial cells (Nishikawa and Baba 1985a, b) and can be used clinically to suppress purulent postsurgical infection (Nishikawa and Baba 1985a, b). Human Papilloma virus contains steroid responsive elements within its promoter regions, which may make it responsive to progesterone and glucocorticoids (Pater et al. 1994). These studies have revealed that sex hormones may play a crucial role during infection, not just through an influence on host immune response but also by directly influencing microbial virulence. The realization that the susceptibility of some women over others to urogential infections may have more to do with hormonal influences than coincidence may help in the clinical diagnosis and treatment of these diseases.

14.3.5   Quorum Sensing Signals

14.3.5.1   Acyl Homoserine Lactones (AHLs)

AHLs are lipophilic compounds secreted by a variety of species of Gram-negative bacteria that participate in intercellular communication by acting as ligands for inducible transcriptional regulatory proteins (Fuqua and Greenberg 2002). AHLs consist of a homoserine lactone ring joined to an acyl side chain that can vary in length, saturation, and side chain modifications. AHLs are synthesized by enzymes of the LuxI family utilizing S-adenosyl methionine and a fatty acid as substrates. The concept of QS is based on the observation that AHLs accumulate and increase as cell numbers expand, so that activation of QS-dependent genes by liganded transcription factors only occurs at elevated cell densities. QS systems in various bacterial species are associated with symbiotic relationships, such as between *Vibrio spp.* and some fish and squid species (reviewed in (Fuqua et al. 2001)), and in regulating virulence and biofilm production in certain bacterial pathogens, such as *P. aeruginosa* (reviewed in Shiner et al. 2005a, b). Biofilms are communities of single or multiple species of microbes adhered to surfaces and surrounded by an extracellular matrix (Hall-Stoodley et al. 2004). Biofilms are ubiquitous in nature and can be found in the oral cavity, in the lungs of cystic fibrosis patients, and on the surfaces of catheters and implanted prosthetic devices (Morris et al. 1999; Singh et al. 2000; Reisner et al. 2005). Biofilms represent a significant challenge for the treatment of infectious diseases as bacteria within biofilms are resistant to standard antibiotic treatments and clearance by the host immune system.

Studies initiated in the late 1990s began to identify apparent physiological effects of AHLs on mammalian cells. These studies focused on one of the two major AHLs produced by *P. aeruginosa*, N-3-oxododecanoyl-homoserine lactone ($3OC_{12}$-HSL), which promoted anti- and proinflammatory effects (Telford et al. 1998; Smith et al. 2001, 2002; Shiner et al. 2006), proapoptotic effects (Tateda et al. 2003; Li et al. 2004; Shiner et al. 2006), and hemodynamic effects (Lawrence et al. 1999; Gardiner et al. 2001). A limited number of studies have also described examples of interkingdom signaling involving AHLs other than $3OC_{12}$-HSL (reviewed in Shiner et al. 2005a, b). It was recently demonstrated that AHLs do not simply function as PAMPs (Kravchenko et al. 2006), suggesting that alternative, endogenous receptors may mediate AHL-dependent signaling in mammalian cells. We will return to the topic of mammalian receptors for interkingdom signals at a later time after discussing other putative interkingdom signals.

14.3.5.2   Autoinducer-2 and Autoinducer-3

Not all autoinducers are acylated homoserine lactones as a furanosyl::borate::diester and an aromatic aminated structure, respectively termed Autoinducer-2 and Autoinducer-3, have been described as compounds responsible for population density dependent signal transduction in bacteria. Although these compounds have not yet been shown to affect host cells, they may have vital roles in bacterial communication. Autoinducer-2 has been linked to interspecies communication (Federle and Bassler 2003) and the bacterial receptor for Autoinducer-3 is activated by the mammalian hormones epinephrine and norepinephrine (Clarke et al. 2006). Whether or not these molecules cause direct effects in host cells remains to be seen.

14.3.5.3 Autoinducing Peptides

In contrast to Gram-negative bacteria, Gram-positive bacteria utilize oligopeptide autoinducers for intracellular communication and quorum sensing. These peptides are termed autoinducing peptides (AIPs) and are 5–17 amino acids in length (Sturme et al. 2002). Autoinducing peptides are utilized by *S. aureus*, *Bacillus subtilis*, *Lactobacillus spp.*, *Streptococcus pneumoniae*, and many others (Sturme et al. 2002). Unlike AHL signaling, the bacterial cell membrane is impermeable to AIPs, necessitating cell-surface oligopeptide transporters to facilitate AIP secretion into the extracellular environment. Detection of AIPs is mediated by two-component sensory-transduction systems, consisting of a membrane-located receptor histidine protein kinase and an intracellular response regulator (Grebe and Stock 1999; Sturme et al. 2007). Upon phosphorylation, the response regulator activates transcription of target genes as well as genes responsible for autoinducer production. AIPs are synthesized as precursor peptides and exported either by ABC transporters or dedicated proteins depending upon the bacterial species (Sturme et al. 2002). Autoinducing precursor peptides may undergo intercellular posttranslational modifications and extracellular proteolytic processing to generate the mature signaling peptides (Sturme et al. 2005).

   AIPs produced by one strain of bacteria may alter the behavior of other bacterial strains. For example, cross-inhibition of *agr* gene expression between *S. aureus* strains and *Staphlococcus epidermis* and *S. lugdunensis* has been reported (Ji et al. 1997; Otto and Gotz 2001). Some peptide autoinducers, such as the antibiotic nisin produced by *L. lactis* during fermentation or the cationic peptide subtilin from *B. subtilis*, also exhibit antimicrobial activity (Schuller et al. 1989; Cheigh et al. 2002). Evidence that AIPs participate in interkingdom signaling is currently lacking; however, it is interesting to note that mammalian cells synthesize petides that resemble AIPs that are called defensins or host defense peptides (HDPs). HDPs are sythesized as precursor peptides and proteolytically processed in a manner similar to AIPs (Sahl et al. 2005). HDPs are generally short oligopetides that are cationic, amphiphilic, and are capable of killing bacteria. The antimicrobial function of HDPs involves membrane depolarization and resembles the antimicrobial activities of AIPs described above. The structural and functional resemblance of AIPs and HDPs, at least in terms of their antimicrobial activities, suggests that these oligo-peptides could also share signaling capabilities and make AIPs attractive candidates for interkingdom signaling molecules.

14.3.5.4 Pseudomonas Quinolone Signal

Quinolones represent another class of bacterial autoinducers that are exemplified by the 2-heptyl-3-hydroxy-4-quinolone compound from *P. aeruginosa* and termed the Pseudomonas quinolone signal (PQS) (Gallagher et al. 2002). PQS can be detected in the sputum of Cystic Fibrosis patients with chronic *P. aeruginosa* infections (Collier et al. 2002) and also in CF patient airways during early colonization (Guina et al. 2003). PQS collaborates with AHLs to regulate the production of virulence factors, including elastase, rhamnolipids, and pyocyanin, and to influence biofilm development (Deziel et al. 2004). PQS biosynthesis involves a "head-to-head" condensation of anthranilic acid and β-keto dodecanoate and proceeds via the intermediate 2-heptyl-4(1*H*)-quinolone (HHQ) (Diggle et al. 2003). The conversion of HHQ to PQS is catalyzed by the LasR-regulated monooxygenase PqsH, linking AHL and PQS signaling (Wade et al. 2005; Xiao et al. 2006).

PQS may serve a dual function, both as a quorum sensing agent for bacteria and as an immunomodulatory agent within the host. Exposure to PQS significantly reduces lymphocyte proliferation in response to the panactivating lectin ConA (Hooi et al. 2004). Furthermore, PQS enhanced interleukin-2 release from T-cells stimulated with anti-CD3/anti-CD28 antibodies (Hooi et al. 2004). PQS also significantly increased TNF-α secretion from LPS-stimulated human leukocytes (Hooi et al. 2004). Thus, PQS may not only be involved in intracellular bacterial communication, but may also confer a survival advantage to *P. aeruginosa* by manipulating the host immune response. At present, the mechanisms by which PQS affect mammalian cell responses are unclear.

14.3.5.5 Cyclic Diketopiperazines

Diketopiperazines are cyclic dipeptides that are secreted by numerous bacterial species and can interfere with QS by acting as AHL antagonists or agonists, depending on the bacterial species (Holden et al. 1999). DKPs share structural similarities with mammalian signaling peptides such as thyotropin-releasing hormone cyclic dipeptides and cyclo-L-His-L-Pro that are synthesized in mammalian cells (Prasad 1995). Some synthetic cyclic dipeptides have antitumor and antiviral activities and also affect heart rate, coronary flow rate, and ventricular pressure (Lucietto et al. 2006). Cyclo(His-Gly) possesses anticoagulant activity as it inhibits platelet aggregation, showing greatest activity against the thrombin-induced platelet aggregation pathway (Lucietto et al. 2006). These observations suggest that further investigations of the biological activities of DKPs in mammalian cells may yield additional evidence for a role in IKS.

14.3.5.6 Farnesol

Quorum sensing is not limited to prokaryotic organisms as a similar process has been observed in some species of yeast. Hornby et al. (2001) described a lipophilic compound in spent medium from the fungus *Candida albicans*, which prevented yeast-to-mycelium transformations and controlled the developmental decision of *C. albicans* between budding yeasts or mycelia. This compound also modulated biofilm formation by *C. albicans* (Ramage et al. 2002) and was identified as the oxygenated lipid farnesol ($C_{16}H_{26}O$), a sesquiterpene alcohol consisting of three isoprene units (Table 14.1). In *C. albicans*, farnesol is produced by dephosphorylation of farnesyl pyrophosphate, which represents a branch point in lipid metabolism (Nickerson et al. 2006). Farnesol is produced constitutively during growth in amounts roughly proportional to cell mass and accumulates at concentrations up to 500 µM in the media under appropriate growth conditions (Nickerson et al. 2006). However, while at least 47 species of fungi have the enzymatic capability to produce farnesol, most reports of farnesol production or response to farnesol are limited to *Candida spp.* (Nickerson et al. 2006). Farnesol also affects the physiology of non-*Candida* species and may mediate antagonism between different species of fungi (Machida and Tanaka 1999; Machida et al. 1999; Nickerson et al. 2006; Semighini et al. 2006)

Farnesol is a particularly attractive candidate as an interkingdom signal as related if not identical compounds are also synthesized in organisms from other kingdoms, including bacteria, insects, and mammals. *C. albicans* is commonly isolated from the sputum of CF patients and coexists with bacteria such as *P. aeruginosa* in the lungs of these patients. As both organisms utilize lipidic signals for intercellular communication, it has been hypothesized that these signals may also cross kingdoms. Farnesol displays structural similarities to AHLs produced by Gram-negative bacteria, and high concentrations of $3OC_{12}$-HSL from *P. aeruginosa* can mimic the effects of farnesol in *C. albicans* (Hogan et al. 2004; McAlester et al. 2008). Thus, interactions with *P. aeruginosa* could theoretically alter *C. albicans* virulence in coinfected tissue by modulating *C. albicans* morphology. Farnesol also inhibits swarming motility in *P. aeruginosa* and decreases PQS synthesis and the production of pyocyanin, indicating that this example of interkingdom signaling occurs in both directions across the prokaryote–eukaryote divide. Farnesol is an intermediate in mevalonate synthesis in mammals, and supraphysiological levels of farnesol are capable of modulating the activity of a member of the nuclear hormone receptor (NHR) superfamily named Farnesoid X Receptor (FXR) or NR1H4 (Cariou and Staels 2007). Thus, farnesol from external sources could influence mammalian cell responses, for example in cases of *C. albicans* infection. Farnesol, and other isoprenoids can inhibit proliferation and induce apoptosis in human and murine cell lines (Haug et al. 1994; Ferrandina et al. 2000; Rioja et al. 2000; Mo and Elson 2004) and enhance virulence of *C. albicans* in murine models of candiasis (Nickerson et al. 2006). Thus, it is intriguing to consider whether interplay between farnesol and AHL signaling could mediate an interkingdom relationship between bacteria and mammals. This theory will be revisited when we consider signal receptor interactions in a later section.

Table 14.1 Evidence for microbial chemicals as interkingdom signals

Structure	Signal	Chemical class	Endogenous function	IK effect	References
Quorum sensing signals					
	Acyl homoserine lactones (3OC$_{12}$-HSL)	Lipid	Virulence, biofilm formation, motility, type III secretion, and more	Apoptosis, immunomodulation, intracellular Ca^{++} signaling	(Shiner et al. 2005a, b; 2006)
YSTCDFIM	Autoinducing Peptides	Small peptide	Virulence, toxin production, genetic competence	Antimicrobial, cell death through pore formation	(Sturme et al. 2002; 2007; Hooi et al. 2004)
	PQS	Phenazine	Virulence, biofilm formation	Antiproliferative, immunomodulation, apoptosis	(Deziel et al. 2004; Hooi et al. 2004)
	Cyclic diketopiperazines	Small peptide	Peptide linkers	Cardiovascular effects, inhibit thrombin, inhibit cancer cells	(Prasad 1995; Lucietto et al. 2006)
	Farnesol	Alcohol derivative	Control fungal dimorphism, inhibit biofilm formation	Inhibition of cell growth, accumulation of reactive oxygen species, disorganize cytoskeleton	(Machida et al. 1999; Hornby et al. 2001; Nickerson et al. 2006)
Bacterial secondary metabolites					
	Pseudomonas pyocyanin	Phenazine	Redox agent/ antibacterial	Apoptosis, tissue destruction, inhibition of respiration, immunomodulation, Ca++ disruption	(Sorensen and Klinger 1987; Wilson et al. 1988; Kanthakumar et al. 1993; Kamath et al. 1995; Usher et al. 2002; Ran et al. 2003; Allen et al. 2005; Look et al. 2005)

Photorhabdus stilbene	Polyketide	Secondary metabolite	Nematode reproductive development	(Joyce et al. 2008)

Fungal secondary metabolites/mycotoxins

Aflatoxin	Polyketide	Unknown	Carcinogen	(Bennett and Klich 2003)
Zearalenone	Aromatic metabolites	Unknown	Endocrine disruptor, estrogenic	(Aucock et al. 1980; Diekman et al. 1992; Massart et al. 2008)
Psilocybin	Amino acid derivative	Unknown	Hallucinogen, serotonin agonist	(Sard et al. 2005)
Ergotamine	Alkaloid	Defense against predation	Vasoconstriction, modulates serotonin/dopamine/NA receptors	(Tfelt-Hansen and Koehler 2008)

Universal signal

GABA	Amino acid derivative	Energy synthesis, metabolite	Attenuates renal vasoconstriction, islet cell paracrine signal, inhibitory neurotransmitter	(Franklin and Wollheim 2004; Shelp et al. 2006)

14.3.6 Secondary Metabolites

Secondary metabolites are compounds produced by an organism that are not required for primary metabolic processes, and in general, their absence does not result in immediate death of the organism. However, these metabolites may have significant effects on other organisms, for example, by acting as toxins. Secondary metabolites are produced by a wide range of microorganisms, including many species of bacteria and fungi, and include several that may be considered as candidate participants in interkingdom signaling.

14.3.6.1 Bacterial Secondary Metabolites as Interkingdom Signals

Pseudomonas Pyocyanin

Many phenazines are produced by bacteria including *Pseudomonas* and *Streptomyces spp*. Phenazines are synthesized via the shikimic acid biosynthesis pathway subsequent to chorismic acid. Subsequent modifications lead to a variety of phenazines with differing biological activities. Phenazines are commonly pigmented and have antibacterial activities, however several phenazines also have negative effects on host cells and are considered virulence factors. The prototypical example is the pyocyanin phenazine (1-hydroxy-5-methylphenazine) produced by more than half of all *P. aeruginosa* clinical isolates. Pyocyanin can be detected in the sputa of patients with *P. aeruginosa* lung infections at concentrations up to 100 μM (Wilson et al. 1988).

Pyocyanin was first extracted in 1860 from wound dressings (Fordos 1860). At neutral pH, pyocyanin has an intense blue color and is responsible for the "green urine" and blue-tinted skin that some *P. aeruginosa* infected patients display (Silvestre and Betlloch 1999). Pyocyanin is involved in redox cycling in bacteria and mammalian liver and epithelial cells. The generation of free-radicals is thought to contribute to its antibacterial properties and it is these reactive oxygen species that mediate many of the effects on host cells as well (Miller et al. 1987; Gardner 1996). The described effects of pyocyanin and its phenazine precursor, phenazine-1-carboxylic acid, are diverse and may affect many different types of cells and potentially contribute to the inflammation and subsequent tissue destruction characteristic of *P. aeruginosa* infections. These effects include the inhibition of human cell respiration, ciliary function and growth of epithelial cells (Sorensen and Klinger 1987; Wilson et al. 1988; Kanthakumar et al. 1993), inhibition of prostacyclin release from lung endothelial cells (Kamath et al. 1995), induction of neutrophil apoptosis (Usher et al. 2002; Allen et al. 2005), induction of chemokines and adhesion molecules (Look et al. 2005), disruption of calcium homeostasis (Denning et al. 1998), and inactivation of human vacuolar ATPases in human airway epithelial cells (Ran et al. 2003).

Photorhabdus Stilbene

Secondary metabolites synthesized by bacteria have primarily been studied and exploited for their antibiotic potential. In fact, over two-thirds of the antibiotics currently in use clinically are produced by *Streptomyces spp.* (Watve et al. 2001). However, the interkingdom signaling potential of secondary metabolites is now becoming a popular area of study. One very interesting example is the polyketide secondary metabolite 3,5-dihydroxy-4-isopropylstilbene (ST), produced by *Photorhabdus luminescens*. *Photorhabdus* is an insect pathogen that also exists in a mutualistic relationship with nematodes (Joyce et al. 2008). *Photorhabdus* colonizes the gut of the juvenile nematode, which lives in the soil and preys on insect larvae. Once the worm finds its larvae prey, it invades and migrates to the insect's hemolymph (combined blood and lymph), where it regurgitates *Photorhabdus*. The bacteria grow quickly and kill the insect within 72 h, after which the nematodes feed and undergo several rounds of reproduction within the cadaver (Joyce et al. 2008). In response to environmental cues, the new juvenile nematodes will emerge from the cadavers carrying *Photorhabdus* in their guts to find new larvae and begin the cycle again. *Photorhabdus* ST, which is chemically similar to the plant antioxidant compound resveratol (see Table 14.1), has antibiotic activity and may serve to protect the nematode and its insect cadaver from intruding microbes in the soil (Hu and Webster 2000). However, ST also acts as a virulence factor by inhibiting components of the insect's innate immune response (Eleftherianos et al. 2007). Most interestingly, there is now evidence that *Photorhabdus* ST is an interkingdom signal that is important for the normal reproductive development of its nematode host (Joyce et al. 2008). The transformation of juvenile nematodes into self-fertilizing hermaphrodites requires an environmental cue(s) from the infected insect hemolymph. The ability of the nematodes to undergo normal reproductive development can be distinguished experimentally by determining the postinfection recovery of juveniles, which then develop into hermaphrodites. Nematodes infected with mutant *Photorhabdus,* not able to synthesize ST, displayed recovery rates 5–15% of those infected with ST-producing *Photorhabdus* (Joyce et al. 2008), indicating that ST is a versatile chemical with important roles in pathogenicity and mutualism.

14.3.6.2 Fungal Secondary Metabolites/Mycotoxins

Aflatoxins

The fungi make and excrete a wide array of secondary metabolites, many of which act as toxins to their mammalian hosts. These secondary metabolites are synthesized in a number of different biosynthetic pathways (Table 14.1). The polyketide metabolites are the products of polymerized acetate, resulting in a fatty acid or a polyketide. Processing of the fatty acid chain by cyclization, lactonization, or formation of thioesters or amides, results in a huge number of possible structures. Polyketide secondary metabolites include orsellinic acid, tetrahydroxynaphthalene (precursor for melanin), statins, fumonisin, and the aflatoxins, which are some of the most toxic compounds on earth. The function of aflatoxins in fungi is unknown however, after consumption of plant material contaminated by *Aspergillus spp.*, toxicosis followed by death can occur in even large animals by almost undetectable quantities. Species' susceptibility varies greatly; however, the liver is the primary target, where cytochrome P450 enzymes convert aflatoxins to the reactive 8,9-epoxide form (Mishra and Das 2003). The reactive aflatoxin epoxide then binds to eukaryotic DNA and proteins, specifically the N^7 position of guanines. Aflatoxin-DNA adducts can result in GC to TA transversions, and it's carcinogenic potency is highly correlated with the extent of aflatoxin–DNA adducts formed in vivo (Bennett and Klich 2003). While aflatoxin appears to affect hosts by a nonspecific mechanism, other mycotoxins exert their function by interacting with specific host endocrine receptors (see below).

Zearalenone

Secondary aromatic metabolites are synthesized through the polyketide or shikimic acid pathways. Zearalenone (ZEA) is one interesting example from this group. ZEA is a phenolic resorcyclic acid lactone that is estrogenic when consumed by animals, primarily swine (Hidy et al. 1977) (Fig. 14.2). It is produced by the fungus *Fusarium roseum*, which grows on grain commonly used to feed livestock. Grains infected with *F. roseum* may exhibit the pink color associated with the production of a pigment simultaneously produced with ZEA. Most often, this mycotoxin is found in corn. However, it is also found in other important crops such as wheat, barley, sorghum, and rye throughout various countries of the world. Generally, the *Fusarium spp.* grow and invade crops in moist, cool conditions. Variations in the incidence of zearalenone production occur with different crop years, types of crop, and geographical area. The most notable effects of zearalenone are the premature mammae development and other estrogenic effects in young female and prepucial enlargement in young male pigs (Aucock et al. 1980). Swine are the most significantly affected species and are considerably more sensitive to ZEA than, for example, rodents and other species such as cattle and poultry. The relative sensitivity of swine to the effects of ZEA may be attributed to differences in its bioconversion (Fig. 14.2). Following hepatic biotransformation of ZEA by subcellular fractions of pig livers alpha-zearelenol is the major metabolite, but in bovine hepatic microsomes beta-zearalenol is the major metabolite (Olsen and Kiessling 1983). The conversion of ZEA into its metabolites α and β zearalenol is catalyzed by 3α and 3 β-hydroxysteroid hydrogenase (Olsen et al. 1981). These enzymes are typically found in the liver and are involved in the synthesis of steroid hormones in mammals, including estradiol and testosterone. One study demonstrated that biotransformation of ZEA occurred in porcine granulosa cells (GCs) expressing predominantly 3α-hydroxysteroid hydrogenase, whereas bovine GCs expressed

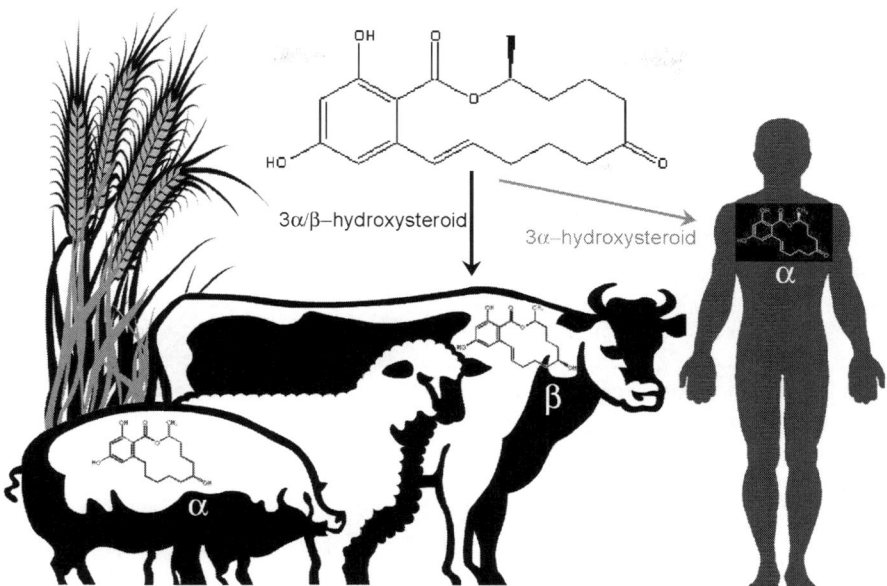

Fig. 14.2 Zearalenone is produced by the fungus *Fusarium roseum*, which grows on grain commonly used to feed livestock. Once consumed, it is bioconverted to primarily alpha-zearelenol in humans and swine or beta-zearalenol in cattle, by 3α and 3β-hydroxysteroid hydrogenase, respectively. Zearalenone and its metabolites disrupt reproductive processes by mimicking estradiol-17β and binding to estrogen receptors. The most notable effects of zearalenone are premature mammae development and other estrogenic effects in young female animals and humans

both the α and β enzyme isoforms (Malekinejad et al. 2006). This could potentially explain the species-related differences in ZEA metabolism and sensitivity.

ZEA and its metabolites, α and β-ZEA disrupt reproductive processes because they mimic estradiol-17β action by binding to estrogen receptors in target cells exerting estrogenic action (Diekman and Green 1992; Nagel et al. 1998; Shier et al. 2001). Furthermore, unlike endogenous hormones, ZEA metabolites tend not to bind carrier proteins (Nagel et al. 1998), allowing easier access to estrogen target sites and a potential increased potency. In 1989, an increased incidence of early breast development in young Hungarian girls was associated with exposure to ZEA (Szuets et al. 1997). Researchers detected patients with serum ZEA levels of 18.9–103 µg/L and suggested that contamination of foodstuffs with ZEA caused early maturation. ZEA was also cited as a possible causative agent for an epidemic of precocious puberty in Puerto Rico from 1978 to 1984 (Schoental 1983). Recently, researchers at the University of Pisa in Italy analyzed the serum of girls affected by central precocious puberty syndrome and serum from healthy female controls to evaluate whether mycotoxin exposure was a triggering factor for premature sexual development (Massart et al. 2008). They discovered a higher growth rate among girls who tested positive for serum ZEA and/or its metabolite α-zearalenol, and suggested possible correlations between environmental mycoestrogen exposure, precocious puberty, and anabolic growth effects of mycotoxins in exposed girls (Massart et al. 2008).

Psilocybin

Secondary metabolites produced via the amino acid pathway are most noted for their antibacterial activities. The β lactam antibiotics, penicillin and cephalosporin, are produced by some *Ascomycota spp.* of fungi and many bacteria, while the antibiotic peptide defensins, produced mainly by vertebrate and invertebrate hosts, are also synthesized by the amino acid pathway. However, some toxins derived from amino acid synthesis affect the mammalian endocrine system by acting as agonists of the mammalian 5-HT_{2A} receptor, which is a subtype of the 5-HT_2 receptor belonging to the serotonin receptor family. One example of these is psilocybin, an indole produced by hundreds of species of fungi (often called "magic mushrooms"), including those of the genus *Psilocybe*. Psilocybin is rapidly absorbed through the lining of the mouth and stomach, and is dephosphorylated to psilocin in acidic conditions (Sard et al. 2005). Psilocin travels through the blood to the brain and binds to 5-HT_{2A} receptors, causing nerve impulse effects, hallucinations, and other "psychedelic" effects (Sard et al. 2005).

Ergotamine

The alkaloids are a broad category of chemicals produced by fungi that are synthesized from several different pathways. Many of these chemicals are thought to be involved in defense and disrupt metabolic processes in herbivores that eat infected grasses, or deter attacking insects. The ergot alkaloids are a collection of alkaloid compounds based on lysergic acid amide, which are produced by fungi in the genus *Claviceps* (Bacon et al. 1986). There are about 50 known species of *Claviceps*, commonly called the ergots, most living in tropical regions. The ergots infect many grasses, including rye, wheat, and barley as part of their life cycle. The fungi are endophytic and cause no known disease symptoms in their hosts. However, the alkaloids produced by the ergots can induce various metabolic disorders in the herbivores that consume infected grasses. For example, one of these alkaloids, ergotamine, possesses structural similarity to several neutotransmitters including serotonin, dopamine, and epinephrine (Tfelt-Hansen et al. 2000). It acts as both an agonist and antagonist to several mammalian receptors including $5\text{-HT}_{1B/1D}$, dopamine, and norepinephrine receptors (Tfelt-Hansen et al. 2000). Ergotamine is used medicinally to treat migraine headaches because it causes constriction of the intercranial extracerebral blood vessels (Tfelt-Hansen and Koehler 2008). It is also used to induce childbirth and prevent postpartum hemorrhage (Akins 1994). In addition to ergotamine, other alkaloids produced by fungi, such as paxilline and loline, reduce the feeding and reproductive rates of insects (Schardl et al. 2007).

14.4   Identifying Receptors for Interkingdom Signals

The essential concept of interkingdom signaling is that signals exchanged between organisms must be interpreted via receptors in the receiving organism. Recent developments have suggested that members of the NHR superfamily may be excellent candidates as mammalian receptors for microbial signals. NHRs are ligand-activated transcriptional regulatory proteins that mediate signaling by lipophilic compounds such as steroids, retinoids, and fatty acids (Aranda and Pascual 2001). We have already mentioned one member of this family, namely the Farnesoid X Receptor (FXR, NR1H4), whose activity is sensitive to supraphysiological levels of farnesol. Our recent studies on the effects of the *P. aeruginosa*-derived AHL, $3OC_{12}$-HSL, on mammalian cells have implicated additional NHRs in interkingdom signaling (Jahoor et al. 2008).

Using a combination of cellular and genetic assays, we have uncovered evidence for two independent signaling pathways that mediate proapoptotic and proinflammatory signaling by $3OC_{12}$-HSL. Proapoptotic signaling is mediated by a calcium-dependent pathway that involves an as yet unknown receptor located at or near the cell membrane (Shiner et al. 2006). However, proinflammatory gene induction was unaffected by inhibitors of the calcium-dependent pathway (Shiner et al. 2006), and our more recent data implicate the peroxisome proliferator activated receptor gamma (PPARγ) NHR as a potential mediator of this signaling event (Jahoor et al. 2008a, b). PPARγ is a member of the Class I subfamily of the NHR superfamily and binds DNA as a heterodimer with retinoid X receptors (RXRs), which belong to the Class II subfamily of NHRs. PPARγ was first identified as a regulator of adipogenesis where it cooperates with members of the CCAAT/enhancer-binding protein family to promote adipocyte differentiation and activate adipocyte-specific genes (Willson et al. 2001). PPARγ was subsequently characterized as an anti-inflammatory mediator that can bind to and block the activity of nuclear factor kappa B (NFκB), a universal regulator of proinflammatory gene expression (Glass and Ogawa 2006). Our results revealed that $3OC_{12}$-HSL functions as a PPARγ antagonist, thereby inhibiting its ability to suppress NFκB activity, leading to enhanced proinflammatory gene expression (Jahoor et al. 2008). Interestingly, PPARγ activity is also regulated by isoprenoids, including farnesol and its derivatives (Takahashi et al. 2002), and other members of the NHR superfamily are targeted by AHLS and farnesol (Takahashi et al. 2002; Jahoor et al. 2008a, b). Considering the functional similarities between NHR and AHL/LuxR signaling, wherein the direct binding of a lipidic ligand alters the conformation and activity of a receptor that functions as a transcriptional regulator, it is interesting to speculate that a broad spectrum of interkingdom signaling might be mediated via NHRs. Current studies to demonstrate direct binding of microbial ligands to mammalian NHRs should assist in defining the scope of this mode of signaling and provide a fertile area for future research into communication and molecular interactions across kingdom boundaries.

A. Jahoor et al.

14.5 Concluding Remarks

The coexistence of prokaryotes and eukaryotes over several millennia may have enabled the evolution of communication systems based on soluble lipidic signaling compounds. One of the best examples of such interactions defined thus far is the effects of AHLs on host cells during pathogenic interactions; however, it is clear that these interactions do not represent true communication as the outcomes appear to be to the benefit of the microbe. It is exciting to consider the possibility that autoinducers, or other microbial products, might participate in bidirectional communication, such as in the gut, and that the messages exchanged might influence the development of one or both organisms. For example, one implication of recent metagenomic studies that showed that gut microbe compositions varied and coevolved with their mammalian hosts (Ley et al. 2008) is that any description of human evolution must consider the genetic makeup of our microbial coinhabitants.

References

Akins, S. 1994. Postpartum hemorrhage. A 90s approach to an age-old problem. J Nurse Midwifery 39: 123S-134S.
Allen, L., D. H. Dockrell, T. Pattery, D. G. Lee, P. Cornelis, P. G. Hellewell and M. K. Whyte 2005. Pyocyanin production by *Pseudomonas aeruginosa* induces neutrophil apoptosis and impairs neutrophil-mediated host defenses in vivo. J Immunol 174: 3643-3649.
Aranda, A. and A. Pascual 2001. Nuclear hormone receptors and gene expression. Physiol Rev 81: 1269-1304.
Aucock, H. W., W. F. Marasas, C. J. Meyer and P. Chalmers 1980. Field outbreaks of hyperoestrogenism (vulvo-vaginitis) in pigs consuming maize infected by *Fusarium graminearum* and contaminated with zearalenone. J S Afr Vet Assoc 51: 163-166.
Bacon, C. W., P. C. Lyons, J. K. Porter and J. D. Robbins 1986. Ergot toxicity from endophyte-infected grasses: a review. Agron J 78: 106-116.
Bennett, J. W. and M. Klich 2003. Mycotoxins. Clin Microbiol Rev 16: 497-516.
Bouche, N., B. Lacombe and H. Fromm 2003. GABA signaling: a conserved and ubiquitous mechanism. Trends Cell Biol 13: 607-10.
Bown, A. W., K. B. Macgregor and B. J. Shelp 2006. Gamma-aminobutyrate: defense against invertebrate pests? Trends Plant Sci 11: 424-7.
Bown, A. W. and B. J. Shelp 1997. The metabolism and functions of [gamma]-aminobutyric acid. Plant Physiol 115: 1-5.
Cariou, B. and B. Staels 2007. FXR: a promising target for the metabolic syndrome? Trends Pharmacol Sci 28: 236-43.
Cheigh, C., I., Choi, H., J., Park, H., Kim, S.,B., Kook, M.,C., Kim, T.,S., Hwang, J.,K. and Y. R. Pyun 2002. Influence of growth conditions on the production of a nisin-like bacteriocin by *Lactococcus lactis* subsp. *lactis* A164 isolated from kimchi. J Biotechnol 95:225-35.
Chevrot, R., Rosen, R., Haudecoeur, E., Cirou, A., Shelp, B., J., Ron, E. and D. Faure 2006. GABA controls the level of quorum-sensing signal in *Agrobacterium tumefaciens*. Proc Natl Acad Sci U S A. 103: 7460-4.
Clarke, M. B., D. T. Hughes, C. Zhu, E. C. Boedeker and V. Sperandio 2006. The QseC sensor kinase: a bacterial adrenergic receptor. Proc Natl Acad Sci U S A 103: 10420-10425.

Collier, D. N., L. Anderson, S. L. McKnight, T. L. Noah, M. Knowles, R. Boucher, U. Schwab, P. Gilligan and E. C. Pesci 2002. A bacterial cell to cell signal in the lungs of cystic fibrosis patients. FEMS Microbiol Lett 215: 41-46.

Cotter, P. D. and C. Hill 2003. Surviving the acid test: responses of gram-positive bacteria to low pH. Microbiol Mol Biol Rev 67: 429-453, table of contents.

de Souza, A. M., Lopez, J.A. 2004. Insulin or insulin-like studies on unicellular organisms: a review. Braz Arch Biol Technol 47: 973-981.

Denning, G. M., L. A. Wollenweber, M. A. Railsback, C. D. Cox, L. L. Stoll and B. E. Britigan 1998. *Pseudomonas* pyocyanin increases interleukin-8 expression by human airway epithelial cells. Infect Immun 66: 5777-5784.

Deziel, E., F. Lepine, S. Milot, J. He, M. N. Mindrinos, R. G. Tompkins and L. G. Rahme 2004. Analysis of *Pseudomonas aeruginosa* 4-hydroxy-2-alkylquinolines (HAQs) reveals a role for 4-hydroxy-2-heptylquinoline in cell-to-cell communication. Proc Natl Acad Sci U S A 101: 1339-1344.

Diekman, M. A. and M. L. Green 1992. Mycotoxins and reproduction in domestic livestock. J Anim Sci 70: 1615-1627.

Diggle, S. P., K. Winzer, et al. 2003. The *Pseudomonas aeruginosa* quinolone signal molecule overcomes the cell density-dependency of the quorum sensing hierarchy, regulates rhl-dependent genes at the onset of stationary phase and can be produced in the absence of LasR. Mol Microbiol 50: 29-43.

Eleftherianos, I., S. Boundy, S. A. Joyce, S. Aslam, J. W. Marshall, R. J. Cox, T. J. Simpson, D. J. Clarke, R. H. ffrench-Constant and S. E. Reynolds. 2007. An antibiotic produced by an insect-pathogenic bacterium suppresses host defenses through phenoloxidase inhibition. Proc Natl Acad Sci U S A 104: 2419-2424.

Federle, M. J. and B. L. Bassler 2003. Interspecies communication in bacteria. J Clin Invest 112: 1291-1299.

Ferrandina, G., P. Filippini, et al. 2000. Growth inhibitory effects and radiosensitization induced by fatty aromatic acids on human cervical cancer cells. Oncol Res 12: 429-40.

Ford, L. C., H. A. Hammill, R. J. DeLange, D. A. Bruckner, F. Suzuki-Chavez, K. L. Mickus and T. B. Lebherz 1987. Determination of estrogen and androgen receptors in *Trichomonas vaginalis* and the effects of antihormones. Am J Obstet Gynecol 156: 1119-21.

Fordos, J. 1860. Recherches sur la matiere colorante des supurations bleues:pyocyanine. C.R. Acad. Sci. 51: 215-217.

Franklin, I. K. and C. B. Wollheim 2004. GABA in the endocrine pancreas: its putative role as an islet cell paracrine-signalling molecule. J Gen Physiol 123: 185-90.

Fulop, A. K. and G. Csaba 1991. Turnover and intranuclear localization of 125I-insulin in Tetrahymena. An autoradiographic study. Acta Morphol Hung 39: 71-77.

Fujimura, S., H. Shimakage, et al. 1999. Effects of GABA on noradrenaline release and vasoconstriction induced by renal nerve stimulation in isolated perfused rat kidney. Br J Pharmacol 127: 109-14.

Fuqua, C. and E. P. Greenberg 2002. Listening in on bacteria: acyl-homoserine lactone signalling. Nat Rev Mol Cell Biol 3: 685-695.

Fuqua, C., M. R. Parsek and E. P. Greenberg 2001. Regulation of gene expression by cell-to-cell communication: acyl-homoserine lactone quorum sensing. Annu Rev Genet 35: 439-468.

Gallagher, L. A., S. L. McKnight, M. S. Kuznetsova, E. C. Pesci and C. Manoil 2002. Functions required for extracellular quinolone signaling by *Pseudomonas aeruginosa*. J Bacteriol 184: 6472-6480.

Gardner, P. R. 1996. Superoxide production by the mycobacterial and pseudomonad quinoid pigments phthiocol and pyocyanine in human lung cells. Arch Biochem Biophys 333: 267-274.

Gardiner, S. M., S. R. Chhabra, et al. 2001. Haemodynamic effects of the bacterial quorum sensing signal molecule, N-(3-oxododecanoyl)-L-homoserine lactone, in conscious, normal and endotoxaemic rats. Br J Pharmacol 133: 1047-54.

Glass, C. K. and S. Ogawa 2006. Combinatorial roles of nuclear receptors in inflammation and immunity. Nat Rev Immunol 6: 44-55.

Grebe, T., W. and J., B. Stock 1999. The histidine protein kinase superfamily. Adv Microb Physiol. 41:139-227.

Guina, T., Purvine, S., O., Yi, E., C., Eng, J., Goodlett, D.,R., Aebersold, R. and S., I. Miller 2003. Quantitative proteomic analysis indicates increased synthesis of a quinolone by *Pseudomonas aeruginosa* isolates from cystic fibrosis airways. Proc Natl Acad Sci U S A 100:2771-6.

Hall-Stoodley, L., J. W. Costerton and P. Stoodley 2004. Bacterial biofilms: from the natural environment to infectious diseases. Nat Rev Microbiol 2: 95-108.

Haug, J. S., C. M. Goldner, et al. 1994. Directed cell killing (apoptosis) in human lymphoblastoid cells incubated in the presence of farnesol: effect of phosphatidylcholine. Biochim Biophys Acta 1223: 133-40.

Hense, B. A., C. Kuttler, J. Muller, M. Rothballer, A. Hartmann and J.-U. Kreft 2007. Does efficiency sensing unify diffusion and quorum sensing? Nat Rev Micro 5: 230-239.

Hidy, P. H., R. S. Baldwin, R. L. Greasham, C. L. Keith and J. R. McMullen 1977. Zearalenone and some derivatives: production and biological activities. Adv Appl Microbiol 22: 59-82.

Higuchi, T., H. Hayashi and K. Abe 1997. Exchange of glutamate and gamma-aminobutyrate in a *Lactobacillus* strain. J Bacteriol 179: 3362-3364.

Hogan, D. A., A. Vik, et al. 2004. A *Pseudomonas aeruginosa* quorum-sensing molecule influences *Candida albicans* morphology. Mol Microbiol 54: 1212-23.

Holden, M. T., S. Ram Chhabra, et al. 1999. Quorum-sensing cross talk: isolation and chemical characterization of cyclic dipeptides from *Pseudomonas aeruginosa* and other gram-negative bacteria. Mol Microbiol 33: 1254-66.

Hooi, D. S., B. W. Bycroft, S. R. Chhabra, P. Williams and D. I. Pritchard 2004. Differential immune modulatory activity of *Pseudomonas aeruginosa* quorum-sensing signal molecules. Infect Immun 72: 6463-6470.

Hooper, L. V. and J. I. Gordon 2001. Commensal host-bacterial relationships in the gut. Science 292: 1115-1118.

Hornby, J. M., E. C. Jensen, A. D. Lisec, J. J. Tasto, B. Jahnke, R. Shoemaker, P. Dussault and K. W. Nickerson 2001. Quorum sensing in the dimorphic fungus *Candida albicans* is mediated by farnesol. Appl Environ Microbiol 67: 2982-2992.

Hu, K. and J. M. Webster 2000. Antibiotic production in relation to bacterial growth and nematode development in Photorhabdus – *Heterorhabditis* infected *Galleria mellonella* larvae. FEMS Microbiol Lett 189: 219-223.

Iwasaki, A. and R. Medzhitov 2004. Toll-like receptor control of the adaptive immune responses. Nat Immunol 5: 987-995.

Jahoor, A., R. Patel, A. Bryan, C. Do, J. Krier, C. Watters, W. Wahli, G. Li, S. C. Williams and K. P. Rumbaugh 2008. Peroxisome proliferator-activated receptors mediate host cell proinflammatory responses to *Pseudomonas aeruginosa* autoinducer. J Bacteriol 190: 4408-4415.

Jakoby, W. B. and J. Fredericks 1959. Pyrrolidine and putrescine metabolism: gamma-aminobutyraldehyde dehydrogenase. J Biol Chem 234: 2145-2150.

Ji, G., Beavis, R. and R., P. Novick 1997. Bacterial interference caused by autoinducing peptide variants. Science 276: 2027-30.

Joyce, S. A., A. O. Brachmann, I. Glazer, L. Lango, G. Schwar, D. J. Clarke and H. B. Bode 2008. Bacterial biosynthesis of a multipotent stilbene. Angew Chem Int Ed Engl 47: 1942-1945.

Kamath, J. M., B. E. Britigan, C. D. Cox and D. M. Shasby 1995. Pyocyanin from *Pseudomonas aeruginosa* inhibits prostacyclin release from endothelial cells. Infect Immun 63: 4921-4923.

Kanthakumar, K., G. Taylor, K. W. Tsang, D. R. Cundell, A. Rutman, S. Smith, P. K. Jeffery, P. J. Cole and R. Wilson 1993. Mechanisms of action of *Pseudomonas aeruginosa* pyocyanin on human ciliary beat in vitro. Infect Immun 61: 2848-2853.

Komatsuzaki, N., Nakamura, T., Kimura, T. and J. Shima 2008. Characterization of glutamate decarboxylase from a high gamma-aminobutyric acid (GABA)-producer, *Lactobacillus paracasei*. Biosci Biotechnol Biochem 72: 278-85.

Kravchenko, V. V., G. F. Kaufmann, J. C. Mathison, D. A. Scott, A. Z. Katz, M. R. Wood, A. B. Brogan, M. Lehmann, J. M. Mee, K. Iwata, Q. Pan, C. Fearns, U. G. Knaus, M. M. Meijler, K. D. Janda and R. J. Ulevitch 2006. N-(3-OXO-ACYL) homoserine lactones signal cell activation through a mechanism distinct from the canonical pathogen-associated molecular pattern recognition receptor pathways. J Biol Chem 281: 28822-28830.

Krnjevic, K. 1984. Some functional consequences of GABA uptake by brain cells. Neurosci Lett 47: 283-287.

Lawrence, R. N., W. R. Dunn, et al. 1999. The *Pseudomonas aeruginosa* quorum-sensing signal molecule, N-(3-oxododecanoyl)-L-homoserine lactone, inhibits porcine arterial smooth muscle contraction. Br J Pharmacol 128: 845-848.

LeRoith, D., J. Shiloach, R. Heffron, C. Rubinovitz, R. Tanenbaum and J. Roth 1985. Insulin-related material in microbes: similarities and differences from mammalian insulins. Can J Biochem Cell Biol 63: 839-849.

Ley, R. E., M. Hamady, et al. 2008. Evolution of mammals and their gut microbes. Science 320: 1647-51.

Li, L., D. Hooi, et al. 2004. Bacterial N-acylhomoserine lactone-induced apoptosis in breast carcinoma cells correlated with down-modulation of STAT3. Oncogene 23: 4894-4902.

Look, D. C., L. L. Stoll, S. A. Romig, A. Humlicek, B. E. Britigan and G. M. Denning 2005. Pyocyanin and its precursor phenazine-1-carboxylic acid increase IL-8 and intercellular adhesion molecule-1 expression in human airway epithelial cells by oxidant-dependent mechanisms. J Immunol 175: 4017-4023.

Lucietto, F. R., P. J. Milne, G. Kilian, C. L. Frost and M. Van De Venter 2006. The biological activity of the histidine-containing diketopiperazines cyclo(His-Ala) and cyclo(His-Gly). Peptides 27: 2706-2714.

MacGregor, K. B., B. J. Shelp, S. Peiris and A. W. Bown 2003. Overexpression of glutamate decarboxylase in transgenic tobacco plants deters feeding by phytophagous insect larvae. J Chem Ecol 29: 2177-2182.

Machida, K. and T. Tanaka 1999. Farnesol-induced generation of reactive oxygen species dependent on mitochondrial transmembrane potential hyperpolarization mediated by F(0)F(1)-ATPase in yeast. FEBS Lett 462: 108-12.

Machida, K., T. Tanaka, Y. Yano, S. Otani and M. Taniguchi 1999. Farnesol-induced growth inhibition in *Saccharomyces cerevisiae* by a cell cycle mechanism. Microbiology 145 (Pt 2): 293-299.

Malekinejad, H., R. Maas-Bakker and J. Fink-Gremmels 2006. Species differences in the hepatic biotransformation of zearalenone. Vet J 172: 96-102.

Massart, F., V. Meucci, G. Saggese and G. Soldani 2008. High growth rate of girls with precocious puberty exposed to estrogenic mycotoxins. J Pediatr 152: 690-695, 695 e1.

Miller, K. M., D. G. Dearborn and R. U. Sorensen 1987. In vitro effect of synthetic pyocyanine on neutrophil superoxide production. Infect Immun 55: 559-563.

Miller, M. B. and B. L. Bassler 2001. Quorum sensing in bacteria. Annu Rev Microbiol 55: 165-199.

Mishra, H. N. and C. Das 2003. A review on biological control and metabolism of aflatoxin. Crit Rev Food Sci Nutr 43: 245-264.

Mo, H. and C. E. Elson 2004. Studies of the isoprenoid-mediated inhibition of mevalonate synthesis applied to cancer chemotherapy and chemoprevention. Exp Biol Med (Maywood) 229: 567-85.

Morris, N. S., D. J. Stickler, et al. 1999. The development of bacterial biofilms on indwelling urethral catheters. World J Urol 17: 345-50.

Morse, D. E., N. Hooker, H. Duncan and L. Jensen 1979. ggr-Aminobutyric acid, a neurotransmitter, induces planktonic abalone larvae to settle and begin metamorphosis. Science 204: 407-410.

Nagel, S. C., F. S. vom Saal and W. V. Welshons 1998. The effective free fraction of estradiol and xenoestrogens in human serum measured by whole cell uptake assays: physiology of delivery modifies estrogenic activity. Proc Soc Exp Biol Med 217: 300-309.

Nickerson, K. W., A. L. Atkin and J. M. Hornby 2006. Quorum sensing in dimorphic fungi: farnesol and beyond. Appl Environ Microbiol 72: 3805-3813.

Nishikawa, Y. and T. Baba 1985. Effects of ovarian hormones on manifestation of purulent endometritis in rat uteruses infected with *Escherichia coli*. Infect Immun 47: 311-317.

Nishikawa, Y. and T. Baba 1985. In vitro adherence of *Escherichia coli* to endometrial epithelial cells of rats and influence of estradiol. Infect Immun 50: 506-509.

Nomura, M., H. Kimoto, Y. Someya, S. Furukawa and I. Suzuki 1998. Production of gamma-aminobutyric acid by cheese starters during cheese ripening. J Dairy Sci 81: 1486-1491.

Olsen, M. and K. H. Kiessling 1983. Species differences in zearalenone-reducing activity in sub-cellular fractions of liver from female domestic animals. Acta Pharmacol Toxicol (Copenh) 52: 287-291.

Olsen, M., H. Pettersson and K. H. Kiessling 1981. Reduction of zearalenone to zearalenol in female rat liver by 3 alpha-hydroxysteroid dehydrogenase. Acta Pharmacol Toxicol (Copenh) 48: 157-161.

Otto, M., Echner, H., Voelter, W. and F. Götz 2001. Pheromone cross-inhibition between *Staphylococcus aureus* and *Staphylococcus epidermidis*. Infect Immun 69:1957-60.

Pater, M. M., R. Mittal and A. Pater 1994. Role of steroid hormones in potentiating transformation of cervical cells by human papillomaviruses. Trends Microbiol 2: 229-234.

Plotkin, B. J., R. J. Roose, Q. Erikson and S. M. Viselli 2003. Effect of androgens and glucocor-ticoids on microbial growth and antimicrobial susceptibility. Curr Microbiol 47: 514-520.

Prasad, C. 1995. Bioactive cyclic dipeptides. Peptides 16: 151-164.

Ramage, G., S. P. Saville, B. L. Wickes and J. L. Lopez-Ribot 2002. Inhibition of *Candida albi-cans* biofilm formation by farnesol, a quorum-sensing molecule. Appl Environ Microbiol 68: 5459-5463.

Ran, H., D. J. Hassett and G. W. Lau 2003. Human targets of *Pseudomonas aeruginosa* pyocya-nin. Proc Natl Acad Sci U S A 100: 14315-14320.

Redfield, R. J. 2002. Is quorum sensing a side effect of diffusion sensing? Trends Microbiol 10: 365-370.

Reisner, A., N. Hoiby, et al. 2005. Microbial pathogenesis and biofilm development. Contrib Microbiol 12: 114-31.

Rioja, A., A. R. Pizzey, et al. 2000. Preferential induction of apoptosis of leukaemic cells by farnesol. FEBS Lett 467: 291-5.

Roberts, R. C. and M. Difiglia 1988. Localization of immunoreactive GABA and enkephalin and NADPH-diaphorase-positive neurons in fetal striatal grafts in the quinolinic-acid-lesioned rat neostriatum. J Comp Neurol 15: 406-21.

Rumbaugh, K. P. 2007. Convergence of hormones and autoinducers at the host/pathogen interface. Anal Bioanal Chem 387: 425-435.

Sahl, H. G., U. Pag, S. Bonness, S. Wagner, N. Antcheva and A. Tossi 2005. Mammalian defensins: structures and mechanism of antibiotic activity. J Leukoc Biol 77: 466-475.

Sard, H., G. Kumaran, C. Morency, B. L. Roth, B. A. Toth, P. He and L. Shuster 2005. SAR of psilocybin analogs: discovery of a selective 5-HT 2C agonist. Bioorg Med Chem Lett 15: 4555-4559.

Schaeffer, A. J., S. K. Amundsen and L. N. Schmidt 1979. Adherence of *Escherichia coli* to human urinary tract epithelial cells. Infect Immun 24: 753-759.

Schardl, C. L., R. B. Grossman, P. Nagabhyru, J. R. Faulkner and U. P. Mallik 2007. Loline alka-loids: Currencies of mutualism. Phytochemistry 68: 980-996.

Schoental, R. 1983. Precocious sexual development in Puerto Rico and oestrogenic mycotoxins (zearalenone). Lancet 1: 537.

Schüller, F., Benz, R., and H., G. Sahl 1989. The peptide antibiotic subtilin acts by formation of voltage-dependent multi-state pores in bacterial and artificial membranes. Eur J Biochem 182: 181-6.

Semighini, C. P., J. M. Hornby, et al. 2006. Farnesol-induced apoptosis in *Aspergillus nidulans* reveals a possible mechanism for antagonistic interactions between fungi. Mol Microbiol 59: 753-64.

Shelp, B. J., A. W. Bown and D. Faure 2006. Extracellular gamma-aminobutyrate mediates com-munication between plants and other organisms. Plant Physiol 142: 1350-1352.

Shier, W. T., A. C. Shier, et al. 2001. Structure-activity relationships for human estrogenic activity in zearalenone mycotoxins. Toxicon 39: 1435-8.

Shiner, E. K., K. P. Rumbaugh and S. C. Williams 2005a. Interkingdom signaling: deciphering the language of acyl homoserine lactones. FEMS Microbiol Rev 29: 935-947.

Shiner, E. K., K. P. Rumbaugh and S. C. Williams 2005b. Interkingdom signaling: deciphering the language of acyl homoserine lactones. FEMS Microbiol Rev 29: 935-947.

Shiner, E. K., D. Terentyev, A. Bryan, S. Sennoune, R. Martinez-Zaguilan, G. Li, S. Gyorke, S. C. Williams and K. P. Rumbaugh 2006. *Pseudomonas aeruginosa* autoinducer modulates host cell responses through calcium signalling. Cell Microbiol 8: 1601-1610.

Silvestre, J. F. and M. I. Betlloch 1999. Cutaneous manifestations due to *Pseudomonas* infection. Int J Dermatol 38: 419-431.

Singh, P. K., A. L. Schaefer, et al. 2000. Quorum-sensing signals indicate that cystic fibrosis lungs are infected with bacterial biofilms. Nature 407: 762-4.

Smith, D. K., T. Kassam, et al. 1992. *Escherichia coli* has two homologous glutamate decarboxylase genes that map to distinct loci. J Bacteriol 174: 5820-6.

Smith, R. S., E. R. Fedyk, et al. 2001. IL-8 production in human lung fibroblasts and epithelial cells activated by the *Pseudomonas* autoinducer N-3-oxododecanoyl homoserine lactone is transcriptionally regulated by NF-kappa B and activator protein-2. J Immunol 167: 366-74.

Smith, R. S., S. G. Harris, et al. 2002. The *Pseudomonas aeruginosa* quorum-sensing molecule N-(3-oxododecanoyl)homoserine lactone contributes to virulence and induces inflammation in vivo. J Bacteriol 184: 1132-9.

Sobel, J. D., J. Schneider, D. Kaye and M. E. Levison 1981. Adherence of bacteria to vaginal epithelial cells at various times in the menstrual cycle. Infect Immun 32: 194-197.

Solomon, P. S. and R. P. Oliver 2001. The nitrogen content of the tomato leaf apoplast increases during infection by *Cladosporium fulvum*. Planta 213: 241-249.

Sonnex, C. 1998. Influence of ovarian hormones on urogenital infection. Sex Transm Infect 74: 11-19.

Sorensen, R. U. and J. D. Klinger 1987. Biological effects of *Pseudomonas aeruginosa* phenazine pigments. Antibiot Chemother 39: 113-124.

Sperandio, V., A. G. Torres, B. Jarvis, J. P. Nataro and J. B. Kaper 2003. Bacteria-host communication: the language of hormones. Proc Natl Acad Sci U S A 100: 8951-8956.

Sturme, M. H., C. Francke, R. J. Siezen, W. M. de Vos and M. Kleerebezem 2007. Making sense of quorum sensing in lactobacilli: a special focus on *Lactobacillus plantarum* WCFS1. Microbiology 153: 3939-3947.

Sturme, M. H., M. Kleerebezem, J. Nakayama, A. D. Akkermans, E. E. Vaugha and W. M. de Vos 2002. Cell to cell communication by autoinducing peptides in gram-positive bacteria. Antonie Van Leeuwenhoek 81: 233-243.

Sturme, M. H., J. Nakayama, D. Molenaar, Y. Murakami, R. Kunugi, T. Fujii, E. E. Vaughan, M. Kleerebezem and W. M. de Vos 2005. An agr-like two-component regulatory system in *Lactobacillus plantarum* is involved in production of a novel cyclic peptide and regulation of adherence. J Bacteriol 187: 5224-5235.

Sugarman, B. and N. Mummaw 1988. The effect of hormones on *Trichomonas vaginalis*. J Gen Microbiol 134: 1623-1628.

Sugarman, B. and N. Mummaw 1990. Oestrogen binding by and effect of oestrogen on trichomonads and bacteria. J Med Microbiol 32: 227-232.

Sweet, R. L., M. Blankfort-Doyle, M. O. Robbie and J. Schacter 1986. The occurrence of chlamydial and gonococcal salpingitis during the menstrual cycle. Jama 255: 2062-2064.

Szuets, P., A. Mesterhazy, G. Falkay and T. Bartok 1997. Early thelarche symptoms in children and their relations to zearalenon contamination in foodstuffs. Cereal Res Commun 25: 429-436.

Takahashi, N., T. Kawada, et al. 2002. Dual action of isoprenols from herbal medicines on both PPARgamma and PPARalpha in 3T3-L1 adipocytes and HepG2 hepatocytes. Febs Lett 514: 315-22.

Tateda, K., Y. Ishii, et al. 2003. The *Pseudomonas aeruginosa* autoinducer N-3-oxododecanoyl homoserine lactone accelerates apoptosis in macrophages and neutrophils. Infect Immun 71: 5785-93.

Telford, G., D. Wheeler, et al. 1998. The *Pseudomonas aeruginosa* quorum-sensing signal molecule N-(3-oxododecanoyl)-L-homoserine lactone has immunomodulatory activity. Infect Immun 66: 36-42.

Tfelt-Hansen, P., P. R. Saxena, C. Dahlof, J. Pascual, M. Lainez, P. Henry, H. Diener, J. Schoenen, M. D. Ferrari and P. J. Goadsby 2000. Ergotamine in the acute treatment of migraine: a review and European consensus. Brain 123 (Pt 1): 9-18.

Tfelt-Hansen, P. C. and P. J. Koehler 2008. History of the use of ergotamine and dihydroergotamine in migraine from 1906 and onward. Cephalalgia 28: 877-886.

Thompson, J. F., J. K. Pollard and F. C. Steward 1953. Investigations of Nitrogen Compounds and Nitrogen Metabolism in Plants. Iii. gamma-Aminobutyric acid in plants, with special reference to the potato tuber and a new procedure for isolating amino acids other than alpha-amino acids. Plant Physiol 28: 401-414.

Ueno, Y., K. Hayakawa, et al. 1997. Purification and characterization of glutamate decarboxylase from *Lactobacillus brevis* IFO 12005. Biosci Biotechnol Biochem 61: 1168-71.

Usher, L. R., R. A. Lawson, I. Geary, C. J. Taylor, C. D. Bingle, G. W. Taylor and M. K. Whyte 2002. Induction of neutrophil apoptosis by the *Pseudomonas aeruginosa* exotoxin pyocyanin: a potential mechanism of persistent infection. J Immunol 168: 1861-1868.

Wade, D. S., M. W. Calfee, et al. 2005. Regulation of *Pseudomonas* quinolone signal synthesis in *Pseudomonas aeruginosa*. J Bacteriol 187: 4372-80.

Watve, M. G., R. Tickoo, M. M. Jog and B. D. Bhole 2001. How many antibiotics are produced by the genus *Streptomyces*? Arch Microbiol 176: 386-390.

Williams, S. C., E. K. Patterson, N. L. Carty, J. A. Griswold, A. N. Hamood and K. P. Rumbaugh 2004. *Pseudomonas aeruginosa* autoinducer enters and functions in mammalian cells. J Bacteriol 186: 2281-2287.

Willson, T. M., M. H. Lambert and S. A. Kliewer 2001. Peroxisome proliferator-activated receptor gamma and metabolic disease. Annu Rev Biochem 70: 341-367.

Wilson, R., D. A. Sykes, D. Watson, A. Rutman, G. W. Taylor and P. J. Cole 1988. Measurement of *Pseudomonas aeruginosa* phenazine pigments in sputum and assessment of their contribution to sputum sol toxicity for respiratory epithelium. Infect Immun 56: 2515-2517.

Xiao, G., J. He, et al. 2006. Mutation analysis of the *Pseudomonas aeruginosa* mvfR and pqsABCDE gene promoters demonstrates complex quorum-sensing circuitry. Microbiology 152: 1679-86.

Yokoyama, S., J. Hiramatsu and K. Hayakawa 2002. Production of gamma-aminobutyric acid from alcohol distillery lees by *Lactobacillus brevis* IFO-12005. J Biosci Bioeng 93: 95-97.

Chapter 15
Mycologic Endocrinology

Karl V. Clemons, Jata Shankar, and David A. Stevens

15.1   Introduction

Hormones serve as regulatory messenger molecules, inducing regulation of gene-expression through receptor-mediated interactions, and the subsequent functional response to the presence of the hormone. Molecules considered to be hormones include steroids, such as 17β-estradiol (E_2), progesterone, and testosterone, and protein or peptide hormones such as insulin, human growth hormone, and luteinizing hormone (LH).

The present understanding of the action of hormones is derived from mammalian systems, where hormones regulate a variety of functional processes in a broad range of tissues. Target cells are affected in a tissue specific manner after binding of the hormone by a specific receptor; the extent of the specific response is determined by the proteins, pathways, and processes with which the receptors interact, as well as the concentration of hormone present. Classically, hormone receptors (e.g., steroid receptors) act as transcriptional regulators in the nucleus, after the hormone–receptor complex moves across the nuclear membrane, where it binds to specific response elements in the DNA (Beato et al. 1996; Funder 1997; Walters and Nemere 2004; Whitfield et al. 1999). However, not all steroid receptors are located intracellularly and some are plasma membrane associated (Hammes 2003; Walters and Nemere 2004). In addition, steroids, and other hormones, can have nongenomic actions, such

K.V. Clemons (✉)
Division of Infectious Diseases, Santa Clara Valley Medical Center,
751 South Bascom Ave, San Jose, CA 95128, USA
e-mail: clemons@cimr.org

J. Shankar
California Institute for Medical Research, 2260 Clove Drive, San Jose, CA 95126, USA
e-mail: jata@standford.edu

D.A. Stevens
Division of Infectious Diseases, Department of Medicine, Santa Clara Valley Medical Center,
Rm. 6C097, 751 S. Bascom Avenue, San Jose, CA 95128-2699, USA
e-mail: stevens@standford.edu

M. Lyte and P.P.E. Freestone (eds.), *Microbial Endocrinology*,
Interkingdom Signaling in Infectious Disease and Health,
DOI 10.1007/978-1-4419-5576-0_15, © Springer Science+Business Media, LLC 2010

as modulation of MAP kinases or tyrosine kinases and regulation of membrane ion channels and G-proteins (Losel and Wehling 2003; Simoncini and Genazzani 2003).

Fungi also utilize hormones (sometimes called pheromones) as messenger molecules that regulate various activities of the organism. Primarily, these molecules are related to the control of sexual reproduction in various fungi, and take the form of novel steroids, peptides, and acid derivatives (Gooday 1974; Gooday and Adams 1993). In addition, fungi produce secondary metabolites, such as zearalenone, which has high estrogenic activity in mammalian cells. A recent study suggested zearalenone and its metabolites, α-zearalanol, β-zearalanol, α-zearalenol, and β-zearalenol, as trigger factors for the development of central precocious puberty and high growth rate of girls exposed to environmental sources of these mycotoxins (Massart et al. 2008). It has long been known that fungi can metabolize, transform, and convert mammalian steroids (including the production of steroid molecules that mammalian cells cannot produce), which has proven useful in the biotechnology production field (Fernades et al. 2003; Mahato and Garai 1997).

The present chapter addresses different aspects of fungal endocrinology, particularly interactions of mammalian hormones with fungi. Two reviews have been written previously addressing in detail the work done in our laboratory and that of our collaborators (Feldman 1988; Stevens 1989). We present an overview of this work, which focused on characterization of hormone receptors in fungi, as well as any physiological effect or functional responses induced by the presence of the hormone in the systems we studied. In addition, we include more recent studies from other investigators addressing functional responses of fungi to mammalian hormones.

15.2   Hormones and Mating in Fungi

Sexual reproduction of several fungi is known to be under the control of hormones or pheromones, which act as chemoattractants and also stimulate a functional response via a specific receptor for the target fungal mating type to become capable of mating with the opposite mating type (e.g., male and female, or + and − strains, or a and α strains) (Bardwell 2005; Gooday 1974; Gooday and Adams 1993). For instance, in the water mould *Allomyces,* swimming female gametes produce sirenin, which is a sesquiterpine that acts as a chemoattractant for the male gametes, causing them to swim toward the female, where eventual contact between the two initiates plasmogamy.

An extensively studied fungal hormone system is that of the oomycete water mould, *Achyla* (Brunt et al. 1990, 1998; Brunt and Silver 1986a, b; Gooday 1974; Gooday and Adams 1993; Riehl and Toft 1984, 1985; Riehl et al. 1984; Silver et al. 1993). This organism produces two steroidal hormones, antheridiol by the female and oogonial by the male that cause the switch from vegetative growth to the formation of mating specific structures. Antheridiol induces the formation of antheridia by the male, which grow toward the female cells and similarly, oogonial induces the formation of oogonial initials by the female cells. The organism has been demonstrated to have a high-affinity binding protein for antheridiol, with a K_d of 7×10^{-10} M and capacity of up to 2,000 fmol/mg of protein (Riehl and Toft 1984; Riehl et al. 1984). Furthermore, this specific binding protein has been demonstrated to be a true receptor, modulating gene expression in the fungus and in particular hsp70 and hsp90 (Brunt et al. 1998, 1990; Brunt and Silver 1986a, b; Silver et al. 1993). Thus, in this system, specific binding of the hormone and subsequent receptor-mediated regulation of gene-expression is similar to the regulatory mechanisms of hormonal interaction in mammalian systems.

Saccharomyces cerevisiae produces two peptide hormones, a- and α-factor, which act on the reciprocal mating type to arrest vegetative growth and enable mating. Interestingly, the α-factor peptide has sequence similar to mammalian gonadotropin-releasing hormone and can also bind to the mammalian receptor, inducing pituitary cells to release luteinizing hormone (Loumaye et al. 1982). Another yeast producing mating type specific peptide hormone is *Rhodosporidium* (Miyakawa et al. 1985). *Tremella* (a basidiomycete) utilizes two pheromones, A-10 and α-13 (isoprenyl peptides), to induce mating of haploid cells (Gooday 1974; Gooday and Adams 1993; Miyakawa et al. 1984). In the zygomycetes, trisporic acids act on both + and − mating types to induce gametangial tropism and subsequent mating, but appear not to be species specific (Gooday and Adams 1993). Overall, it is likely that sexual reproduction of most fungi is regulated by the production of mating type hormones.

A reciprocal mating pheromone system has been described in the human pathogen *Candida albicans*. Naturally occurring a and α clinical isolates have been identified and can undergo mating. For these cells to become mating competent, they must first undergo transition from the white phase to the opaque phase. This well-described phenotypic switching system involves changes in the expression of many genes, including many involved in mating (Bennett et al. 2003). Cells of the mating type respond to the α pheromone by producing long polarized projections that can result in the mating process. During this process, transcription of many genes is induced. Of note, several of the genes encode surface and secreted proteins implicated in *C. albicans* virulence.

15.3   Interaction of Fungal Hormones with Plants, and Plant Hormones with Fungi

Host-microbial interactions are also important in the plant kingdom, and these interactions affecting fungal infection may be mediated by hormones (Gogala 1991; Prusty et al. 2004). However, for the remainder of the chapter we focus on mammalian-fungal hormonal interactions.

15.4   Interaction of Mammalian Hormones with Fungi

In addition to the action of fungal-produced hormones, several pathogenic fungi are known to interact and respond to various mammalian hormones. These interactions can influence the growth, and even the pathogenesis of the organism. In the following sections, we will examine some of these interactions and effects on specific organisms, as well as potential implications for pathogenesis.

15.4.1 Hormone Influence on Growth of Fungi

Early studies with the interactions of fungi and hormones focused on the growth inhibitory effects of high concentrations (e.g., $>10^{-3}$M) of the hormones. However, in some instances mammalian hormones are stimulatory to the growth of the organisms (Stevens 1989). Dermatophytic fungi (i.e., *Trichophyton*, *Microsporum*, and *Epidermophyton*) have been demonstrated to be inhibited by the presence of various steroids including androgens and progesterone (Brasch and Flader 1996; Capek and Simek 1971; Stevens 1989). Furthermore, hydroxylation of progesterone reduces the inhibitory activity (Capek and Simek 1971). The inhibition of *Trichophyton rubrum* and *Epidermophyton floccosum* by androgens has been suggested as a reason that these organisms do not cause tinea capitis (Brasch and Flader 1996; Brasch and Gottkehaskamp 1992). In addition, a dematiaceous fungus, *Phialophora verrucosa* is inhibited by the mammalian hormones progesterone and testosterone (Hernandez-Hernandez et al. 1995).

Interestingly, some fungi have been shown to have stimulated growth in the presence of mammalian hormones. For instance, two studies have reported the stimulation of the growth of *C. albicans* by dexamethasone (Gupta et al. 1982; Khosla et al. 1978) and another reported the stimulation of adherence to buccal cells by dexamethasone, but inhibition by corticosterone (Ghannoum and Elteen 1987). Estrogens have also been reported to stimulate the growth of *C. albicans* (Gujjar et al. 1997; Zhang et al. 2000) and promote spherule growth and endospores release in *Coccidioides* (Powell et al. 1983). Furthermore, hydrocortisone has been shown to stimulate the growth of *Aspergillus fumigatus* and *Aspergillus flavus* (Ng et al. 1994). Overall, results showing growth stimulation by mammalian hormones in these studies have implications for enhancing or promoting the pathogenesis of these organisms.

15.5 Specific Steroid Binding Proteins in Fungi

As can be noted from the investigations noted earlier, in some organisms, specific receptors for endogenous hormones have been demonstrated and various fungi interact with mammalian hormones. This raises the question of whether the fungi respond to the nonendogenous mammalian hormones via the binding of the hormone to a fungal protein, which acts as a receptor, or do fungi have specific receptors for mammalian hormones? These questions have been addressed for a number of fungal pathogens and are presented in the following sections.

15.5.1 Paracoccidioides brasiliensis

P. brasiliensis is a thermally dimorphic fungus, which causes the most prevalent systemic mycosis (paracoccidioidomycosis) in Latin America. The fungus exists in soil as a filamentous form, which transforms into a yeast form when it invades mammalian hosts or by a temperature switch from room temperature to 37°C, in vitro. Infection is initiated after inhalation of conidia or mycelial fragments by the mammalian host. These mycelial propagules further differentiate into the yeast form of the organism, which is found in the tissues. The transition of mycelial to yeast form is crucial for the establishment of disease. Epidemiologically, clinical disease is more common in adult men than women despite equal frequencies of exposure to this fungus, and it has been estimated that females are about 13–70 times less likely as males to develop clinical disease (Brummer et al. 1993). Furthermore, the development of clinical disease is equal in males and prepubertal or postmenopausal females. Thus, this led to the speculation that human sex hormones might have an effect on progression of this disease.

Using a microculture system, morphological transformation of mycelia to yeast form of clinical isolates of *P. brasiliensis* was tested in the presence of various steroids, including 17β-estradiol (E_2), testosterone, tamoxifen and 17α-estradiol, and the nonsteroidal estrogen, diethylstilbesterol (DES), at concentrations ranging from 2×10^{-10} to 2×10^{-6} M to span relevant physiological and pharmacological concentrations (Restrepo et al. 1984). Mycelial to yeast transformation was dose-dependently inhibited by E_2 to 71% (2×10^{-10}), 33% (2×10^{-8}) and 19% (2×10^{-6} M) of the transformation of control cultures. In addition, DES at these concentrations resulted in 85% (2×10^{-10}), 54% (2×10^{-8}) and 37% (2×10^{-6} M) inhibition when compared to control. Testosterone, tamoxifen, and 17α-estradiol were inactive. Inhibition of the morphological transformation occurs only in one direction and none of the tested compounds blocked the transformation of yeast to mycelia nor did they affect yeast growth or budding. Similarly, the transformation of conidia to yeast was shown to be blocked by E_2 (Salazar et al. 1988). Thus, inhibition of transformation of conidia or mycelia fragments to yeast form is a biologically specific phenomenon with E_2 effective and its stereoisomer 17α-estradiol ineffective.

To identify a possible steroid hormone receptor in the cytosol of *P. brasiliensis*, yeast cells were grown and disrupted with glass beads to produce a cytosolic extract. Specific binding activity was examined by addition of [³H]-E_2 and 500-fold molar excess of unlabeled E_2 to the cytosol to perform conventional steroid binding assays (Loose et al. 1983b). Bound hormone was separated from free hormone using centrifuged Sephadex microcolumns. In a single point study using 130 nM [³H]-E_2 specific E_2 binding was demonstrated in the range of 200 fmol/mg of cytosol protein. Some specific binding was detected with [³H]-progesterone (60 fmol/ mg protein) and DES (42 fmol/mg protein); only trace binding was detected with dihydrotestosterone and testosterone, and no binding was seen with corticosterone. Progesterone binding was completely blocked by E_2, suggesting the presence of a single estrogen binding site with some cross reactivity for other steroids. However, the presence of a separate progesterone or DES binding site has not been completely excluded.

More extensive binding studies with cytosol and [³H]-E_2 done at 0°C indicated that binding was maximal after about 90 min of incubation (Loose et al. 1983b). The dissociation rate was determined and the off-rate time for half of the bound hormone to be released was 29 min. Binding characteristics at equilibrium showed binding to be saturable with low nonspecific binding. Scatchard plots suggested a single class of noninteracting binding sites, with a dissociation constant (K_d) of 8.5 nM and a binding capacity (N_{max}) of 210 fmol/mg protein, indicative of high-affinity low capacity binding. The specificity of the binding was assessed in competition assays, where unlabelled hormones compete for specific binding with [^{3}H]-E_2. Competition for bound E_2 was maximal with E_2, whereas the related estrogens, estrone, and estriol, had only about 25% of the affinity of E_2. DES was a weak competitor, in contrast to its affinity for the mammalian E_2 receptor. Thermal stability studies at 0, 37, and 56°C for over 30 min showed a second binding protein, with lower affinity at 37°C. However at 56°C, only the high capacity $(N_{max}$ 1700–2600 fmol/mg protein) binding protein was detected. Treatments with DNase, RNase, and phospholipase A2 had little effect on the [^{3}H]-E_2 binding, whereas binding activity was inhibited by trypsin and reduced by N-ethylmaleimide. These results together suggested specific binding was due to a protein containing sulfhydryl groups. Liquid chromatographic (HPLC) studies indicated that the binding protein has a relative molecular mass of 60,000 Da and sucrose gradient centrifugation indicated a sedimentation coefficient of 4.4S.

Because the functional response of inhibition of form transformation was that of inhibiting the mycelial form, it was important to determine whether the mycelial form also contained a specific binding protein for E_2, and we were able to devise ways to work with the mycelial form of *P. brasiliensis* for binding assays (Stover et al. 1986). A binding protein was discovered in this form, in many isolates of the fungus, as had been shown for the yeast form. The K_d and N_{max} were 13 nM and 78 fmol/mg protein, respectively. In addition, a second low-affinity $(K_d$ 150 nM) high-capacity $(N_{max}$ 3,000–4,500 fmol/mg protein) binding protein was demonstrated in the yeast form. In competition studies, DES was a potent competitor for E_2 with the mycelial binder, correlating better with its functional effect of inhibiting the mycelium-to-yeast transformation described earlier.

To examine the cellular response of *P. brasiliensis* to the presence of E_2, we followed temporal protein expression during mycelial to yeast transformation using [^{35}S]-methionine incorporation and 1-D SDS-PAGE (Clemons et al. 1989a). E_2 altered protein expression in *P. brasiliensis* in vitro when shifted to the temperature that permits mycelium-to-yeast transformation, blocking the appearance of proteins associated with transformation or yeast and maintaining a more mycelial protein profile until later time points, where de novo protein expression was virtually shut down (Clemons et al. 1989a). In addition, exposure of mycelial cultures to E_2 without a switch in temperature induced the uptake of labeled methionine (see Figs. 15.1 and 15.2).

In light of the demonstration of specific E_2 binding and effects on morphological form transition and protein synthesis, we performed in vivo studies in a murine model of pulmonary infection to determine whether E_2 could alter the pathogenesis of *P. brasiliensis* and explain the observed epidemiologic data of resistance of females. In the initial studies we followed morphologic transformation of conidia

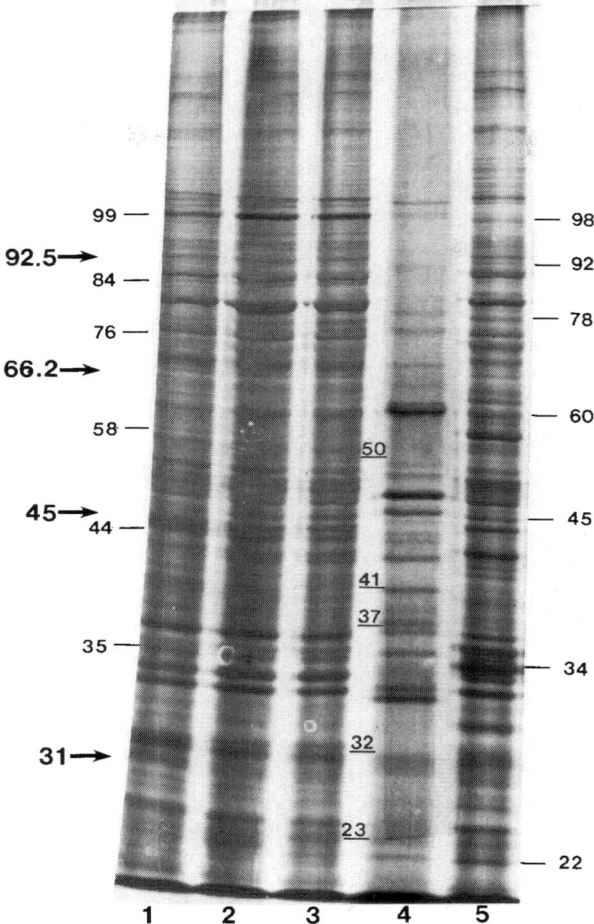

Fig. 15.1 Comparison of proteins from the cytosol fractions of E$_2$-treated *P. brasiliensis* undergoing M to Y transition with M and Y controls. Proteins were resolved through 9% SDS-PAGE and silver stained. Lane 1, M-cells incubated solely at 25°C, treated with E$_2$ for 24 h; lanes 2, 3, and 4, M-cells after 24, 72, and 120 h of E$_2$ treatment at 37°C, respectively. Lane 5, Y-cells incubated at 37°C only. Lanes 1–4 represent M cultures treated with E$_2$ (2.6 × 10^{-7} M), and lane 5 represents the Y control. Note the maintenance of the M-form profile at 24 and 72 h (lanes 2 and 3) and the decreased total number of bands by 120 h (lane 4), which demonstrates little similarity to the Y-form profile (lane 5). For reference, the molecular mass in kDa of the transition bands (between lanes 3 and 4) as well as selected M-specific (near left) and Y-specific (near right) bands are indicated. Molecular masses in kDa of standards are indicated on the far left. Reprinted with permission from the Society for General Microbiology, United Kingdom from reference (Clemons et al. 1989a)

after pulmonary instillation. Conidial transformation into yeast was not impeded in male mice, with budding yeast forms present as early as 48 h. In contrast, female mice had progressively fewer cells in the bronchoalveolar lavage fluid and no

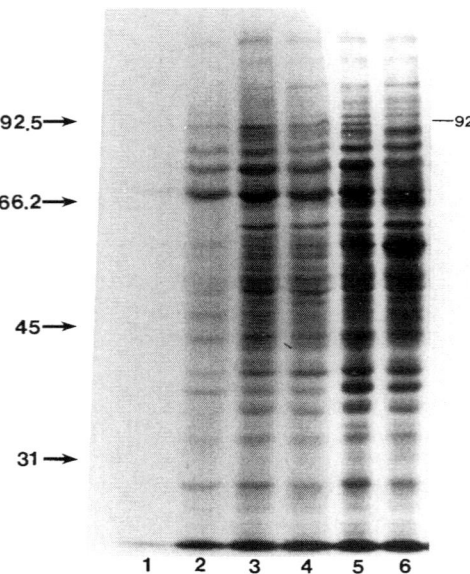

$92.5\longrightarrow$ -92

$66.2\longrightarrow$

$45\longrightarrow$

$31\longrightarrow$

1   2   3   4   5   6

Fig. 15.2 Comparison of [^{35}S]methionine-labeled proteins in cytosol fractions of E$_2$-treated and untreated *P. brasiliensis* undergoing M to Y transition. Labeling was done for 2 h in the absence of radioinert methionine prior to disruption of organisms and extraction of cellular proteins. Lanes were loaded with equal counts of [^{35}S]-labeled proteins, electrophoresed and processed for fluorography. Lane assignments: lane 1 , M control; 2, M E$_2$,-treated; 3 and 5, M controls grown at 37°C for 24 h, and 72 h, respectively. Lanes 4 and 6 are M treated with E$_2$ (2.6×10^{-7} M) grown at 37°C for 24 h, and 72 h, respectively. Note the effect of E$_2$ on label incorporation of M-form (lane 2) as compared to untreated control (lane 1). Absence of the 92 kDa Y-specific band in lane 6 is indicated. The fluorogram was intentionally over-exposed to enhance the bands in lane 2. Molecular masses in kDa of standards are indicated on the left. Reprinted with permission from the Society for General Microbiology, United Kingdom from reference (Clemons et al. 1989a)

budding yeasts were observed (Aristizabal et al. 1998). In addition, no CFU were recovered from the lungs of female mice after 2–6 weeks of infection, whereas $\log_{10}$ 2 or more CFU were recovered from male mice (Aristizabal et al. 1998). Thus, in female mice, transformation of the conidia was severely impeded, with the mice able to clear detectable infection (see Fig. 15.3). Furthermore, we found that castrated male mice reconstituted with high-doses of E$_2$ initially inhibited conidial transformation to yeast and subsequent proliferation. However, these mice had recurrence of disease with progression later in infection, which we speculate may have been due to the immunoregulatory effects of the high-doses of E$_2$ administered to those animals. In contrast, castrated females reconstituted with testosterone were unable to restrict disease (Aristizabal et al. 2002). Taken together, these studies support the hypothesis that the interaction of the organism with E$_2$ contributes to the resistance of females to this infection.

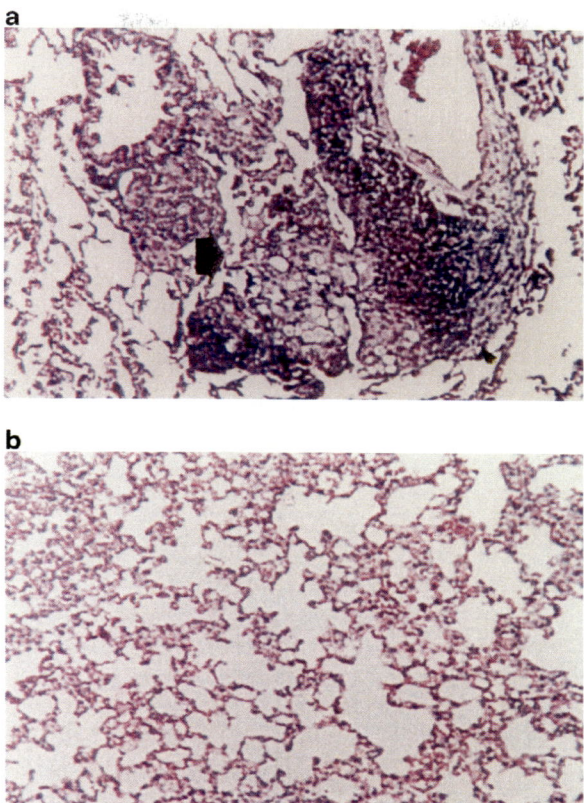

Fig. 15.3 Histopathology of the lungs in normal mice 4 weeks after infection with conidia of *P. brasiliensis* (H&E, X100). (**a**) Normal males (NM) showing an intense chronic inflammatory reaction, also with granuloma formation and presence of yeast cells. (**b**) Normal females (NF) showing a slight inflammatory reaction, with no yeast cells present. Reprinted with permission of the publisher (Taylor & Francis Ltd, http://www.tandf.co.uk/journals) from reference (Aristizabal et al. 2002)

15.5.2 C. albicans

C. albicans is an opportunistic dimorphic fungal pathogen of medical importance. Studies by Loose et al. (Loose and Feldman 1982; Loose et al. 1981, 1983a), into the evolution of hormone-receptors, represents a classical example of mammalian steroid hormone interaction with the fungus. In those studies, *C. albicans* was shown to have a protein capable of specific corticosteroid-binding (CBP) that exhibits high affinity for corticosterone and progesterone. Specific corticosterone binding was found in the cytosol of *C. albicans* with a K_d 6.3 nM and binding capacity N_{max} 650 fmol/mg of protein. Specific binding was demonstrated to be due to a protein with apparent molecular mass of 43 kD (Loose and Feldman 1982; Loose et al. 1981). Interestingly, lipid extracts of the cell pellet or of culture filtrate displaced specific [^{3}H]-corticosterone from CBP, and are presumed to contain an endogenous ligand (Loose et al. 1981). Thus, CBP exhibits properties consistent with a receptor molecule. It is stereo-specific, extremely selective with affinity for corticosterone (~7 nM) that is equivalent to that of mammalian glucocorticoid receptors (Loose et al. 1981).

CBP has been found in both serotypes A and B of *C. albicans* and a survey of various species of *Candida* indicated that the binding protein appears ubiquitous within the genus (Loose et al. 1983a) and was confirmed in other studies (Powell and Drutz 1983). Although the binding parameters differed between species, Scatchard plots were linear, indicating a single class of binding sites in each. It was shown that corticosterone can enter an intact yeast cell and bind to the protein. *Candida* growth, yeast to mycelia conversion of *Candida* or glucose oxidation was not affected by the addition of different steroids (corticosterone, progesterone and dexamethasone) over a range of 1×10^{-6} to 2×10^{-10} M (Loose et al. 1983a).

To determine the relationship of CBP from *C. albicans* to the mammalian hormone receptors, the CBP gene has been cloned and expressed. It revealed an open reading frame of 1,467 bp that encodes a protein with a molecular weight of 44,545 Da; the expressed protein has the properties of the native CBP. Sequence comparison of CBP gene to members of mammalian steroid-thyroid-retinoic acid receptor gene superfamily showed that CBP is unrelated to these hormone receptors (Malloy et al. 1993). Interestingly, in *S. cerevisiae*, the *FMS1* (fenpropimorph multicopy suppressor gene 1) yeast gene shows a protein identity of 35% with *C. albicans* CBP (Joets et al. 1996).

C. albicans has also been shown to have an estrogen-binding protein (EBP) that displays high affinity for estradiol and estrone (Powell et al. 1984; Skowronski and Feldman 1989). Specific binding was found for E_2 with a K_d of about 6×10^{-8} M and N_{max} of 400–13,000 fmol/mg of protein depending on the study and the strain of *C. albicans* tested (Powell et al. 1984; Skowronski and Feldman 1989). Furthermore, the abundance of EBP was found to be significantly higher in early log phase growth (Skowronski and Feldman 1989). Binding was saturable and other estrogens, estrone and estriol, were the best competitors and unlike the situation of the human estrogen receptor, tamoxifen does not bind to EBP in *C. albicans* (Powell et al. 1984; Skowronski and Feldman 1989).

Cloning of the EBP gene revealed an open reading frame of 1,221 bp that encodes a protein with 407 amino acids and having a molecular mass of 46,073 Da, the estimated size of EBP (Madani et al. 1994). The expressed gene showed high affinity with binding for estradiol and a competitive profile comparable to wild-type *C. albicans* EBP. Sequence comparison showed that EBP shares a 46% amino acid identity with the old yellow enzyme, an oxidoreductase from *S. cerevisiae*, but, as anticipated, was unrelated to the human estrogen receptor. Expressed protein exhibited oxidoreductase activity and showed inhibition by the treatment of E_2 in vitro (Madani et al. 1994).

C. albicans has also been shown to interact with the human peptide hormones luteinizing hormone (LH) and human chorionic gonadotropin (Bramley et al. 1990, 1991). Both low and high affinity binding sites have been found for these peptides, and the binding activity has been demonstrated in microsomes and cytosol preparations. Furthermore, LH was found to stimulate germ tube formation (Kinsman et al. 1988), and adenylate cyclase activity (Williams et al. 1990). In addition, other investigators have reported a human chorionic gonadotropin-like protein in extracts of *C. albicans* that was a potent stimulator of germ tube formation in the presence of serum (Caticha et al. 1993), which is suggestive that this protein is an endogenous ligand and regulator of germ tube formation.

15.5.3   S. cerevisiae

Similar to *C. albicans, S. cerevisiae* has been demonstrated to have an EBP (Burshell et al. 1984; Feldman et al. 1982), as does *P. brasiliensis* as discussed earlier. The studies in *S. cerevisiae* showed it to have a 60–70 kDa protein that specifically bound E_2 with a K_d 1 nM with a binding capacity of 2,000–4,000 fmol/mg of protein. Interestingly, *S. cerevisiae* was thought to possess an endogenous estrogen ligand, but carefully performed studies showed the presence of estrone in the components of culture media (e.g., molasses and Bacto-Peptone) that could be enzymatically converted to E_2 by the organism (Feldman 1988). The question of whether the EBP in *S. cerevisiae, Candida,* and *P. brasiliensis* are similar in sequence or activity is unknown.

15.5.4   Trichophyton and Microsporum

As noted earlier, the growth of dermatophytic fungi is inhibited by various steroid hormones. In addition, epidemiologic data are suggestive that men exhibit dermatophytic infections more frequently than women. Thus, we undertook a series of studies to investigate possible interactions with mammalian steroids.

Initial steroid binding studies were done with *Trichophyton mentagrophytes*. A specific binding of progesterone was demonstrated in the cytosol of the organism with a K_d 95 nM and binding capacity of close to 5,000 fmol/mg of protein. Both deoxycorticosterone and dihydrotestosterone were strong competitors for progesterone binding, whereas other steroids had minimal binding activity (Schar et al. 1986). Furthermore, progesterone inhibited the growth of the organism, as did deoxycorticosterone and dihydrotestosterone, all in the same rank-order as for binding (Schar et al. 1986). Further studies showed that the binding was due to a protein and that similar to *T. mentagrophytes, Microsporum canis* had a progesterone binding protein similar in affinity and high capacity, whereas the progesterone binding protein in *T. rubrum* was of higher affinity (i.e., K_d 16 nM) and lower capacity (i.e., 196 fmol/mg protein) (Clemons et al. 1988). Growth inhibition was demonstrated similar to the initial study, but this inhibition appeared to be a delay in growth from which the organisms escaped. Because *Trichophyton* had been reported previously to be able to metabolize progesterone (Capek and Simek 1971; Stevens 1989), we examined whether this might be an explanation for the escape from growth inhibition (Clemons et al. 1989b). Those results indicated that progesterone was metabolized to more polar and less growth inhibitory compounds including 15α-hydroxyprogesterone, 1-dehydroprogesterone, 11α-hydroxyprogesterone, and 1-dehydro-hydroxyprogesterone (Clemons et al. 1989b). Thus, metabolism of progesterone is the likely mechanism for escape from the growth inhibition.

15.5.5   *Coccidioides*

The frequency of coccidioidal disease has been linked to gender, with males exhibiting clinical disease more frequently (Drutz and Huppert 1983). However, pregnancy, especially in the third trimester has been considered a risk-factor for the development of serious life-threatening disease. Studies by Powell et al. (Powell and Drutz 1984) showed that *Coccidioides* had a high affinity receptor for E_2 (i.e. K_d 21 nM, N_{max} 1,500 fmol/mg protein) and a low affinity receptor for testosterone (i.e., K_d 190 nM, N_{max} 39,000 fmol/mg protein). Similarly, they found a progestin binding protein in the organism (Powell et al. 1983). As noted above, E_2 at physiological concentrations achieved during pregnancy (i.e., 10^{-8} M) stimulated endosporulation and release of endospores by the organism, which could promote pathogenesis during pregnancy (Powell et al. 1983).

15.5.6   *Rhizopus nigricans*

In the saprophytic fungus *R. nigricans*, steroids such as progesterone are toxic at high concentrations, inhibiting its growth (Jeraj et al. 2005). Steroid binding sites with high affinity for progesterone (K_d 40±14 nM determined by binding, and K_d 71±22 nM determined by displacement studies) and lower affinity for 21-hydroxyprogesterone or testosterone were demonstrated; no affinity was detected for 17β-estradiol, onapristone and K-naphthoflavone in the enriched plasma membrane fraction of this fungus (Lenasi et al. 2002). Further investigation into the functional response to steroid showed that progesterone induced about 30% activation of G-proteins over basal level, as determined by GTPase activity (EC50=32±8 nM) and by the guanosine 5P-O-(3-thiotriphosphate) (GTPγS) binding rate (EC50=61±21 nM), in the membrane fraction. The affinity of receptors for progesterone was substantially decreased in the presence of GTPγS and cholera toxin (Lenasi et al. 2002). Furthermore, the intracellular level of cAMP decreased in the presence of steroids, suggestive of a possible role for cAMP signaling in the response of *R. nigricans* to steroids. Growth analysis, in the presence of cAMP increasing agents, indicates a role for cAMP in fungal growth inhibition by steroids (Jeraj et al. 2005).

15.6 Functional Responses of Fungi to Hormones

As noted in several of the studies already, when a fungus harboring one of these steroid-binding proteins is exposed to cognate hormone, direct physiological responses have been elicited. In addition to the effects on fungal growth, other physiological effects of mammalian hormones on fungal physiology have been described, such as calcium uptake (Berdicevesky and Silbermann 1982) and oxygen consumption (Shah 1981).

15.6.1 C. albicans

Although E_2 blocks the dimorphic conversion of the mycelia form to the invasive yeast form in *P. brasiliensis*, in *C. albicans,* E_2 directly stimulates the dimorphic transition from yeast to the hyphal form (Cheng et al. 2006; White and Larsen 1997), suggesting that different fungi respond to the same hormone in a different way.

For *C. albicans,* genome-wide inventory of up- or down-regulated genes in the presence of E_2 and progesterone has been reported (Banerjee et al. 2007; Cheng et al. 2006). Several genes, including methionine synthase (Burt et al. 1999), *HSP90* (Burt et al. 2003), CDR1 (Cheng et al. 2006; Krishnamurthy et al. 1998; Zhang et al. 2000) and *PDR16* (Cheng et al. 2006) have been shown to be induced following E_2-treatment. Progesterone also enhanced the expression of multidrug resistance (MDR) genes belonging to ATP Binding Cassette (*CDR1* and *CDR2*) super-family of multidrug transporters and several genes associated with hyphal induction (Banerjee et al. 2007). Transcripts of E_2-treated *C. albicans* cells in the Cheng et al. (2006), study showed an increased expression of *Candida* drug resistance genes, *CDR1* and *CDR2*, across several strain-E_2 concentration-time point combinations, suggesting that these genes are the most responsive to estrogen exposure. Expression of *CDR1* and *CDR2*, which encode multidrug transporters of the ABC family, are upregulated in the presence of E_2 at concentrations ranging from 10^{-3} to 10^{-9} M. The drug efflux pump *CDR1* has been implicated in resistance to the azole antifungal drugs commonly used in clinical practice (Albertson et al. 1996; Sheehan et al. 1999). In a detailed study, it has been indicated that the promoter region of the *CDR1* gene contains steroid responsive elements (SRE) and a drug response element (DRE) (de Micheli et al. 2002; Karnani et al. 2004). The transcription factor *TAC1* binds to the DRE in response to E_2 and induces *CDR1* expression (Coste et al. 2004). Another AP-1 recognition element was noted upstream of the multidrug resistance gene *CDR1* in *C. albicans*, which has been shown to mediate *CDR1* induction in response to various types of drugs, including some azoles (Puri et al. 1999). The DRE consensus sequence is 5′ CGGA (A/T) ATCGGATATTTTTTTT 3′. Although this consensus sequence is found uniquely upstream of *CDR1* and *CDR2*, a more redundant form of the sequence, 5′ WCGGWWWWWCGGWWW 3′ (W is A or T), is found in the promoters of *IFU5*, *RTA3*, and *HSP12* (Coste et al. 2004; de Micheli et al. 2002). Further investigation on SRE indicated that it consists of two distinct elements, viz. SRE1 and SRE2. Although SRE1 responds only to progesterone, SRE2 responded to both progesterone and E$_2$. Both SRE1 and SRE2 were specific for steroids, as they did not respond to drugs, such as cycloheximide, miconazole, and terbinafine (Karnani et al. 2004). Analysis of *C. albicans* strain DSY654, which lacks the *CDR1* and *CDR2* coding sequences, showed a significantly decreased number of germ tube-forming cells in the presence of E$_2$. *PDR16* was the most highly upregulated gene in strain DSY654 under these growth conditions (Cheng et al. 2006). It is conceivable that the components of the phospholipid and sterol metabolic pathways may interact to affect *C. albicans* germ tube formation and length. Thus, increased *CDR1* expression is an adaptation response to the environment inside a human host, aimed at enhancing the organism's tolerance for steroids or drugs used to treat fungal infection.

The neuroendocrine hormone, serotonin, has also been shown to modulate several virulence properties of *C. albicans* in in vitro studies (Mayr et al. 2005). The presence of serotonin (5-hydroxytryptamine) reduces hyphal elongation, phospholipase activity, and production of secreted aspartyl protease. Interestingly, a serotonin reuptake inhibitor, sertraline, also had a similar effect on *C. albicans*, reducing the same virulence properties and reduced fungal viability (Lass-Florl et al. 2003).

15.6.2   S. cerevisiae

Banerjee et al. (2004), carried out a genome-wide expression profile of human steroid progesterone response (1 mM progesterone exposure for 30 min) in *S. cerevisiae*. These studies revealed that the most highly upregulated genes included two ABC transporters *PDR5* (11-fold), a close homolog of *CDR1* of *C. albicans*, and *SNQ2* (fivefold), both of which are known to be involved in pleiotropic drug resistance. Three other consistently upregulated genes were *ICT1*, *OYE2*, and *YAL061W*, which according to the *Saccharomyces* genome database (SGD), do not have a known function assigned to them. Among these, *OYE2* and *YAL061W* belong to the category of oxidoreductase/dehydrogenase genes (SGD), the homologs of which have been found to be highly overexpressed in *Candida* populations treated with inhibitory concentrations of antifungal drugs (Cowen et al. 2002). The consistently down-regulated genes *MTD1* (3.5-fold) and *ADE17* (threefold) are both involved in de novo purine biosynthesis.

In *S. cerevisiae*, which lacks endogenous estrogen receptor (ER), it appears that the protein essential to estrogen receptor function is conserved among eukaryotes to such an extent that introduction of human ER into this organism is sufficient for faithful reconstitution of estrogen signaling within these cells (Knoblauch and Garabedian 1999). This transgenic expression in fungi has proven to be very valuable in understanding hormone signaling and receptor function in mammalian systems. In addition to the human-ER, a number of accessory proteins are apparently required to efficiently transduce the steroid hormone signal. In the absence of E_2, the human-ER, like other steroid receptors, is complexed with Hsp90 and other molecular chaperone components, including an immunophilin, and p23. This Hsp90-based chaperone complex is thought to repress the transcriptional regulatory activities of human ER, while maintaining the receptor in a conformation that is competent for high-affinity steroid binding. However, a role for p23 in ER signal transduction has not been demonstrated. Using a mutant human ER (G400V) with decreased hormone-binding capacity as a substrate in a dosage suppression screen in yeast cells (*S. cerevisiae*), a yeast homologue of the human p23 protein (yhp23) as a positive regulator of ER function was identified. Overexpression of yhp23 in yeast cells increases ER transcriptional activation by increasing E_2 binding. Importantly, the magnitude of the effect of yhp23 on ER transcriptional activation was inversely proportional to the concentration of both ER and E_2 in the cell. Under conditions of high ER expression, ER transcriptional activity is largely independent of yhp23, whereas at low levels of ER expression, ER transcriptional activation is primarily dependent on yhp23. The same relationship holds for E_2 levels.

15.6.3   Aspergillus spp.

In addition to the enhancement of growth by glucocorticoids, several species of *Aspergillus* respond to serotonin. In vitro, serotonin in concentrations ranging from 14 to 235 mM has antifungal properties for conidia and hyphae (Lass-Florl et al. 2002). Interestingly, serotonin receptor antagonists did not have any effect on growth (Lass-Florl et al. 2002). However, similar to data for *C. albicans*, sertaline inhibited the growth of *Aspergillus,* and the authors suggested that the mechanism may be that of an interaction with a fungal transporter system (Lass-Florl et al. 2001).

15.6.4   Cryptococcus

Little is known about possible interactions of host hormones with species of *Cryptococcus*. *Cryptococcus neoformans* var *grubii* and *Cryptococcus neoformans* var. *neoformans* are primary etiologic agents of fungal meningitis in immunocompromised patients (e.g., AIDS), whereas *Cryptococcus gattii* causes meningitis in immunocompetent individuals (Casadevall and Perfect 1998). Dopamine is a neuroendocrine hormone produced by the *tinea nigra* of the brain. *Cryptococcus* is able to utilize dopamine in the production of melanin, which is considered a virulence factor for this organism (Polacheck et al. 1990) and may also be related to the tropism of these fungi for the central nervous system.

15.7   Effect of Hormonal Interactions on Antifungal Therapy

We have previously discussed the interactions between hormones and drug-resistance genes in *Candida* and *Saccharomyces*. Early studies had shown that ketoconazole was a competitive stereo-specific binder for the CBP in *C. albicans*, and acted as an antagonist against mammalian glucocorticoid receptors (Stover et al. 1983). This binding did not appear related to the antifungal activity of the drug, and there appeared no corticoid effect on the activity. Others have reported inhibitory (Hogl and Raab 1980, 1982) or stimulatory (Ramondenc et al. 1998, 2001) interactions of steroids on azoles. It has been speculated that the presence of an estrogen-like 1,2 diarylethane moiety in azoles may relate to their antifungal activity, and synthesis of azole derivatives emphasizing this moiety produced compounds with antifungal activity (Massa et al. 1992).

The observation of gynecomastia in some of our patients led to our discovery that azole drugs block steroidogenesis in mammalian cells and in man because of the interference with specific P450 steroidogenic enzymes (Stevens 1985). This has led to the utility of these agents in the therapy of hypercortisolemic states and in androgen blockade in prostatic cancer.

15.8   Future Directions

As is evident from the information presented in this chapter, the field of fungal endocrinology encompasses a number of different areas and shows that these simple eukaryotic organisms can indeed have true hormone receptor interactions. Additional studies are needed in the search for endogenous hormones that regulate mating of higher fungi, such as the dermatophytes, or *Coccidioides*. The influence and interactions of mammalian hormones on pathogenic fungi remains an area for study that has thus far given indications that host hormones can positively or negatively influence the pathogenesis of a fungal infection. For example, hormonal influences on *C. albicans* have been implicated in the pathogenesis of candidal vaginitis (Tarry et al. 2005). The question of whether these interactions occur through a fungal receptor-mediated process has begun to be addressed for *C. albicans*, where genomic expression studies have demonstrated upregulation of genes in the organism related to drug resistance, for example. With the advent of genomics tools, such as microarrays and high throughput sequencing technologies, as well as improvements in proteomic technologies, additional studies are now possible to examine whether the hormonal influence on growth of a fungus is receptor-mediated and what cellular pathways are involved. That is to say, how does estrogen stimulate *C. albicans* or *Coccidioides* or how does progesterone inhibit the growth of *Trichophyton*? The alternative explanations for fungal pathogenesis include hormonal effects on host mucosa, as is important in vaginal candidiasis, and possibly oral candidiasis (Junqueira et al. 2005), and hormonal effects on the host immune response (Marriott and Huet-Hudson 2006). There is an extensive literature on hormonal effects on the mucosa and on the immune response, which could profoundly affect the host-fungal pathogen interplay, and we have only cited some key articles here.

Of particular interest to our laboratory over many years has been the dramatic effect that E_2 has on the inhibition of mycelial to yeast transformation of *P. brasiliensis*. We are in the initial stages of examining the mechanism of this block at the genomic level with the use of a random-shear genomic DNA microarray we have developed (Monteiro et al. 2009). Examination of the temporal events during form transformation in the presence of E_2 are currently underway and have indicated alteration of gene expression as early as 4 h after the addition of E_2 to transforming cultures (Shankar et al., unpublished data). Studies such as this will clarify not only the hormonal interactions, but on the basic biology of the organism and the mechanisms by which morphologic change might be regulated. Similarly, does an organism like *P. brasiliensis* make an endogenous hormone that acts as a regulator of morphology?

Lastly are the questions of what other pathogenic fungi show interactions with the host hormonal milieu and what are the responses of the organisms to the hormones? The elucidation of a second mammalian estrogen receptor (Gustafsson 1999), with quite different ligand-binding domains, may necessitate reexamination of the putative estrogen receptor in fungi, particularly with receptor-specific ligands. Thus, we believe that fungal endocrinology is still in its infancy as a field of research and that much remains to be done to gain a full understanding of these interactions.

Acknowledgement We thank Dr. L. Crapo, Chief of the Division of Endocrinology, Department of Medicine, Santa Clara Valley Medical Center for careful review of the manuscript.

References

Albertson, G. D., Niimi, M., Cannon, R. D. and Jenkinson, H. F. 1996. Multiple efflux mechanisms are involved in *Candida albicans* fluconazole resistance. Antimicrob Agents Chemother 40: 2835–2841.

Aristizabal, B. H., Clemons, K. V., Cock, A. M., Restrepo, A. and Stevens, D. A. 2002. Experimental *Paracoccidioides brasiliensis* infection in mice: influence of the hormonal status of the host on tissue responses. Med Mycol 40: 169–178.

Aristizabal, B. H., Clemons, K. V., Stevens, D. A. and Restrepo, A. 1998. Morphological transition of *Paracoccidioides brasiliensis* conidia to yeast cells: in vivo inhibition in females. Infect Immun 66: 5587–5591.

Banerjee, D., Martin, N., Nandi, S., Shukla, S., Dominguez, A., Mukhopadhyay, G. and Prasad, R. 2007. A genome-wide steroid response study of the major human fungal pathogen *Candida albicans*. Mycopathologia 164: 1–17.

Banerjee, D., Pillai, B., Karnani, N., Mukhopadhyay, G. and Prasad, R. 2004. Genome-wide expression profile of steroid response in *Saccharomyces cerevisiae*. Biochem Biophys Res Commun 317: 406–413.

Bardwell, L. 2005. A walk-through of the yeast mating pheromone response pathway. Peptides 26: 339–350.

Beato, M., Chavez, S. and Truss, M. 1996. Transcriptional regulation by steroid hormones. Steroids 61: 240–251.

Bennett, R. J., Uhl, M. A., Miller, M. G. and Johnson, A. D. 2003. Identification and characterization of a *Candida albicans* mating pheromone. Mol Cell Biol 23: 8189–8201.

Berdicevsky, I. and Silbermann, M. 1982. Effect of glucocorticoid hormones on calcium uptake and morphology of *Candida albicans*. Cell Biol Int Rep 6: 783–790.

Bramley, T. A., Menzies, G. S., Williams, R. J., Adams, D. J. and Kinsman, O. S. 1990. Specific, high-affinity binding sites for human luteinizing hormone (hLH) and human chorionic gonadotrophin (hCG) in *Candida* species. Biochem Biophys Res Commun 167: 1050–1056.

Bramley, T. A., Menzies, G. S., Williams, R. J., Kinsman, O. S. and Adams, D. J. 1991. Binding sites for LH in *Candida albicans*: comparison with the mammalian corpus luteum LH receptor. J Endocrinol 130: 177–190.

Brasch, J. and Flader, S. 1996. Human androgenic steroids affect growth of dermatophytes in vitro. Mycoses 39: 387–392.

Brasch, J. and Gottkehaskamp, D. 1992. The effect of selected human steroid hormones upon the growth of dermatophytes with different adaptation to man. Mycopathologia 120: 87-92.

Brummer, E., Castaneda, E. and Restrepo, A. 1993. Paracoccidioidomycosis: an update. Clin Microbiol Rev 6: 89–117.

Brunt, S. A., Borkar, M. and Silver, J. C. 1998. Regulation of hsp90 and hsp70 genes during antheridiol-induced hyphal branching in the oomycete *Achlya ambisexualis*. Fungal Genet Biol 24: 310–324.

Brunt, S. A., Riehl, R. and Silver, J. C. 1990. Steroid hormone regulation of the *Achlya ambisexualis* 85-kilodalton heat shock protein, a component of the *Achlya* steroid receptor complex. Mol Cell Biol 10: 273–281.

Brunt, S. A. and Silver, J. C. 1986a. Cellular localization of steroid hormone-regulated proteins during sexual development in *Achlya*. Exp Cell Res 165: 306–319.

Brunt, S. A. and Silver, J. C. 1986b. Steroid hormone-induced changes in secreted proteins in the filamentous fungus *Achlya*. Exp Cell Res 163: 22–34.

Burshell, A., Stathis, P. A., Do, Y., Miller, S. C. and Feldman, D. 1984. Characterization of an estrogen-binding protein in the yeast *Saccharomyces cerevisiae*. J Biol Chem 259: 3450–3456.

Burt, E. T., Daly, R., Hoganson, D., Tsirulnikov, Y., Essmann, M. and Larsen, B. 2003. Isolation and partial characterization of Hsp90 from *Candida albicans*. Ann Clin Lab Sci 33: 86–93.

Burt, E. T., O'Connor, C. and Larsen, B. 1999. Isolation and identification of a 92-kDa stress induced protein from *Candida albicans*. Mycopathologia 147: 13–20.

Capek, A. and Simek, A. 1971. Antimicrobial agents. IX. Effect of steroids on dermatophytes. Folia Microbiol (Praha) 16: 299–302.

Casadevall, A. and Perfect, J. R. 1998. *Cryptococcus neoformans*. Washington, D.C., ASM Press.

Caticha, O., Li, Y., Griffin, J., Winge, D. and Odell, W. D. 1993. Characterization of a human chorionic gonadotropin-like protein from *Candida albicans*. Endocrinology 132: 667–673.

Cheng, G., Yeater, K. M. and Hoyer, L. L. 2006. Cellular and molecular biology of *Candida albicans* estrogen response. Eukaryot Cell 5: 180–191.

Clemons, K. V., Feldman, D. and Stevens, D. A. 1989a. Influence of oestradiol on protein expression and methionine utilization during morphogenesis of *Paracoccidioides brasiliensis*. J Gen Microbiol 135: 1607–1617.

Clemons, K. V., Schar, G., Stover, E. P., Feldman, D. and Stevens, D. A. 1988. Dermatophyte–hormone relationships: characterization of progesterone-binding specificity and growth inhibition in the genera *Trichophyton* and *Microsporum*. J Clin Microbiol 26: 2110–2115.

Clemons, K. V., Stover, E. P., Schar, G., Stathis, P. A., Chan, K., Tokes, L., Stevens, D. A. and Feldman, D. 1989b. Steroid metabolism as a mechanism of escape from progesterone-mediated growth inhibition in *Trichophyton mentagrophytes*. J Biol Chem 264: 11186–11192.

Coste, A. T., Karababa, M., Ischer, F., Bille, J. and Sanglard, D. 2004. TAC1, transcriptional activator of CDR genes, is a new transcription factor involved in the regulation of *Candida albicans* ABC transporters CDR1 and CDR2. Eukaryot Cell 3: 1639–1652.

Cowen, L. E., Nantel, A., Whiteway, M. S., Thomas, D. Y., Tessier, D. C., Kohn, L. M. and Anderson, J. B. 2002. Population genomics of drug resistance in *Candida albicans*. Proc Natl Acad Sci USA 99: 9284–9289.

de Micheli, M., Bille, J., Schueller, C. and Sanglard, D. 2002. A common drug-responsive element mediates the upregulation of the *Candida albicans* ABC transporters CDR1 and CDR2, two genes involved in antifungal drug resistance. Mol Microbiol 43: 1197–1214.

Drutz, D. J. and Huppert, M. 1983. Coccidioidomycosis: factors affecting the host-parasite interaction. J Infect Dis 147: 372–390.

Feldman, D. 1988. Evidence for the presence of steroid hormone receptors in fungi. *Steroid Hormone Action*. G. Ringold (ed.). p 169–176, Alan R. Liss, Inc., New York.

Feldman, D., Do, Y., Burshell, A., Stathis, P. and Loose, D. S. 1982. An estrogen-binding protein and endogenous ligand in *Saccharomyces cerevisiae*: possible hormone receptor system. Science 218: 297–298.

Fernades, P., Cruz A., Angelova B., Pincheiro H. M. and Cabral, J. M. S. 2003. Microbial conversion of steroid compounds: recent developments. Enzyme Microb Technol 32: 688–705.

Funder, J. W. 1997. Glucocorticoid and mineralocorticoid receptors: biology and clinical relevance. Annu Rev Med 48: 231–240.

Ghannoum, M. A. and Elteen, K. A. 1987. Effect of growth of *Candida* spp. in the presence of various glucocorticoids on the adherence to human buccal epithelial cells. Mycopathologia 98: 171–178.

Gogala, N. 1991. Regulation of mycorrhizal infection by hormonal factors produced by hosts and fungi. Experientia 47: 331–340.

Gooday, G. W. 1974. Fungal sex hormones. Annu Rev Biochem 43: 35–87.

Gooday, G. W. and Adams, D. J. 1993. Sex hormones and fungi. Adv Microb Physiol 34: 69–145.

Gujjar, P. R., Finucane, M. and Larsen, B. 1997. The effect of estradiol on *Candida albicans* growth. Ann Clin Lab Sci 27: 151–156.

Gupta, P. N., Ichhpujani, R. L., Bhatia, R., Arora, D. R. and Chugh, T. D. 1982. Effect of dexamethasone on in vitro growth of *Candida albicans*. J Commun Dis 14: 259–262.

Gustafsson, J. A. 1999. Estrogen receptor beta – a new dimension in estrogen mechanism of action. J Endocrinol 163: 379–383.

Hammes, S. R. 2003. The further redefining of steroid-mediated signaling. Proc Natl Acad Sci USA 100: 2168–2170.

Hernandez-Hernandez, F., De Bievre, C., Camacho-Arroyo, I., Cerbon, M. A., Dupont, B. and Lopez-Martinez, R. 1995. Sex hormone effects on *Phialophora verrucosa* in vitro and characterization of progesterone receptors. J Med Vet Mycol 33: 235–239.

Hogl, F. and Raab, W. 1980. The influence of steroids on the antifungal and antibacterial activities of imidazole derivatives. Mykosen 23: 426–439.

Hogl, F. and Raab, W. 1982. The influence of glucocorticoids on the membrane disrupting activity of antifungal imidazole derivatives. Mykosen 25: 315–320.

Jeraj, N., Lenasi, H. and Breskvar, K. 2005. The involvement of cAMP in the growth inhibition of filamentous fungus *Rhizopus nigricans* by steroids. FEMS Microbiol Lett 242: 147–154.

Joets, J., Pousset, D., Marcireau, C. and Karst, F. 1996. Characterization of the *Saccharomyces cerevisiae* FMS1 gene related to *Candida albicans* corticosteroid-binding protein 1. Curr Genet 30: 115–120.

Junqueira, J. C., Colombo, C. E., Martins Jda, S., Koga Ito, C. Y., Carvalho, Y. R. and Jorge, A. O. 2005. Experimental candidosis and recovery of *Candida albicans* from the oral cavity of ovariectomized rats. Microbiol Immunol 49: 199–207.

Karnani, N., Gaur, N. A., Jha, S., Puri, N., Krishnamurthy, S., Goswami, S. K., Mukhopadhyay, G. and Prasad, R. 2004. SRE1 and SRE2 are two specific steroid-responsive modules of *Candida* drug resistance gene 1 (CDR1) promoter. Yeast 21: 219–239.

Khosla, P. K., Chawla, K. S., Prakash, P. and Mahajan, V. M. 1978. The effect of dexamethason and oxytetracyclin on the growth of *Candida albicans*. Mykosen 21: 342–348.

Kinsman, O. S., Pitblado, K. and Coulson, C. J. 1988. Effect of mammalian steroid hormones and luteinizing hormone on the germination of *Candida albicans* and implications for vaginal candidosis. Mycoses 31: 617–626.

Knoblauch, R. and Garabedian, M. J. 1999. Role for Hsp90-associated cochaperone p23 in estrogen receptor signal transduction. Mol Cell Biol 19: 3748–3759.

Krishnamurthy, S., Gupta, V., Prasad, R. and Panwar, S. L. 1998. Expression of CDR1, a multidrug resistance gene of *Candida albicans*: transcriptional activation by heat shock, drugs and human steroid hormones. FEMS Microbiol Lett 160: 191–197.

Lass-Florl, C., Dierich, M. P., Fuchs, D., Semenitz, E., Jenewein, I. and Ledochowski, M. 2001. Antifungal properties of selective serotonin reuptake inhibitors against *Aspergillus* species in vitro. J Antimicrob Chemother 48: 775–779.

Lass-Florl, C., Ledochowski, M., Fuchs, D., Speth, C., Kacani, L., Dierich, M. P., Fuchs, A. and Wurzner, R. 2003. Interaction of sertraline with *Candida* species selectively attenuates fungal virulence in vitro. FEMS Immunol Med Microbiol 35: 11–15.

Lass-Florl, C., Wiedauer, B., Mayr, A., Kirchmair, M., Jenewein, I., Ledochowski, M. and Dierich, M. P. 2002. Antifungal properties of 5-hydroxytryptamine (serotonin) against *Aspergillus* spp. in vitro. Int J Med Microbiol 291: 655–657.

Lenasi, H., Bavec, A. and Zorko, M. 2002. Membrane-bound progesterone receptors coupled to G proteins in the fungus *Rhizopus nigricans*. FEMS Microbiol Lett 213: 97–101.

Loose, D. S. and Feldman, D. 1982. Characterization of a unique corticosterone-binding protein in *Candida albicans*. J Biol Chem 257: 4925–4930.

Loose, D. S., Schurman, D. J. and Feldman, D. 1981. A corticosteroid binding protein and endogenous ligand in *C. albicans* indicating a possible steroid-receptor system. Nature 293: 477–479.

Loose, D. S., Stevens, D. A., Schurman, D. J. and Feldman, D. 1983a. Distribution of a corticosteroid-binding protein in *Candida* and other fungal genera. J Gen Microbiol 129: 2379–2385.

Loose, D. S., Stover, E. P., Restrepo, A., Stevens, D. A. and Feldman, D. 1983b. Estradiol binds to a receptor-like cytosol binding protein and initiates a biological response in *Paracoccidioides brasiliensis*. Proc Natl Acad Sci USA 80: 7659–7663.

Losel, R. and Wehling, M. 2003. Nongenomic actions of steroid hormones. Nat Rev Mol Cell Biol 4: 46–56.

Loumaye, E., Thorner, J. and Catt, K. J. 1982. Yeast mating pheromone activates mammalian gonadotrophs: evolutionary conservation of a reproductive hormone? Science 218: 1323–1325.

Madani, N. D., Malloy, P. J., Rodriguez-Pombo, P., Krishnan, A. V. and Feldman, D. 1994. *Candida albicans* estrogen-binding protein gene encodes an oxidoreductase that is inhibited by estradiol. Proc Natl Acad Sci USA 91: 922–926.

Mahato, S. B. and Garai, S. 1997. Advances in microbial steroid biotransformation. Steroids 62: 332–345.

Malloy, P. J., Zhao, X., Madani, N. D. and Feldman, D. 1993. Cloning and expression of the gene from *Candida albicans* that encodes a high-affinity corticosteroid-binding protein. Proc Natl Acad Sci USA 90: 1902–1906.

Marriott, I. and Huet-Hudson, Y. M. 2006. Sexual dimorphism in innate immune responses to infectious organisms. Immunol Res 34: 177–192.

Massa, S., Di Santo, R., Retico, A., Artico, M., Simonetti, N., Fabrizi, G. and Lamba, D. 1992. Antifungal agents I. Synthesis and antifungal activities of estrogen-like imidazole and triazole derivatives. Eur J Med Chem 27: 495–502.

Massart, F., Meucci, V., Saggese, G. and Soldani, G. 2008. High growth rate of girls with precocious puberty exposed to estrogenic mycotoxins. J Pediatr 152: 690–695.

Mayr, A., Hinterberger, G., Dierich, M. P. and Lass-Florl, C. 2005. Interaction of serotonin with *Candida albicans* selectively attenuates fungal virulence in vitro. Int J Antimicrob Agents 26: 335–337.

Miyakawa, T., Kadota, T., Okubo, Y., Hatano, T., Tsuchiya, E. and Fukui, S. 1984. Mating pheromone-induced alteration of cell surface proteins in the heterobasidiomycetous yeast *Tremella mesenterica*. J Bacteriol 158: 814–819.

Miyakawa, T., Kaji, M., Yasutake, T., Jeong, Y. K., Tsuchiya, E. and Fukui, S. 1985. Involvement of protein sulfhydryls in the trigger reaction of rhodotorucine A, a farnesyl peptide mating pheromone of *Rhodosporidium toruloides*. J Bacteriol 162: 294–299.

Monteiro, J. P., Clemons, K. V., Mirels, L. F., Coller, J. A., Wu, T. D., Shankar, J., Lopes, C. R. and Stevens, D. A. 2009. Genomic DNA microarray comparison of gene expression patterns in *Paracoccidioides brasiliensis* mycelia and yeasts in vitro. Microbiology 155: 2795–2808.

Ng, T. T., Robson, G. D. and Denning, D. W. 1994. Hydrocortisone-enhanced growth of *Aspergillus* spp.: implications for pathogenesis. Microbiology 140: 2475–2479.

Polacheck, I., Platt, Y. and Aronovitch, J. 1990. Catecholamines and virulence of *Cryptococcus neoformans*. Infect Immun 58: 2919–2922.

Powell, B. and Drutz, D. 1984. Identification of a high-affinity binder for estradiol and a low-affinity binder for testosterone in *Coccidioides immitis*. Infect Immun 45: 784–786.

Powell, B. L. and Drutz, D. J. 1983. Confirmation of corticosterone and progesterone binding activity in *Candida albicans*. J Infect Dis 147: 359.

Powell, B. L., Drutz, D. J., Huppert, M. and Sun, S. H. 1983. Relationship of progesterone- and estradiol-binding proteins in *Coccidioides immitis* to coccidioidal dissemination in pregnancy. Infect Immun 40: 478–485.

Powell, B. L., Frey, C. L. and Drutz, D. J. 1984. Identification of a 17β-estradiol binding portein in *Candida albicans* and *Candida* (*Torulopsis*) *glabrata*. Exp Mycol 8: 304–313.

Prusty, R., Grisafi, P. and Fink, G. R. 2004. The plant hormone indoleacetic acid induces invasive growth in *Saccharomyces cerevisiae*. Proc Natl Acad Sci USA 101: 4153–4157.

Puri, N., Krishnamurthy, S., Habib, S., Hasnain, S. E., Goswami, S. K. and Prasad, R. 1999. CDR1, a multidrug resistance gene from *Candida albicans*, contains multiple regulatory domains in its promoter and the distal AP-1 element mediates its induction by miconazole. FEMS Microbiol Lett 180: 213–219.

Ramondenc, I., Pinel, C., Ambroise-Thomas, P. and Grillot, R. 1998. Does hydrocortisone modify the in vitro susceptibility of *Aspergillus fumigatus* to itraconazole and amphotericin B? Med Mycol 36: 69–73.

Ramondenc, I., Pinel, C., Parat, S., Ambroise-Thomas, P. and Grillot, R. 2001. Hydrocortisone, prednisolone and dexamethasone act on *Aspergillus fumigatus* in vitro susceptibility to itraconazole. Microbios 104: 17–26.

Restrepo, A., Salazar, M. E., Cano, L. E., Stover, E. P., Feldman, D. and Stevens, D. A. 1984. Estrogens inhibit mycelium-to-yeast transformation in the fungus *Paracoccidioides brasiliensis*: implications for resistance of females to paracoccidioidomycosis. Infect Immun 46: 346–353.

Riehl, R. M. and Toft, D. O. 1984. Analysis of the steroid receptor of *Achlya ambisexualis*. J Biol Chem 259: 15324–15330.

Riehl, R. M. and Toft, D. O. 1985. Effect of culture medium composition on pheromone receptor levels in *Achlya ambisexualis*. J Steroid Biochem 23: 483–489.

Riehl, R. M., Toft, D. O., Meyer, M. D., Carlson, G. L. and McMorris, T. C. 1984. Detection of a pheromone-binding protein in the aquatic fungus *Achlya ambisexualis*. Exp Cell Res 153: 544–549.

Salazar, M. E., Restrepo, A. and Stevens, D. A. 1988. Inhibition by estrogens of conidium-to-yeast conversion in the fungus *Paracoccidioides brasiliensis*. Infect Immun 56: 711–713.

Schar, G., Stover, E. P., Clemons, K. V., Feldman, D. and Stevens, D. A. 1986. Progesterone binding and inhibition of growth in *Trichophyton mentagrophytes*. Infect Immun 52: 763–767.

Shah, K. 1981. Steroid effects O_2 consumption and susceptibility to an azole. Indian J Med Res 73: 965–969.

Sheehan, D. J., Hitchcock, C. A. and Sibley, C. M. 1999. Current and emerging azole antifungal agents. Clin Microbiol Rev 12: 40–79.

Silver, J. C., Brunt, S. A., Kyriakopoulou, G., Borkar, M. and Nazarian-Armavil, V. 1993. Regulation of two different hsp70 transcript populations in steroid hormone-induced fungal development. Dev Genet 14: 6–14.

Simoncini, T. and Genazzani, A. R. 2003. Non-genomic actions of sex steroid hormones. Eur J Endocrinol 148: 281–292.

Skowronski, R. and Feldman, D. 1989. Characterization of an estrogen-binding protein in the yeast *Candida albicans*. Endocrinol 124: 1965–1972.

Stevens, D. A. 1985. Ketoconazole metamorphosis. An antimicrobial becomes an endocrine drug. Arch Intern Med 145: 813–815.

Stevens, D. A. 1989. The interface of mycology and endocrinology. J Med Vet Mycol 27: 133–140.

Stover, E. P., Loose, D. S., Stevens, D. A. and Feldman, D. 1983. Ketoconazole binds to the intra-cellular corticosteroid-binding protein in *Candida albicans*. Biochem Biophys Res Commun 117: 43–50.

Stover, E. P., Schar, G., Clemons, K. V., Stevens, D. A. and Feldman, D. 1986. Estradiol-binding proteins from mycelial and yeast-form cultures of *Paracoccidioides brasiliensis*. Infect Immun 51: 199–203.

Tarry, W., Fisher, M., Shen, S. and Mawhinney, M. 2005. *Candida albicans*: the estrogen target for vaginal colonization. J Surg Res 129: 278–282.

Walters, M. R. and Nemere, I. 2004. Receptors for steroid hormones: membrane-associated and nuclear forms. Cell Mol Life Sci 61: 2309–2321.

White, S. and Larsen, B. 1997. *Candida albicans* morphogenesis is influenced by estrogen. Cell Mol Life Sci 53: 744–749.

Whitfield, G. K., Jurutka, P. W., Haussler, C. A. and Haussler, M. R. 1999. Steroid hormone receptors: evolution, ligands, and molecular basis of biologic function. J Cell Biochem Suppl 32–33: 110–122.

Williams, R. J., Dickinson, K., Kinsman, O. S., Bramley, T. A., Menzies, G. S. and Adams, D. J. 1990. Receptor-mediated elevation of adenylate cyclase by luteinizing hormone in *Candida albicans*. J Gen Microbiol 136: 2143–2148.

Zhang, X., Essmann, M., Burt, E. T. and Larsen, B. 2000. Estrogen effects on *Candida albicans*: a potential virulence-regulating mechanism. J Infect Dis 181: 1441–1446.

Chapter 16
Experimental Design Considerations for In Vitro Microbial Endocrinology Investigations

Richard D. Haigh

16.1 Introduction

Microbial endocrinology when first conceived was defined simply as the ability of microorganisms to recognise and exploit the host's hormonal output for their own growth and pathogenesis (Lyte 1993). It was proposed that microorganisms would use the presence of hormones as an indicator that they were in the host, and they would then initiate the expression of their virulence determinants, in a manner somewhat similar to their known responses to host factors such as temperature, iron restriction or pH. In the past two decades, there have been many examples of microbial endocrinology documented, for example: transcriptional responses to the steroid hormones estrogen and progesterone in *Candida albicans* (Zhang et al. 2000; Banerjee et al. 2007); increase in virulence in *Pseudomonas aeruginosa* due to a quorum-signalling response to the human kappa-opioid peptide dynorphin (Zaborina et al. 2007); and, by far the largest grouping, transcriptional and/or growth responses of more than 50 gram-positive and gram-negative bacterial species to the catecholamine neuroendocrine hormones norepinephrine (NE), epinephrine (Epi) and dopamine (Dop) (Freestone et al. 2008). In this chapter, we consider only those aspects of experimental design which pertain to bacterial responses to these catecholamine neurohormones.

Though there was much circumstantial evidence in the literature for the role of neuroendocrine catecholamines in promoting bacterial growth and virulence (see Lyte 2004 and Chap. 1 for a comprehensive review), it was not until Lyte and Ernst's (1992) in vitro work, using NE to promote the growth of *Escherichia coli*, it was appreciated that the hormones might interact directly with the bacteria rather than merely having a role in immunosuppression. Initial hypotheses of bacterial "adrenergic" receptors were not supported by experiments using adrenergic antagonists

R.D. Haigh (✉)
Department of Genetics, University of Leicester,
University Road, Leicester, Leicestershire, LE1 7RH, UK
e-mail: rxh@le.ac.uk

M. Lyte and P.P.E Freestone (eds.), *Microbial Endocrinology*,
Interkingdom Signaling in Infectious Disease and Health,
DOI 10.1007/978-1-4419-5576-0_16, © Springer Science+Business Media, LLC 2010

(Lyte and Ernst 1993), but equally the action of NE did not appear to be obviously nutritional. Subsequent work has demonstrated that NE promotes bacterial growth in serum-containing media by interacting with host iron binding proteins, such as transferrin and lactoferrin, thereby decreasing their affinity for iron, which can then be obtained more readily by any of the multiple bacterial species (Freestone et al. 1999, 2000). The mechanism by which the bacteria obtain the iron appears to be based very much upon their cognate iron uptake systems, for example, *E. coli* and *Salmonella* spp. have both been shown to require the genes of their *ent* operons, which are required for the synthesis of catechol siderophores enterobactin and salmochelin (Burton et al. 2002; Freestone et al. 2003; Bearson et al. 2008; Methner et al. 2008). However, bacterial species which are known not to have endogenous siderophores also show growth responses to neurohormone catecholamines, for example, *Listeria monocytogenes* (Coulanges et al. 1998), *Staphylococcus epidermidis* (Neal et al. 2001) and *Campylobacter jejuni* (Cogan et al. 2007); it has been proposed that in these bacteria the iron might be obtained by the action of a ferric reductase activity and subsequent ferrous iron uptake, or even simply by direct uptake of a catecholamine-iron complex.

Early microbial endocrinology studies identified NE-related induction of bacterial virulence determinants, such as, the *Pseudomonas aeruginosa* PA-1 host binding protein (Alverdy et al. 2000), the K99 pilus adhesin of enterotoxigenic *E. coli* (Lyte et al. 1997a, b) and Shiga-like toxin in *E. coli* O157:H7 (Lyte et al. 1996). However, in these experiments, it was always unclear if the induction of virulence determinants was due to a specific interaction of NE with the bacteria or if it was part of the general process of growth induction caused by alleviation of the serum-induced iron restriction. More recent work has however identified a family of two component regulators (QseBC and QseEF), present in *E .coli* and some other gram negative bacterial species, whose cognate receptors appear to specifically recognise NE and Epi and whose response regulators modulate virulence determinant expression (Sperandio et al. 2002; Clarke et al. 2006; Reading et al. 2007, see also Chap. 12). It has also been shown that the bacterial ligand for the Qse receptors is a quorum sensing autoinducer, AI-3, which in enterohemorrhagic *E. coli* (EHEC) is responsible for the activation of transcription of motility genes and virulence determinant genes within the LEE (*L*ocus of *e*nterocyte *e*ffacement) pathogenicity island (Sperandio et al. 2003). It has been proposed that a combination of signals from AI-3, produced by EHEC and other gut flora, and host catecholamine hormones released at the mucosal surface are responsible, via signalling events at the Qse receptors, for the spatial and temporal coordination of EHEC infection (Clarke and Sperandio 2005; Kaper and Sperandio 2005; Walters and Sperandio 2006). Norepinephrine has also been observed to regulate the expression of the outer surface protein OspA of *Borrelia burgdoferi* (Scheckelhoff et al. 2007) and to alter transcription of the catechol siderophore receptor BfeA of *Bordetella brochiseptica* (Anderson and Armstrong 2006). Furthermore, there is now evidence from Freestone et al. (2007a) that adrenergic and dopaminergic antagonists can specifically block the induction of gram negative bacterial growth by norepinephrine, epinephrine or dopamine but that this blockade is not at the level of iron uptake;

these data strongly suggest that the process of catecholamine induction of bacterial growth may also involve neurohormone signalling events.

The aim of this chapter is to discuss the methodological variables to consider when designing in vitro experiments that aim either to characterise bacterial responses to neurohormone catecholamines or to identify the genes and molecules that are responsible for such responses. As the data which are produced by such studies will inform future in vivo work, it is of crucial importance to consider whether the in vitro conditions used are relevant to the disease models which will ultimately be employed. In the next three sections, I consider the most important of the experimental variables which exist, and then in the remaining sections I look at examples of the commonalities and differences in the results obtained by research groups using different experimental approaches.

16.2 Choice of Growth Medium

It has generally been observed that propagation of bacteria in traditional rich micro-biological media, which typically allow good growth, frequently does not promote the strong expression of virulence determinants. Indeed, many bacteria will only express their adhesins, invasins or toxins when grown under the same temperature, pH, nutritional or atmospheric conditions that they would face on entry into the host. Consequently, when undertaking microbial endocrinology experiments, where the effects of neuroendocrine stress hormones on the growth and virulence of bacteria are being investigated, it would seem obvious that it is essential to try to mimic, as far as possible, the stressful environments in which the bacteria actually encounters those hormones in the host. Indeed, a failure to provide such an in vivo like environment may likely result in erroneous conclusions as to the influence those neurohormones may have on bacterial growth and/or virulence. A possible example of this is the work of Straub et al. (2006) who used Luria broth as a test medium for analysing the effect of norepinephrine on bacteria, and thereby its potential role in a model of bacteremia. Based upon the lack of any obvious effect of the catecholamine in this media, they concluded that norepinephrine was not directly interacting with the bacterial species investigated; however, considering the ever increasing number of reports of catecholamine–bacterial interactions (Freestone et al. 2008) it seems very premature to rule out such a role without further studies using culture media, which might more accurately reflect the in vivo conditions.

For their initial microbial endocrinology work Lyte and Ernst (1992, 1993) developed serum-SAPI, a minimal salts medium supplemented with 30% adult bovine serum that would provide host-like conditions for their experiments. Due to its limited nutrient availability, iron restriction and the presence of host defence proteins, such as antibodies and complement, serum-SAPI presents a highly stressful bacteriostatic environment. Indeed most bacteria grow very poorly in serum-SAPI without supplementation (Lyte and Ernst 1992, 1993; Lyte 2004). Many subsequent analyses of neuroendocrine catecholamine hormone interactions with bacteria have used serum-SAPI as their medium of choice; this primarily involves those studies concerning growth induction (Lyte and Ernst 1992, 1993; Freestone et al. 1999, 2000, 2002, 2003, 2007a, b; Kinney et al. 1999, 2000; Neal et al. 2001; Belay et al. 2003; O'Donnell et al. 2006; Nakano et al. 2007b) but also includes some studies on virulence factor expression (Lyte et al. 1996, 1997a, b; Nakano et al. 2007a) or global transcriptional changes (Dowd 2007; Bearson et al. 2008). The growth inhibition inherent to serum-SAPI is primarily due to the iron restriction imposed by the serum iron-binding protein transferrin as evidenced by the fact that addition of excess of ferric iron will alleviate the growth repression; however, addition of catecholamine hormones can also promote growth by allowing the bacteria access to the iron bound up in transferrin (Freestone et al. 2000, 2003; Anderson and Armstrong 2008). Indeed, this stimulation in growth with catecholamines can be very large with up to 5 log orders difference in the numbers of bacteria with some species when compared with unsupplemented control cultures (Lyte and Ernst 1992, 1993, Lyte et al. 1996; Freestone et al. 1999, 2002; Neal et al. 2001). Some researchers have used modifications of serum-SAPI for work with fastidious oral pathogens (Roberts et al. 2002) and other research groups, working with bacteria with complex nutritional requirements, have used media composed of rich microbiological medias (e.g. Tryptic Soy Broth, Mueller Hinton Broth, Stainer-Scholte medium), which were then treated with chelators to remove iron, prior to supplementation with serum to render them bacteriostatic (Coulanges et al. 1997; Anderson and Armstrong 2006; Cogan et al. 2007).

In contrast with those researchers investigating catecholamine induced growth, many studies looking at catecholamine modulation of virulence determinant expression, particularly those using enteropathogenic and enterhemorrhagic *E. coli* (Sperandio et al. 2003; Walters and Sperandio 2006; Kendall et al. 2007), have used non-restrictive growth media which do not contain serum. In these studies, the researchers have typically grown their bacteria in cell culture media, such as Dulbecco's Modified Eagle medium (DMEM), which have been previously demonstrated to induce high-level expression of *E. coli* virulence determinants. Whilst the use of such growth media may make analysis of changes in virulence determinant expression much easier, it must be kept in mind that these media are not representative of the conditions which the bacteria will encounter within the host.

16.3 Importance of Bacterial Inoculum Size

Integral to the design of the original microbial endocrinology experiments was a commitment to use a bacterial inoculum that accurately reflected the infectious dose encountered in vivo. Therefore, the starting inocula used were low, typically these were around 10^1–10^2 CFU/ml, which was intended to reflect the numbers of pathogenic bacteria likely to be present at the start of an infection (Lyte and Ernst 1992, 1993; Freestone et al. 1999, 2002, 2003, 2007a, b, c). This use of low bacterial numbers is in direct contrast to the majority of in vitro bacterial virulence studies which generally utilise many log orders higher inocula in their experiments. Indeed, several microbial endocrinology-related studies, mainly for technical reasons, have examined responses to catecholamines using high cell density cultures (around 10^8 CFU/ml); the authors have then used the data obtained in these studies to make inferences to in vivo scenarios, where the infecting bacteria are likely to be in much lower numbers (Sperandio et al. 2003; Vlisidou et al. 2004; Walters and Sperandio 2006), but are such extrapolations really valid? It has been observed that at low cell densities (10^2 CFU/ml), *Yersinia enterocolitica* shows no growth response to epinephrine, and furthermore epinephrine can actually antagonize *Yersinia* responses to other catecholamines (Fig. 16.1, Freestone et al. 2007a); however, the same *Y. enterocolitica* when analysed at high cell density (10^8 CFU/ml) are able to use a similar concentration of epinephrine to specifically acquire iron from transferrin (Freestone et al. 2007a). As it has already been determined that bacterial growth induction in low density cultures is mainly due to catecholamine-assisted iron uptake (Freestone et al. 2000), this failure of the *Y. enterocolitica* low density cultures to respond to epinephrine is very difficult to explain unless we can hypothesise that bacteria may in fact use the same catecholamines in different ways in a manner which is dependent upon their population density.

Additional experiments have demonstrated that the population density of a bacterial culture can greatly influence the apparent specificity of catecholamine responsiveness. Freestone et al. (2007a) used serum-SAPI medium to examine the growth responses of *E. coli* O157:H7, *S. enterica* and *Y. enterocolitica* to various catecholamines over an 8-log dilution curve (Fig. 16.2). It can be seen that the effect of catecholamines on bacterial growth in a serum-based medium becomes evident at low ($<10^4$ CFU/ml) cell densities with the greatest differences observed at very low ($<10^2$ CFU/ml) cell densities. Although we see in Fig. 16.2 that *E. coli* O157:H7 and *S. enterica* are both able to respond to epinephrine at very low population densities, which reflect the bacterial numbers that are likely to be present at the initial stages of an infection, it is apparent that there is an order of catecholamine preference evident, with growth responses to norepinephrine and dopamine being at least a log-fold greater than those of epinephrine. It seems obvious based upon these observations and the *Y. enterocolitica* epinephrine response data (Freestone et al. 2007a) that caution should be exercised before assuming that conclusions reached from in vitro observations of microbe–host catecholamine responses in high density cultures can be directly extrapolated to the low population numbers typically present during the initial stage of an infection (Tarr and Neill 2001).

It must also be appreciated that when using host-like media such as serum-SAPI, the use of large numbers of bacteria in the inoculum can lead to problems with evaluating a response to a specific neuroendocrine hormone due to the cell mass overwhelming the media's bacteriostatic potential. For example, Belay et al. 2003 reported differential effects of catecholamines on the growth of several gram negative bacterial species which appeared to be in contrast to those of previous publications (Lyte and Ernst 1992, 1993). In these experiments, the initial bacterial inoculum was so large that no lag phase occurred in the serum-based medium and maximal growth was achieved by both control and catecholamine-supplemented

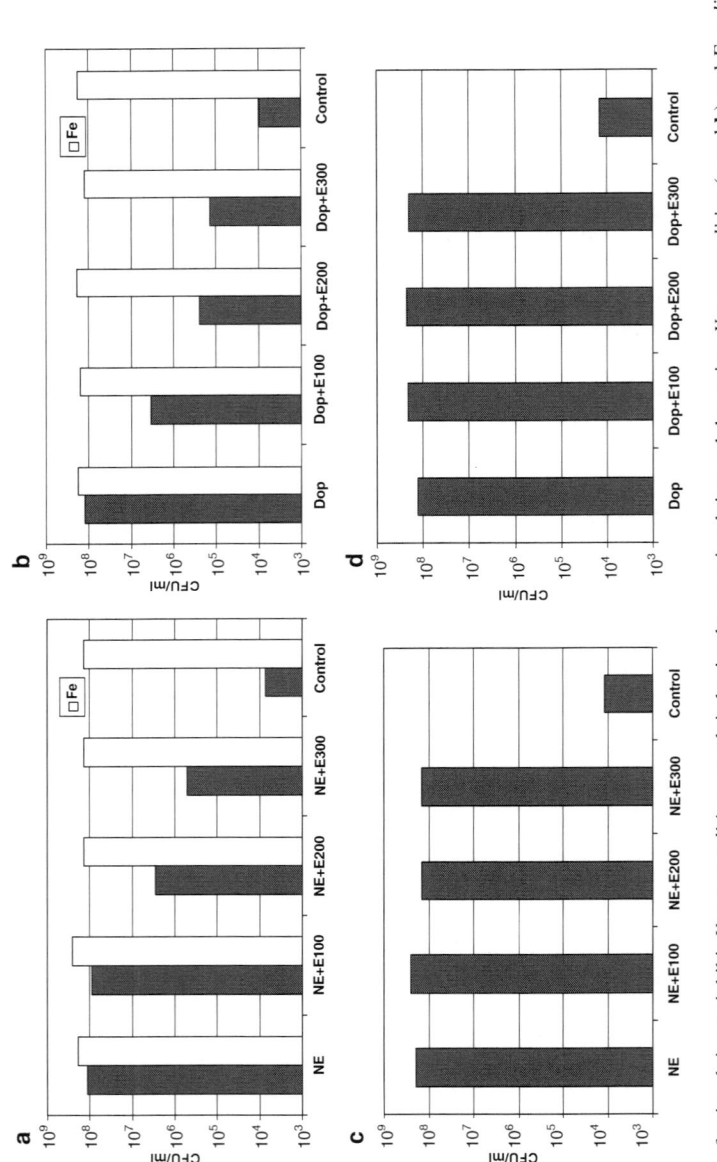

Fig. 16.1 The ability of epinephrine to inhibit *Y. enterocolitica* growth induction by norepinephrine and dopamine. *Y. enterocolitica* (**a** and **b**) and *E. coli* O157:H7 (**c** and **d**) were inoculated at approximately 10^2 CFU/ml into duplicate 1 ml aliquots of serum-SAPI containing the combination of catecholamines shown, and incubated for either 40 h (*Y. enterocolitica*) or 18 h (*E. coli* O157:H7), and enumerated for growth (CFU/ml) as per standard technique (Freestone et al. 2000). The results shown are representative data from four separate experiments; data points typically showed variation of no more than 3%. *Black bar* catecholamine/catecholamine combination only, *light grey bar* catecholamine combinations plus 100 μM Fe(NO₃)₃; Key: *NE* norepinephrine, *Dop* dopamine, and *Epi* epinephrine, (*black bar*). *NE* 50 μM NE, *NE + E100* 50 μM NE plus 100 μM Epi, *NE + E200* 50 μM NE plus 200 μM Epi, *NE + E300* 50 μM NE plus 300 μM Epi, *Dop* 50 μM Dop, *Dop + E100* 50 μM Dop plus 100 μM Epi, *D + E200* 50 μM Dop plus 200 μM Epi, *Dop + E300* 50 μM Dop plus 300 μM Epi. The results shown are representative data from three separate experiments; data points typically showed variation of no more than 3%. This figure was taken with permission from Freestone et al. (2007a)

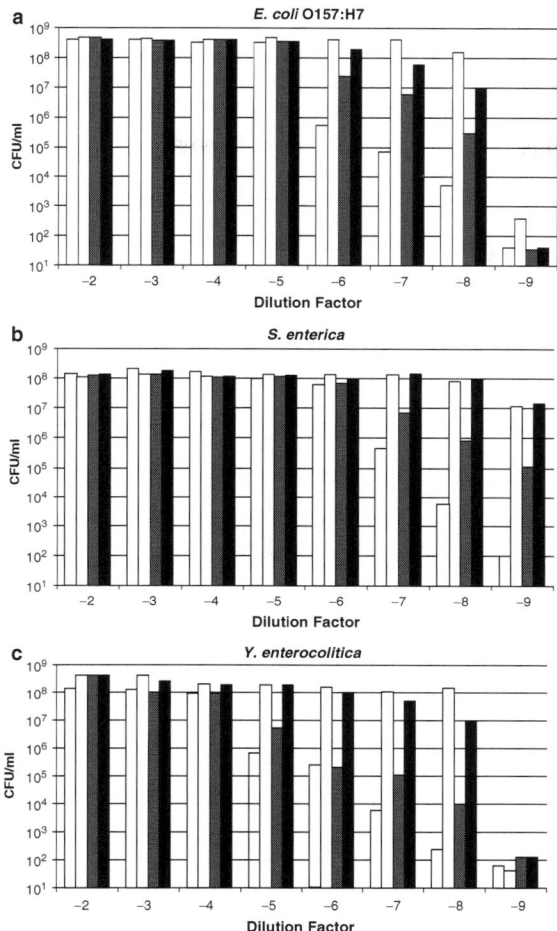

Fig. 16.2 Bacterial population density influences catecholamine specificity. Histograms (**a**)–(**c**) shows the growth response of varying inoculum size on the specificity of growth response to catecholamines norepinephrine (NE), epinephrine (Epi) and dopamine (Dop). Cultures were diluted in tenfold steps into serum-SAPI medium, incubated for either 18 h (*E. coli* O157:H7 and *S. enterica*) or 40 h (*Y. enterocolitica*), and enumerated for growth (CFU/ml) as described in Freestone et al. (2007a). The inoculum size of the *E. coli* culture was 5×10^8 CFU/ml. The results shown are representative data from four separate experiments; individual data points showed variation of no more than 5%. *White bar* no additions (*control*), *light grey bar* 50 μM NE, *black bar* 100 μM Epi, *diagonal hatch* 50 μM Dop. This figure was taken with permission from Freestone et al. (2007a)

cultures in less than 12 h (Belay et al. 2003). However, when the authors later conducted similar experiments using much lower inocula (O'Donnell et al. 2006), they observed a lag phase that is more characteristic of this bacteriostatic medium

(Lyte and Ernst 1992, 1993; Freestone et al. 1999, 2002, 2003, 2007a, b, c); they also reported an increase in bacterial response to norepinephrine which had not been observed in the previous work. It is also paramount that, given the inherent variability in bacterial culture viability, the optical density measurements which are typically used to estimate CFU/ml for culture initiation should always be confirmed by performing plate counts to determine the numbers of bacteria that were actually in the inocula used.

16.4 Choice of Neuroendocrine Hormone and Appropriate Concentration

As would seem evident, the correct concentration of catecholamine neurohormone to use in an in vitro experiment should be the level that the bacteria are likely to encounter within their host. This has two provisos; firstly, the assumption that we can accurately determine the hormone concentrations at sites within the body, and secondly the assumption that bacterial pathogens will occupy just a single site during the course of an infection, and therefore will not be exposed to varying hormone concentrations. It is therefore important to realise that the concentrations of neuroendocrine hormones seen clinically are derived from fluid-based samples, such as plasma or urine, which are easy to obtain from a subject. However, the majority of the body's neurohormones are actually found in the tissue target where they act, and therefore it must be understood that the values obtained from circulatory specimens may give gross underestimates of (often several log orders lower than) the true levels in tissues or at mucosal surfaces where bacteria will interact with them (Leinhardt et al. 1993). Of course, this element of uncertainty will influence experimental design, most notably the choice of concentration(s) at which the specific neuroendocrine hormone under study should be tested (for further discussion, see Lyte 2004). Where the concentration of the neurohormone at the site of action in the body is unknown, or where it is predicted to be very variable, it would be prudent to perform dose responses analyses over a reasonably wide range of neurohormone concentrations. A typical dose response profile of catecholamine growth responsiveness for *E. coli* O157:H7 and *Y. enterocolitica* inoculated into serum-SAPI medium is shown in Fig. 16.3. As previously observed, epinephrine had little effect on growth of *Y. enterocolitica*, and on a concentration-dependent basis was seen to be less potent at stimulating growth of *E. coli* O157:H7 than either norepinephrine or dopamine (see also Fig. 16.3). Previous catecholamine dose response analyses for bacterial species such as *E. coli*, *Salmonella*, *Yersinia*, and the coagulase-negative staphylococci have suggested that initial test concentrations of 50 μM norepinephrine or dopamine, and 100 μM epinephrine (Fig. 16.3) would be suitable. Furthermore, we see that at very high concentrations of catecholamine (>200 μM) the effects upon growth can begin to become inhibitory (Fig. 16.3).

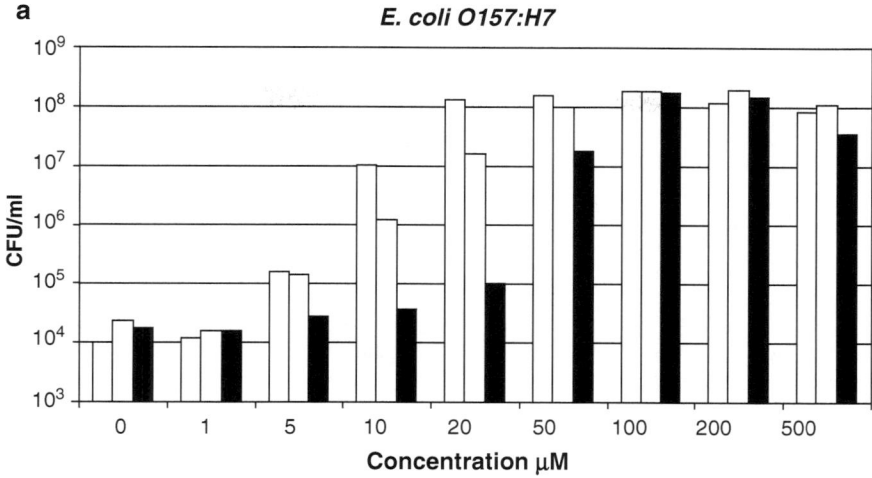

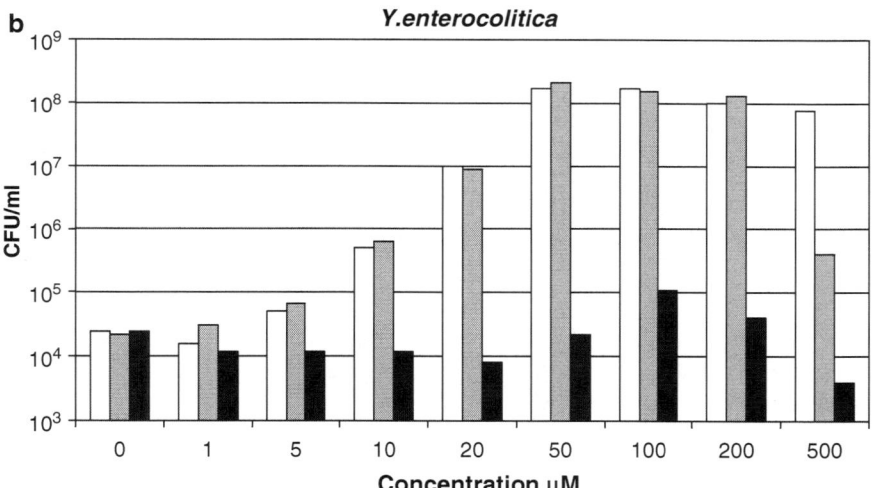

Fig. 16.3 Dose-response effects of catecholamines on *E. coli* O157:H7 and *Y. enterocolitica*. *E. coli* O157:H7 and *Y. enterocolitica* were inoculated at approximately 10^2 CFU/ml into duplicate 1 ml aliquots of serum-SAPI containing the concentrations of the catecholamines shown and incubated for either 40 h (*Y. enterocolitica*) or 18 h (*E. coli* O157:H7), and enumerated for growth (CFU/ml) as described in Freestone et al. (2007a). The results shown for norepinephrine (*NE*), (*grey bar*), dopamine (*Dop*), (*white bar*) and epinephrine (*Epi*), (*black bar*), are representative data from two separate experiments; data points showed variation of less than 3%. This figure was taken with permission from Freestone et al. (2007a)

Catecholamines induce bacterial growth by providing access to host sequestered iron which in iron limited media, such as serum or blood, can then lead to clinically important effects (Coulanges et al. 1998; Freestone et al. 2000, 2002, 2003, 2007a, b, c; Neal et al. 2001, Lyte et al. 2003, Cogan et al. 2007). However, catecholamines can also chelate iron and therefore given the role that iron, and in particular iron restriction, plays in the regulation of virulence factor expression, it is important to consider what the physiological effects of high concentrations of neurohormone would be in non-host-like media where iron is not limited. This is highlighted by the fact that some investigations of the effects of catecholamines on bacteria virulence have used catecholamine concentrations in the mM range. For example, Vlisidou et al. (2004) used 5 mM norepinephrine in Luria broth to investigate the effects of catecholamines on *E. coli* O157:H7 adherence to gut tissues. Though Luria broth is a relatively iron rich culture medium the addition of a 5 mM concentration of an iron chelator, such as norepinephrine, will undoubtedly result in the medium becoming more Fe limited. The possibility that iron limitation may have affected the observed results could be investigated by incubating cultures in iron-depleted Luria broth (made, for example, by Chelex pretreatment) or by the direct addition to test cultures of a ferric iron chelator such as dipyridyl or desferal. The chelation of iron by catecholamines in media containing free iron can also lead to the generation of oxygen-derived free radicals, a cell damaging process that has been implicated in the development of cancers or neurodegenerative diseases such as Parkinson's (Borisenko et al. 2000); therefore, particularly when using high catecholamine concentrations, it is important to consider the potential influence of oxidative stress on bacterial gene expression levels.

16.5 Catecholamine-Induced Growth of Bacteria for Use in Animal Models

Due to the bacteriostatic nature of serum or plasma-SAPI, and the poor growth of control cultures, creation of equivalent conditions between test and control cultures can prove extremely difficult. This challenge is well exemplified in a recent study by Toscano et al. (2007) who showed that by in vitro pre-treatment of *Salmonella typhimurium* with norepinephrine prior to infection of the pathogen into young pigs they could markedly increase its virulence. Examination of the tissue distribution of *Salmonella* 24 h post-challenge revealed that the norepinephrine treated bacteria were present in greater numbers and more widely distributed in the gut tissues than the control bacteria. Toscano et al. (2007) is in itself an interesting study that has implications for food safety, given that widespread nature of catecholamines and related compounds in human foods (Freestone et al. 2007c); however, there are also methodological issues within the investigation that highlight key experimental design challenges for the study of microbial endocrinology. The test medium for pre-incubation with stress neuroendocrine hormone for the Toscano et al. study was serum-SAPI supplemented with 2 mM norepinephrine, but due to poor bacterial growth in non-catecholamine supplemented serum-SAPI, the control cultures were grown in Luria broth, a rich microbiological media. Despite the significant differences in the behaviour of test and control bacteria in the animal model, it is still very difficult to definitively conclude that these effects are due to treatment with the stress neuroendocrine hormone. Because of the extreme difference between the test and control pre-incubation media composition, it is possible that the enhanced virulence seen is partially due to the host factors in the serum component of the test culture media. A further control using serum-SAPI not supplemented with catecholamine but inoculated with high numbers of bacteria, to circumvent the media's bacteriostatic nature, could have addressed this methodological concern. It would also have been informative had the study investigated the effects of addition of 2 mM norepinephrine to Luria broth cultures on dissemination of *Salmonella*; this is especially pertinent since Vlisidou et al. (2004) have shown that 5 mM norepinephrine in Luria broth markedly enhanced both the attachment and enterotoxicity of *E. coli* O157:H7 in a calf ileal loop model of virulence.

16.6 Analysis of Global Bacterial Gene Regulation by Catecholamines

In the last 3 years, there have been a number of studies published utilising microarray technologies to investigate the ability of NE and Epi to influence bacterial gene expression. Five of these studies have been in the well studied pathogens *E. coli* O157:H7 (Bansal et al. 2007; Dowd 2007; Kendall et al. 2007) and *Salmonella enteric* serovar Typhimurium (Bearson et al. 2008; Karavolos et al. 2008), and the most recent has used the porcine pathogen *Mycoplasma hyopneumoniae* (Oneal et al. 2008). Fundamental to microarray work is the necessity of harvesting a minimum number of bacterial cells (approximately 10^8–10^9) in order to obtain enough RNA for the experimental procedure; this necessity relates to all samples including the untreated controls. Thus, even before we start to plan an experiment, it becomes obvious that it will be impossible to design a microarray analysis-based protocol using the standard SAPI-serum media and methodology of Lyte and Ernst (1992); this is simply because the control samples will not achieve a cell density of greater than 10^5 CFU/ml, and therefore the volumes of culture required to obtain sufficient RNA would be infeasible to use. Each of the groups whose microarray analyses are detailed in Table 16.1 has approached this problem in methodology from a slightly different angle, often using media and conditions derived from their previous work, though two of the groups (Dowd 2007; Bearson et al. 2008) have chosen to use serum-SAPI medium albeit with much higher than normal initial inoculums of bacteria in order to circumvent its bacteriostatic nature.

Examination of the *E. coli* microarray studies shows a huge disparity in their results; the numbers and types of genes whose transcription is altered by similar catecholamine exposure are significantly different, and contradictory gene expression profiles are seen between the studies (Table 16.1). If we initially consider just the Dowd (2007) and Bansal et al. (2007) studies we see that, although their initial inocula (10^6 CFU/ml), catecholamine concentration (50 µM) and catecholamine exposure times were similar, the culture media used, i.e. Serum-SAPI and Luria broth plus glucose respectively, are very different. Addition of NE to serum-SAPI would be predicted to make iron relatively more available to bacteria whilst in media rich in free iron, such as Luria broth, NE's iron chelating properties would result in relative iron restriction. As such, it is not surprising to see that Dowd (2007) reports catecholamine exposure resulting in transcriptional repression of some iron uptake genes (e.g. *fepA*, *entCD*) whilst Bansal et al. (2007) reports only the activation of iron repressed genes. Despite the similar catecholamine concentrations and exposure times, it is difficult to make an accurate comparison of the first two studies with that of Kendall et al. (2007) because the latter has only examined the effects of catecholamines upon an *E. coli* O157:H7 strain, VS94, which is mutant in the *luxS* gene. Kendall et al. (2007) reported much larger numbers of genes (5214) being differentially regulated in response to Epi when compared with those observed by Bansal et al. (970); however, these differences may simply be the result of the inherent differences in either the media used, DMEM vs LUB, or the genetic background of the strains, *luxS* vs wildtype. Interestingly, both Dowd (2007) and Kendall et al. (2007) report transcriptional up regulation of *E. coli* virulence determinants (i.e. genes of the LEE operons and shiga toxin genes) in response to catecholamine, whereas Bansal et al. (2007) does not, thereby further underlining the importance of using a host-like culture media when looking at the expression of pathogenicity loci.

Though the catecholamine concentrations and the exposure times used were somewhat different between the two *Salmonella* microarray analyses, we can see a similar pattern of contradictory gene expression data akin to that seen previously in *E. coli*; again these disparities appear to be linked to differences in the culture media used (Table 16.1). The addition of catecholamine to Luria broth cultures of *Salmonella* resulted in transcriptional activation of iron uptake genes, e.g. *feoAB*, *entE*, *fhuC*, (Karavolos et al. 2008), whereas in Serum-SAPI grown cultures under similar catecholamine conditions almost all of the known iron repressed genes were seen to be down regulated (Bearson et al. 2008). Furthermore, Bearson et al. reported the down regulation of genes of the Type III secretion system of *Salmonella* pathogenicity island 2, whereas the Karavalos et al. study found no obvious differential regulation of any virulence associated genes. Intriguingly, the numbers of genes differentially regulated in *Salmonella* is approximately tenfold smaller in either study than that reported for the full *E. coli* microarrays; Karavolos et al. (2008) report only 25 genes differentially regulated and Bearson et al. showed only 88 genes down regulated (they do not comment on, nor give numbers of, genes up-regulated). Even considering the experimental design differences between the *Salmonella* and *E. coli* studies, it is obvious, from the vast disparity in the numbers and nature of the regulated genes that these two species have quite different holistic responses to catecholamine exposure despite their very similar growth responses. This observation will have serious implications for those researchers hoping to extrapolate their microbial endocrinology findings between even closely related species.

The microarray study examining the effects of NE upon *Mycoplasma hyopneumoniae* gene expression (Oneal et al. 2008) also raises another interesting point with respect to experimental methodology. As *M. hyopneumoniae* is an obligate intracellular parasite and has minimal metabolic enzymes, the study required the

Table 16.1 Comparison of studies of global bacterial gene regulation by catecholamines

Study	Organism	Medium used	Initial inoculums (cfu/ml)[a]	Exposure time (hours)	Catecholamines used (concentration)	Number of genes alternately regulated	Genes of interest
Dowd 2007	E. coli O157:H7 EDL933	Serum-SAPI	2.0×10^6	5	NE (50 μM)	Up: 57[b] Down: 44[b]	LEE genes, stx1, stx2 Iron uptake genes
Bansal et al. 2007	E. coli O157:H7 EDL933	LUB+0.2% glucose	10^6	7	NE (50 μM)	938	Up: Iron uptake genes Down: AI-2 uptake system
					Epi (50 μM)	970	Up: Iron uptake genes Down: AI-2 uptake system
Kendall et al. 2007	E. coli O157:H7 VS94 (86-24 luxS)	DMEM	10^7	ND[d]	Epi (50 μM)	Up: 2,367 Down: 2,837	LEE genes, flagella biosynthesis, stx2
Bearson et al. 2008	S. typhmurium BSX8	Serum-SAPI	10^7	ND[d]	NE (2 mM)	Up: Data not shown Down: 88	Iron uptake genes, TTSS
Karavolos et al. 2008	S. typhmurium SL1344	LUB	5×10^8	0.5	Epi (50 μM)	Up: 14 Down: 11	Iron uptake genes Polymixin B resistance genes, invF
Oneal et al. 2008	Mycoplasma hyopneumoniae strain 232	Friis medium[c]	ND[d]	4	NE (100 μM)	Up: 53 Down: 31	Protein expression, i.e. ribosomal proteins. General metabolism

[a]For some studies initial inoculums have been approximated based upon the initial OD measurements given
[b] The microarray in this study was partial and only contained 610 genes
[c]Friis (1975)
[d]Experimental details in these studies give the starting points and end points of growth as early and late log phase

use of a rich microbiological medium, Friis medium (Friis 1975), which contains approximately 18% heat-treated porcine serum. It was observed that addition of 100 μM NE to the *M. hyopneumoniae* cultures resulted in the differential expression of 84 genes (Table 16.1); however, there was also an unexpected reduction in the growth rate and a doubling of the time required to reach exponential phase. It is still unclear exactly how *Mycoplasma* species obtain their iron in the host; however, since none of the *M. hyopneumoniae* genes previously shown to be associated with iron restriction (Madsen et al. 2006) were similarly differentially regulated in this study, it seems unlikely that the NE induced growth defect was related to iron availability. Similarly, based on previous work by the same group examining the effect of hydrogen peroxide on *M. hyopneumoniae* (Schafer et al. 2007), there is no evidence from the NE-induced gene expression profile of a role for oxidative stress in the NE induced growth inhibition. Thus, it appears that there must be another property of NE, of which we are currently unaware, by which it can inhibit the growth of some bacterial species.

16.7   Concluding Remarks

A major methodological concept in the design of Lyte's initial microbial endocrinology experiments (Lyte 1993) was that it was vital to provide the bacteria with in vitro conditions similar to those encountered in the host environment in which to test their reactions to mammalian hormones. Therefore, most of the in vitro studies examining catecholamine-induced growth effects have been conducted in nutrient poor, serum-supplemented media using very low inocula of bacteria to mimic the conditions during the onset of an infection. In contrast, the majority of studies which have examined the effect of catecholamines upon virulence determinant expression and global gene regulation have used more traditional nutrient rich microbiological or tissue culture media, and have used high numbers of bacteria, which are not representative of the host environment. Many of these studies have been very successful, for example, reports from Sperandio's laboratory have detailed the identification and characterisation of specific catecholamine receptors in *E. coli* linked to the expression of known virulence genes (Sperandio et al. 2002; Clarke et al. 2006; Reading et al. 2007); however, there still remain concerns that the inferences made from such in vitro work will not translate well to the in vivo situation (Freestone and Lyte 2008). One major concern which must be addressed in future microbial endocrinology studies is the use of high density bacterial cultures for in vitro gene regulation experiments; this unfortunate methodological necessity is now particularly worrying because of the link established between the catecholamine signalling and quorum sensing systems in *E. coli* (Kendall and Sperandio 2007; Hughes and Sperandio 2008). If bacterial autoinducers like AI3, which are produced at high levels in high density cultures, can substitute for the neurohormones which would be encountered in the host, then how can we be sure that the catecholamine-induced gene regulation we see these in vitro experiments is typical of the low cell density conditions which occur in infection? For the technical reasons detailed previously, microarray analyses will not be able to address this problem, and new methodologies must be sought to examine catecholamine-induced global gene regulation at low bacterial cell densities.

References

Alverdy, J., Holbrook, C., Rocha, F., Seiden, L., Wu, R. L., Musch, M., Chang, E., Ohman, D., and Suh, S. 2000. Gut-derived sepsis occurs when the right pathogen with the right virulence genes meets the right host: Evidence for in vivo virulence expression in *Pseudomonas aeruginosa*. Ann. Surg. 232:480-489.

Anderson, M.T., and Armstrong, S.K. 2006. The Bordetella Bfe system: Growth and transcriptional response to siderophores, catechols, and neuroendocrine catecholamines. J. Bact.188:5731-5740.

Anderson, M. T., and Armstrong, S. K. 2008. Norepinephrine mediates acquisition of transferrin-iron on *Bordetella brontiseptica*. J. Bact. 190:3940-3947.

Banerjee, D., Martin, N., Nandi, S., Shukla, S., Dominguez, A., Mukhopadhyay, G., and Prasad, R. 2007. A genome-wide steroid response study of the major human fungal pathogen *Candida albicans*. Mycopathologia. 164:1-17.

Bansal, T., Englert, D., Lee, J., Hegde, M., Wood, T. K., and Jayaraman, A. 2007. Differential effects of epinephrine, norepinephrine, and indole on *Escherichia coli* O157:H7 chemotaxis, colonization, and gene expression. Infect. Immun. 75:4597-4607.

Bearson, B. L., Bearson, S. M., Uthe, J. J., Dowd, S. E., Houghton, J. O., Lee, I., Toscano, M. J., and Lay Jr., D. C. 2008. Iron regulated genes of *Salmonella enterica* serovar Typhimurium in response to norepinephrine and the requirement of *fepDGC* for norepinephrine-enhanced growth. Microbes Infect. 10:807-816.

Belay, T., Aviles, H., Vance, M., Fountain, K., Sonnenfeld, G. 2003. Catecholamines and in vitro growth of pathogenic bacteria: Enhancement of growth varies greatly among bacterial species. Life Sci. 73:1527-1535.

Borisenko, G., Kagan, A., Hsia, C., and Schor, N. F. 2000. Interaction between 6-Hydroxydopamine and Transferrin: "Let My Iron Go". *Biochemistry*. 39:3392-3400.

Burton, C. L., Chhabra, S. R., Swift, S., Baldwin, T. J., Withers, H., Hill, S. J., and Williams, P. 2002. The growth response of *Escherichia coli* to neurotransmitters and related catecholamine drugs requires a functional enterobactin biosynthesis and uptake system. Infect. Immun. 70:5913-5923.

Clarke, M. B., and Sperandio, V. 2005. Events at the host-microbial interface of the gastrointestinal tract III. Cell-to-cell signaling among microbial flora, host, and pathogens: there is a whole lot of talking going on. Am. J. Physiol. Gastrointest. Liver Physiol. 288: G1105-G1109.

Clarke, M. B., Hughes, D. T., Zhu, C., Boedeker, E. C., and Sperandio, V. 2006. The QseC sensor kinase: A bacterial adrenergic receptor. Proc. Natl. Acad. Sci. U. S. A. 103:10420-10425.

Cogan, T. A., Thomas, A. O., Rees, L. E., Taylor, A. H., Jepson, M. A., Williams, P. H., Ketley, J., and Humphrey. T. J. 2007. Norepinephrine increases the pathogenic potential of *Campylobacter jejuni*. Gut 56:1060-1065.

Coulanges, V., Andre, P., Ziegler, O., Buchheit, L., and Vidon, D. J. 1997. Utilization of iron-catecholamine complexes involving ferric reductase activity in *Listeria monocytogenes*. Infect Immun. 65:2778-2785.

Coulanges, V., Andre, P., Vidon, D. J-M. (1998) Effect of siderophores, catecholamines, and catechol compounds on *Listeria* spp. growth in iron-complexed medium. Biochem. Biophys. Res. Comm. 249:526-530.

Dowd, S. E. 2007. *Escherichia coli* O157:H7 gene expression in the presence of catecholamine norepinephrine. FEMS Microbiol. Lett. 273:214-223.

Freestone, P. P., and Lyte, M. (2008) Microbial endocrinology: experimental design issues in the study of interkingdom signalling in infectious disease. Adv. Appl. Microbiol. 64:75-105.

Freestone, P. P., Haigh, R. D., Williams, P. H., and Lyte, M. 1999. Stimulation of bacterial growth by heat-stable, norepinephrine-induced autoinducers. FEMS Microbiol. Lett. 172:53-60.

Freestone, P. P., Lyte, M., Neal, C. P., Maggs, A. F., Haigh, R. D., and Williams, P. H. 2000. The mammalian neuroendocrine hormone norepinephrine supplies iron for bacterial growth in the presence of transferrin or lactoferrin. J. Bacteriol. 182:6091-6098.

Freestone, P. P., Williams, P. H., Haigh, R. D., Maggs, A. F., Neal, C. P. and Lyte, M. 2002. Growth stimulation of intestinal commensal *Escherichia coli* by catecholamines: A possible contributory factor in trauma-induced sepsis. Shock 18:465-470.

Freestone, P. P., Haigh, R. D., Williams, P. H., and Lyte, M. 2003. Involvement of enterobactin in norepinephrine-mediated iron supply from transferrin to enterohaemorrhagic *Escherichia coli*. FEMS Microbiol. Lett. 222:39-43.

Freestone, P. P., Haigh, R. D., and Lyte, M. 2007a. Blockade of catecholamine-induced growth by adrenergic and dopaminergic receptor antagonists in *Escherichia coli* O157:H7, *Salmonella enterica* and *Yersinia enterocolitica*. BMC Microbiol. 7:8.

Freestone, P. P., Haigh, R. D., and Lyte, M. 2007b. Specificity of catecholamine-induced growth in *Escherichia coli* O157:H7, *Salmonella enterica* and *Yersinia enterocolitica*. FEMS Microbiol. Lett. 269:221-228.

Freestone, P. P., Walton, N. J., Haigh, R. D., and Lyte, M. 2007c. Influence of dietary catechols on the growth of enteropathogenic bacteria. Int. J. Food Microbiol. 119:159-169.

Freestone, P. P., Sandrini, S. M., Haigh, R. D., and Lyte, M. 2008. Microbial endocrinology: how stress influences susceptibility to infection. Trends Microbiol. 16:55-64.

Friis, N. F. 1975. Some recommendations concerning primary isolation of *Mycoplasma hyopneumoniae* and *Mycoplasma flocculare*, a survey. Nord. Vetmed. 27:337-339.

Hughes, D. T., and Sperandio, V. 2008. Inter-kingdom signalling: communication between bacteria and their hosts. Nat. Rev. Microbiol. 6:111-120.

Kaper, J. B., and Sperandio, V. 2005. Bacterial cell-to-cell signalling in the gastrointestinal tract. Infect. Immun. 73:3197-3209.

Karavolos, M. H., Spencer, H., Bulmer, D. M., Thompson, A., Winzer, K., Williams, P., Hinton, J. C., and Khan, C. M. 2008. Adrenaline modulates the global transcriptional profile of *Salmonella* revealing a role in the antimicrobial peptide and oxidative stress resistance responses. BMC Genomics. 9:458.

Kendall, M. M., and Sperandio, V. 2007 Quorum sensing by enteric pathogens. Gastrointestinal infections. Curr. Opin. Gastroenterol. 23:10-15.

Kendall, M. M., Rasko, D. A., and Sperandio, V. 2007. Global effects of the cell-to-cell signaling molecules autoinducer-2, autoinducer-3, and epinephrine in a *luxS* mutant of enterohemorrhagic *Escherichia coli*. Infect. Immun. 75:4875-4884.

Kinney, K.S., Austin, C.E., Morton, D.S., Sonnenfeld, G. 1999. Catecholamine enhancement of *Aeromonas hydrophila* growth. Micro. Pathogen. 26:85-91.

Kinney, K. S., Austin, C. E., Morton, D. S., and Sonnenfeld, G. 2000. Norepinephrine as a growth stimulating factor in bacteria – mechanistic studies. Life Sci. 67:3075-3085.

Leinhardt, D. J., Arnold, J., Shipley, K. A., Mughal, M. M., Little, R. A., and Irving, M. H. 1993. Plasma NE concentrations do not accurately reflect sympathetic nervous system activity in human sepsis. Am. J. Physiol. 265:E284-E288.

Lyte, M. 1993 The role of microbial endocrinology in infectious disease. J. Endocrinol. 137: 343-345.

Lyte, M. 2004. Microbial endocrinology and infectious disease in the 21st century. Trends Microbiol. 12:14-20.

Lyte, M., and Ernst, S. 1992. Catecholamine-induced growth of gram negative bacteria. Life Sci. 50:203-212.

Lyte, M., Ernst, S. 1993. Alpha and beta adrenergic receptor involvement in catecholamine-induced growth of gram-negative bacteria. Biochem. Biophys. Res. Comm. 190:447-452.

Lyte, M., Arulanandam, B. P., and Frank, C. D. 1996. Production of Shiga-like toxins by *Escherichia coli* O157:H7 can be influenced by the neuroendocrine hormone norepinephrine. J. Lab. Clin. Med. 128:392-398.

Lyte, M., Arulanandam, B., Nguyen, K., Frank, C., Erickson, A., and Francis, D. 1997a. Norepinephrine induced growth and expression of virulence associated factors in enterotoxigenic and enterohemorrhagic strains of *Escherichia coli*. Adv. Exp. Med. Biol. 412:331-339.

Lyte, M., Erickson, A. K., Arulanandam, B. P., Frank, C. D., Crawford, M. A., and Francis, D. H. 1997b. Norepinephrine-induced expression of the K99 pilus adhesin of enterotoxigenic *Escherichia coli*. Biochem. Biophys. Res. Commun. 232:682-686.

Lyte, M., Freestone, P. P., Neal, C. P., Olson, B. A., Haigh R. D., Bayston, R., Williams, P. H. 2003. Stimulation of *Staphylococcus epidermidis* growth and biofilm formation by catecholamine inotropes. Lancet 361:130-135.

Madsen, M. L., Nettleton, D, Thacker, E. L., and Minion, F. C. 2006 Transcriptional profiling of *Mycoplasma hyopneumoniae* during iron depletion using microarrays. Microbiology 152:937-944.

Methner, U., Rabsch, W., Reissbrodt, R., and Williams, P. H. 2008. Effect of norepinephrine on colonisation and systemic spread of *Salmonella enterica* in infected animals: Role of catecholate siderophore precursors and degradation products. Int. J. Med. Microbiol. 298:429-439.

Nakano, M., Takahashi, A., Sakai, Y., and Nakaya, Y. 2007a. Modulation of pathogenicity with norepinephrine related to the type III secretion system of *Vibrio parahaemolyticus*. J. Infect. Dis. 195:1353-1360.

Nakano, M., Takahashi, A., Sakai, Y., Kawano, M., Harada, N., Mawatari, K., and Nakaya, Y. 2007b. Catecholamine-induced stimulation of growth in *Vibrio* species. Lett. Appl. Microbiol. 44:649-653.

Neal, C. P., Freestone, P. P. E., Maggs, A. F., Haigh, R. D., Williams, P. H., Lyte, M. 2001. Catecholamine inotropes as growth factors for *Staphylococcus epidermidis* and other coagulase-negative staphylococci. FEMS Microbiol. Lett. 194:163-169.

O'Donnell, P. M., Aviles, H., Lyte, M., and Sonnenfeld, G. 2006. Enhancement of in vitro growth of pathogenic bacteria by norepinephrine: Importance of inoculum density and role of transferrin. Appl. Environ. Microbiol. 72:5097-5099.

Oneal, M. J., Schafer, E. R., Madsen, M. L., and Minion, F. C. 2008. Global transcriptional analysis of *Mycoplasma hyopneumoniae* following exposure to norepinephrine. Microbiology 154:2581-2588.

Reading, N. C., Torres, A. G., Kendall, M. M., Hughes, D. T., Yamamoto, K., and Sperandio, V. 2007. A novel two-component signaling system that activates transcription of an enterohemorrhagic *Escherichia coli* effector involved in remodeling of host actin. J. Bacteriol. 189:2468-2476.

Roberts, A., Matthews, J. B., Socransky, S. S., Freestone, P. P., Williams, P. H., and Chapple, I. L. 2002. Stress and the periodontal diseases: Effects of catecholamines on the growth of periodontal bacteria in vitro. Oral Microbiol. Immunol. 17:296-303.

Schafer, E. R., Oneal, M. J., Madsen, M. L., and Minion, C. F. 2007 Global transcriptional analysis of *Mycoplasma hyopneumoniae* following exposure to hydrogen peroxide. Microbiology 153:3785-3790.

Scheckelhoff, M. R., Telford, S. R., Wesley, M., and Hu, L. T. 2007. *Borrelia burgdorferi* intercepts host hormonal signals to regulate expression of outer surface protein A. Proc. Natl. Acad. Sci. USA. 104:7247–7252.

Sperandio, V., Torres, A. G., and Kaper, J. B. 2002. Quorum sensing *Escherichia coli* regulators B and C (QseBC): A novel two-component regulatory system involved in the regulation of flagella and motility by quorum sensing in *E. coli*. Mol. Microbiol. 43:809-821.

Sperandio, V., Torres, A. G., Jarvis, B., Nataro, J. P., and Kaper, J. B. 2003. Bacteria-host communication: the language of hormones. Proc. Natl. Acad. Sci. U. S. A. 100:8951-8956.

Straub, R. H., Wiest, R., Strauch, U. G., Harle, P., and Scholmerich, J. 2006. The role of the sympathetic nervous system in intestinal inflammation. Gut 55:1640-1649.

Tarr, P. I., and Neill, M. A., 2001. *Escherichia coli* O157:H7, Gastroenterol. Clin. North Am. 30:735-751.

Toscano, M. J., Stabel, T. J., Bearson, S. M. D., Bearson, B. L. and Lay, D. C. 2007. Cultivation of *Salmonella enterica* serovar Typhimurium in a norepinephrine-containing medium alters *in vivo* tissue prevalence in swine. J. Exp. Anim. Sci. 43: 329-338.

Vlisidou, I., Lyte, M., van Diemen, P. M., Hawes, P., Monaghan, P., Wallis, T. S., and Stevens, M. P. 2004. The neuroendocrine stress hormone norepinephrine augments *Escherichia coli* O157:H7-induced enteritis and adherence in a bovine ligated ileal loop model of infection. Infect. Immun. 72:5446-5451.

Walters, M., and Sperandio, V. 2006. Autoinducer 3 and epinephrine signaling in the kinetics of locus of enterocyte effacement gene expression in enterohemorrhagic *Escherichia coli*. Infect. Immun. 74:5445-5455.

Zaborina, O., Lepine, F., Xiao, G., Valuckaite, V., Chen, Y., Li, T., Ciancio, M., Zaborin, A., Petrof, E. O., Turner, J. R., Rahme, L. G., Chang, E., and Alverdy, J. C. 2007. Dynorphin activates quorum sensing quinolone signaling in *Pseudomonas aeruginosa*. PLoS Pathog. 3:e35.

Zhang, X., Essmann, M., Burt, E. T., and Larsen, B. 2000. Estrogen effects on *Candida albicans*: a potential virulence-regulating mechanism. J. Infect. Dis. 181:1441-1446.

Index